On Growth and Form

W. Wilson
from H E Stanley
7/88

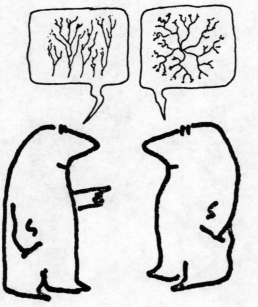

On Growth and Form
Fractal and Non-Fractal Patterns in Physics

edited by

H. Eugene Stanley
Boston University
Boston, Massachusetts
USA

Nicole Ostrowsky
University of Nice
Nice, France

1986 **MARTINUS NIJHOFF PUBLISHERS**
a member of the KLUWER ACADEMIC PUBLISHERS GROUP
BOSTON / DORDRECHT / LANCASTER

Distributors

for the United States and Canada: Kluwer Academic Publishers, 190 Old Derby Street, Hingham, MA 02043, USA
for the UK and Ireland: Kluwer Academic Publishers, MTP Press Limited, Falcon House, Queen Square, Lancaster LA1 1RN, UK
for all other countries: Kluwer Academic Publishers Group, Distribution Center, P.O. Box 322, 3300 AH Dordrecht, The Netherlands

Library of Congress Cataloging in Publication Data

```
Main entry under title:

On growth and form.

  (NATO ASI series. Series E, Applied sciences ; no.
100)
  Proceedings of a course offered as part of the
NATO ASI series, during the period 26 June-6 July,
1985, at the Cargèse Summer School in Cargèse, Corsica.
  "Published in cooperation with NATO Scientific
Affairs Division."
  1. Matter--Congresses.  2. Growth--Congresses.
3. Random walks (Mathematics)--Congresses.  I. Stanley,
H. Eugene (Harry Eugene), 1941-      .  II. Ostrowsky,
Nicole, 1943-      .  III. NATO Advanced Study Institute
(1985 : Cargèse Summer School)  IV. Series.
QC170.O56  1985          539'.1            85-18825
```

ISBN 0-89838-850-3

Copyright

© 1986 by Martinus Nijhoff Publishers, Dordrecht.

All rights reserved. No part of this publication may be reproduced, stored in a retrieval system, or transmitted in any form or by any means, mechanical, photocopying, recording, or otherwise, without the prior written permission of the publishers,
Martinus Nijhoff Publishers, P.O. Box 163, 3300 AD Dordrecht,
The Netherlands

PRINTED IN THE NETHERLANDS

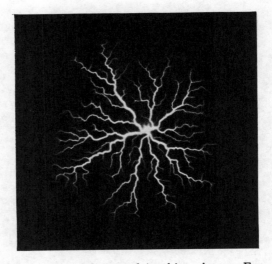

Two of the many growing fractal structures discussed in this volume. Experimental details are given in the book, and in the original references: J. Nittmann, G. Daccord and H. E. Stanley, Nature **314**, 141 (1985) and L. Niemeyer, L. Pietronero and A. J. Wiesmann, Physical Review Letters **52**, 1033 (1984). We thank the authors for permission to display these photographs.

ON GROWTH AND FORM: Fractal and Non-Fractal Patterns in Physics

PREFACE

Within the past few years, remarkable progress has occurred in understanding the structure of a variety of random "forms" and the fashion in which these forms "grow." Examples range from fractal viscous fingering, a topic from fluid mechanics, to dielectric breakdown, a topic from electromagnetism. Photographs of these two phenomena are shown on the frontispiece of this book.

The striking similarity of these two structures suggests that their might be some unifying elementary physical principle underlying both. The remarkable discovery that forms the basis of "ON GROWTH AND FORM" is that both phenomena can be related to the properties of a simple random walk—indeed, even the quantitative equations from fluid mechanics and electromagnetism can be formally related to random walk equations.

Not surprisingly, there has been a tremendous upsurge of interest in this opportunity to unify a large number of diverse physical phenomena, ranging from physics and materials science to chemistry and biology. There has yet to appear a single book on the subject of these new approaches to growth and form. Therefore we organized a formal course on this topic, choosing lecturers known as much for their clarity of presentation as for their seminal contributions to this new field. This course was offered as part of the NATO ASI series, during the period 26 June–6 July 1985, at the Institut d'Etudes Scientifiques de Cargèse (M. Lévy, Director) in Corsica, France. All eleven course lecturers have prepared pedagogical accounts of their presentations.

In addition to a large number of outstanding students, the course was attended by several active researchers who delivered seminars on their topic of interest. Short summaries of these seminars are also included.

To our pleasant surprise, the school was characterized by a degree of *élan* higher than any previous school in memory. Not only were the lectures of uniformly high quality, but many new scientific results were presented for the first time. A spacious library contained preprints from all the participants, with a functioning photocopy machine to permit *permanent* records of these new ideas. Also set up in the library were various on-line computer simulations, video demonstrations of several experiments that could not be transported to Corsica, and a 25-minute pedagogical film "Aggregation" by M. Kolb (produced by ZEAM, Frei Universität Berlin). Copies of the film and its video counterpart are available and nicely complement much of the material in this volume.

Verbal intercourse was stimulated by communal meals on long tables overlooking the Mediterranean, by long discussions on the beaches during the early afternoon break, and by long evenings at cafe tables in town. Not surprisingly, genuine tears were in the eyes of many as the school drew to a close.

Our sincere thanks are due to many. First and foremost, to those lecturers who put forth extra effort to make their presentations models of clarity and acumen, to those students whose frequent questions opened new directions of pursuit, to Karine Ostrowsky, Serge Ostrowsky, Françoise Boon, Chantal Ariano and Marie-France Hanseler for 24-hour days of managerial help, to Dan Ostrowsky, Avraham Simievic and Joseph Antoine Ariano for preparing a genuine *mechoui* for 130 people for our Saturday night banquet, to the citizens of Cargèse for putting up with the same number of "invading continentals" with good cheer and, most importantly, to all 130 whose lightness of heart created that sort of relaxed atmosphere conducive to intellectual stimulation. Without the generous support of NATO, the Office of Naval Research, the National Science Foundation, Exxon, SOHIO, and E. I. duPont de Nemours, this opening chapter of a new field could not have unfolded. Without the inspiration of D'Arcy Thompson, we would not have the title or the spirit of this course.

Finally, it is a pleasure to thank and to record our immense debt to those who transformed this book from written lecture notes into a finished reality. Jerry Morrow began this process by producing TEX copy from those lecturers who did not come prepared with camera-ready manuscripts and re-typing the manuscripts of those who did. Then came the arduous tasks of computer debugging–and of proofreading–which were kindly shared among a number of colleagues, including Robin Blumberg Selinger, Ken Janes, Jim Miller, Ann Price and Peter Reynolds. Betsy E. Kadanoff, Avraham Simievic, and Idahlia Stanley kindly supplied most of the photographs of participants that adorn the otherwise wasted space at the end of chapters. The multi-national offices of Martinus Nijhoff massaged the final product, and we wish to express our deep appreciation to its Managing Director Arne Visser and its staff member Henny A.M.P. Hoogervorst for the three-fold honor of selecting our work to appear as No. 100 in their NATO ASI series on Applied Sciences, for publishing simultaneously a hardbound and an inexpensive paperback edition, and for setting a world record for rapid publishing time. We also wish to thank all the other efficient and cheerful members of the Martinus Nijhoff staff—most especially Dick Dissel, Gerard Aleven and Derek Middleton in Dordrecht and Andrea Howard and Paul Chambers in Boston.

H. Eugene Stanley
Nicole Ostrowsky
Cargèse, 6 July 1985

CONTENTS

PART A. "THE COURSE"

Growth: An Introduction 3
   Hans J. Herrmann
Form: An Introduction to Self-Similarity and Fractal Behavior 21
   H. Eugene Stanley
Scale-Invariant Diffusive Growth 54
   Thomas A. Witten, III
DLA in the Real World 69
   Robin C. Ball
Percolation and Cluster Size Distribution 79
   Dietrich Stauffer
Scaling Properties of the Probability Distribution
for Growth Sites 101
   Antonio Coniglio
Computer Simulation of Growth and Aggregation Processes 111
   Paul Meakin
Rate Equation Approach to Aggregation Phenomena 136
   François Leyvraz
Experimental Methods for Studying Fractal Aggregates 145
   José Teixeira
On the Rheology of Random Matter 163
   Etienne Guyon
Development, Growth, and Form in Living Systems 174
   Jean-Pierre Boon and Alain Noullez

PART B. "THE SEMINARS"

Aggregation of Colloidal Silica 187
   David Cannell and Claude Aubert
Dynamics of Fractals 198
   Dale W. Schaefer, James E. Martin and Alan J. Hurd
Fractal Viscous Fingers: Experimental Results 203
   Gerard Daccord, Johann Nittmann and H. Eugene Stanley
Wetting Induced Aggregation 211
   Daniel Beysens, Coggio Houessou and Françoise Perrot
Light Scattering from Aggregating Systems:
Static, Dynamic (QELS) and Number Fluctuations 218
   John G. Rarity and Peter N. Pusey

Flocculation and Gelation in Cluster Aggregation 222
   Max Kolb, Robert Botet, Rémi Jullien and Hans J. Herrmann
Branched Polymers 227
   Mohamed Daoud
Dynamics of Aggregation Processes 231
   Fereydoon Family
Fractal Properties of Clusters during Spinodal Decomposition 237
   Rashmi C. Desai and Alan R. Denton
Kinetic Gelation 244
   David P. Landau
Dendritic Growth by Monte Carlo 249
   Janos Kertész, Jenö Szép and József Cserti
Flow through Porous Materials 254
   Jorge Willemsen
Crack Propagation and Onset of Failure 260
   Sara Solla
The Theta Point 263
   Naeem Jan, Antonio Coniglio, Imtiaz Majid and H. Eugene Stanley
Field Theories of Walks and Epidemics 265
   Luca Peliti
Transport Exponents in Percolation 273
   Stephane Roux and Etienne Guyon
Non-Universal Critical Exponents for Transport in Percolating Systems 278
   Pabitra N. Sen, James N. Roberts and Bertrand I. Halperin
Lévy Walks Versus Lévy Flights 279
   Michael F. Shlesinger and Joseph Klafter
Growth Perimeters Generated by a Kinetic Walk: Butterflies,
Ants and Caterpillars 284
   Alla E. Margolina
Asymptotic Shape of Eden Clusters 288
   Deepak Dhar
Occupation Probability Scaling in DLA 293
   Leonid A. Turkevich
Fractal Singularities in a Measure and "How to Measure Singularities
on a Fractal" 299
   Leo P. Kadanoff
List of Participants 303
Index 305

PART A

. .

"THE COURSE"

Nicole Ostrowsky

Gene Stanley

GROWTH: AN INTRODUCTION

Hans J. Herrmann*

Service de Physique Théorique, CEN Saclay
91191 Gif-sur-Yvette Cedex, France

Nature provides us with an almost infinite multitude of shapes and forms. Yet, even today, we know very little about how these shapes are created.

One *can* make some crude observations. For example, the shape of any object tends to depend strongly on the kind of process from which it originated. Man-made objects for example, are usually asymmetric to the eye, but when one looks in more detail one finds often that they possess symmetries that are exact. On the other hand, naturally-formed objects often tend to possess apparent symmetries (see, e.g., Fig. 1, the so-called Lichtenberg figures). Looking closer, however, one finds that they are random. In some sense it seems almost paradoxical that random growth processes give rise to objects with inherent symmetry properties. In fact it is still a mystery and a source of intense frustration to those working in the field of artificial intelligence and pattern recognition as to how the human eye discovers the global patterns and self symmetries of these intrinsically random natural objects.

We shall build some theoretical concepts towards solving this apparent paradox. Two questions naturally arise: (1) How can we quantify the form of these naturally-grown objects? (2) How can we obtain these forms as end products of a growth process?

There are several characteristics that a naturally-grown object can possess. There can be a characteristic length, e.g., in iron dendrites or ice crystals on glass. This is quite surprising since it does not seem evident why one characteristic length is selected. There are other objects that possess no characteristic length scale at all, but are *self-similar* over many length scales. Such objects

---

* Notes taken by Ashvin Chhabra, Mason Laboratory, Yale University, New Haven, CT 06520

are called fractals. Examples of such objects are soot particles, or gels at the gel-point.

In many of the systems in nature, growth is a very complex process, involving the optimization of many variables. For example, the leaves of a growing plant tend to maximize their surface area and minimize distance from the branches in order to maximize food production and minimize food transportation. Unfortunately, we have not yet reached the stage where we can predict the patterns and forms of the objects on the basis of these optimizations. Therefore we shall concentrate only on extremely simple, somewhat idealized irreversible growth models where the binding energy is taken to be much greater than the thermal energy. The growth models that we will present here can be represented as Cellular Automata, where time is discrete and the growth rules are local. Cellular Automata are easy to simulate on the computer and they lead to many interesting and beautiful patterns. We shall see that there are very few exact results in this field and almost all of our knowledge comes from fairly extensive computer simulations.

Fig. 1: Example for a self-similar form of nature: Lichtenberg figure produced through the breakdown of an electrical charge that had been injected into plexiglas previously [F. Schwörer in P. Brix, Phys. Bl. **41**, 141 (1985)].

SURVEY OF SOME RECENT MODELS

We shall briefly describe some of the important models, which shall appear repeatedly in the course of this school.

Eden model. This model was proposed in 1961 by Eden[1] in an attempt to describe the growth of tumors. A site on the lattice (e.g., a square lattice) is randomly chosen and occupied. The empty nearest neighbors (NN) of the occupied site are labelled as growth sites. In the next time step, a growth site is randomly chosen and occupied. The NN of the newly occupied site are added on to the list of growth sites. This growth process is repeated several thousand times until we have a large cluster of occupied sites (see Fig. 2a).

There are three variants of this model, depending on how we choose a growth site. Looking at Fig. 2a, we see that the choice of growth sites will be different, depending on whether one chooses randomly (i) any external surface site, (ii) an empty neighbor of a randomly-chosen internal surface site, or (iii) any empty site at a bond on the boundary. Each of these definitions leads to slightly different probability distributions for the configurations generated, thus emphasizing the need for rules to be precisely defined telling us both when and where to grow. Fortunately, in this model, all three variants give the same result, namely that the cluster generated is compact.

Diffusion-limited aggregation (DLA). The DLA growth model proposed by Witten and Sander[2] is an example of how totally random motion can give rise to beautiful self-similar clusters. We start with a square lattice and occupy a site with a seed particle. A particle is then released from the perimeter of a large circle whose center coincides with the seed particle. The particle executes a random walk until it either leaves the circle or reaches a neighboring site of the seed particle. In the latter case it becomes a part of the growing cluster. This process is repeated several thousand times until a large cluster is formed. Although the growth process is deceptively simple, it gives rise to ramified, self-similar structures (see Fig. 2b) as opposed to the Eden model. The crucial reason for the richness of form generated here comes from the fact that the growth rule is non-local. One must emphasize however that non-locality is not a necessary condition for producing such ramified self-similar clusters.

Epidemics. As in the Eden model, we start with a single seed and consider all its neighboring sites as a part of the living surface. We then randomly choose a site of the living surface and do one of the following: (a) occupy the site with probability p and make all its new neighbors part of the living surface or (b) kill the surface site forever with probability $1-p$. On a square lattice this model can be used to describe the growth of epidemics. For example, in a plantation of regularly placed trees (square lattice) the dead sites would correspond to immune trees and the living ones to trees that can transmit the disease.

The sole parameter in this model is the probability p. By varying p we can get structures of various kinds. Fig. 2c shows the sort of clusters we get for $p_c = 0.59273$ (p_c in the percolation threshold on a square lattice). Notice that these clusters are self-similar, possessing holes of all sizes.

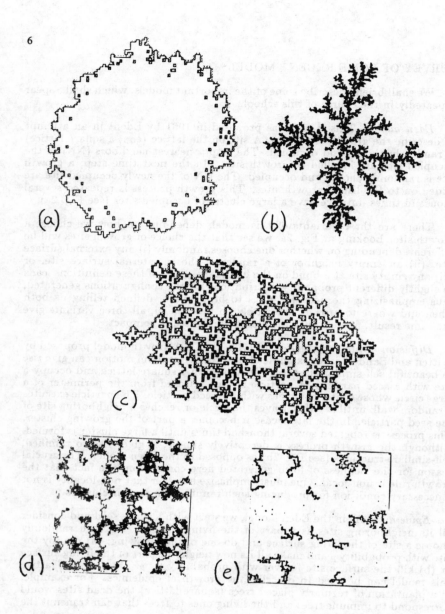

Fig. 2: (a) Eden cluster of 1500 occupied sites. (b) DLA cluster of 11260 particles on a square lattice [P. Meakin, Phys. Rev. A **27**, 604 (1983)]. (c) Epidemic cluster of 1800 sites at $p_c = 0.59273$. (d) Mole's labyrinth close to the threshold at which the infinite cluster appears.[3] (e) Typical configurations of clustering of clusters.[4]

So far we have only described growth models that start with a single seed. Many models have been proposed that start with several seeds and therefore give rise to multiple clusters. We briefly discuss two of them.

Mole's labyrinth. Several seeds (known as moles) are randomly placed on a lattice.[3] At each time step a mole is randomly chosen and moved to a random neighboring site. The traces left by the random walks of these moles define the clusters so formed. Clusters grow and merge, forming various complicated structures until in addition to many finite clusters an infinite cluster (spanning the entire lattice) is formed (see Fig. 2d). At the time that this infinite cluster first appears, it is self-similar in structure, possessing holes of all sizes.

Clustering of clusters. This model was first proposed by Kolb, Jullian and Botet[4] and Meakin.[5] Particles are placed randomly on a lattice and at every time step moved around randomly. If two particles become nearest-neighbors they stick together thus becoming a two-site cluster and so on.

In this model, both particles and clusters move about randomly until most of them aggregate, thus forming larger and larger self-similar structures (see Fig. 2e). The mass of the system remains constant throughout the growth process, and the self-similar structures arise from the random rearrangement of an initially random configuration of the particles of the system.

This last point brings out an important aspect that is worth emphasizing. In all the above growth models, clusters are formed *from very simple and random* growth rules. Yet instead of getting uninteresting or random blobs we find clusters either (i) with a single characteristic length or (ii) with no characteristic length scale but with an intricate self-similarity. The question that quite naturally arises is how are clusters with special properties formed from apparently random rules?

A promising theoretical thrust in this direction is to link these growth processes to critical phenomena. The motivation for this can be summarized

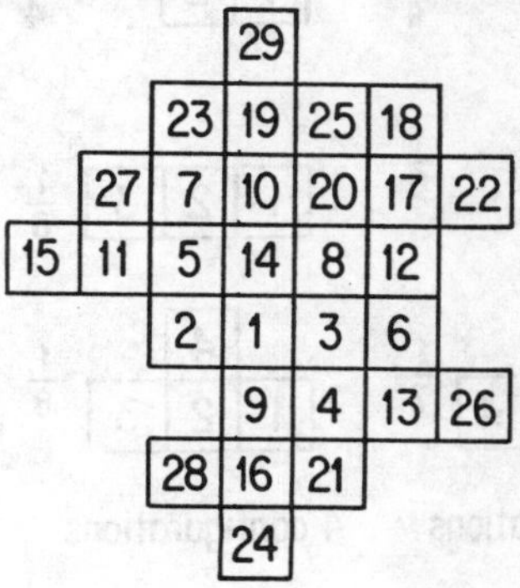

Fig. 3: Eden cluster. The numbers on the sites mark the time at which these sites were occupied.

as follows. At the critical point (in a phase transition) the correlation length is infinite. For structures in the lattice this would mean that they are similar on any length scale. Thus the concept of self-similarity appears quite naturally in a phase transition. On the other hand, the formation of an infinite cluster in gelation phenomena leads to the appearance of new macroscopic properties and can also be viewed as a phase transition. We shall now make an analogy of cluster growth with critical phenomena.

ON THE STATISTICS OF THE EDEN MODEL

In order to be able to calculate statistical averages, the first thing we would like to know is a "partition function" or "generating function" for growth models. This, however, is not a simple problem.

Consider the Eden model. Any Eden cluster has a definite "history"—an order in which its sites grew. We can thus uniquely describe its history by putting on each site an integer, namely the time when the site was occupied (see Fig. 3). It is evident that not all numberings on a cluster are admissable due to the nearest-neighbor growth constraint.

The number of possible histories that a round Eden cluster can have is large: $N! \sim N^N$. On the contrary, ramified clusters have a substantially smaller number of possible histories; e.g., a linear cluster has 2^N possible histories. This statistical effect tends to make Eden clusters rather compact.

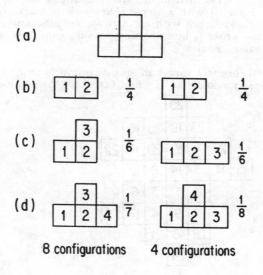

Fig. 4: Calculation of the statistical weight of the geometrical shape (a). The stages (b), (c) and (d) are the different growth steps.

The above argument roughly illustrates the idea that the probability for a certain cluster of fixed size to occur depends both on its shape and history. For example, let us regard all the possibilities of constructing a cluster of the shape shown in Fig. 4a. We show in Figs. 4b, 4c and 4d two different histories each time giving the probability to get this configuration from the preceeding one. So for this example, the total weight is 1/168 for the left history and 1/192 for the right history, i.e., the weights are different for different histories. As the left configuration appears eight times and the right four times the shape given in Fig. 4a has a weight of $8/168 + 4/192 = 23/336$ for the Eden model. It is therefore clear that the calculation of the weight of a given Eden cluster is quite complicated, and that the weight depends as well on the history as on the shape. The second assertion is illustrated in Fig. 5 where we show all the different shapes of $N = 4$ occupied sites with their respective weight. For large N the problem of figuring out the probability of occurence of a cluster of a given shape becomes hopelessly complicated. We shall return to this problem later.

LATTICE ANIMALS

In order to see how one usually treats geometrical models, let us look at a completely different model, namely a classic static problem. The relation to critical phenomena will be elucidated.

Let us assume that all clusters of the same size have equal weights. This model is known as the lattice animal model (see, e.g., Ref. 6). Remember now that we are no longer dealing with a growth model but with static clusters.

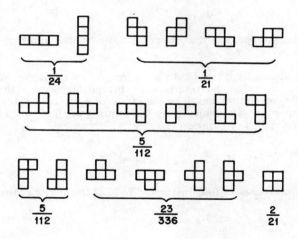

Fig. 5: All 19 shapes of size $N = 4$ with the respective weight if one grows them according to the Eden model growth rule.

Let Ω_N denote the number of different shapes of size N (i.e., lattice animals) that can occur. Then one can assume that asymptotically

$$\Omega_N \sim \mu^N N^{\gamma-1}, \tag{1}$$

where μ is an "effective coordination number."

A priori, the animals of different size N have no relation to each other. We restrict ourselves to a family of size distributions by defining for the ensemble of lattice animals of all sizes a grand-canonical partition function

$$Z(x) = \sum_N \Omega_N x^N \sim \sum_N \mu^N N^{\gamma-1} x^N, \tag{2}$$

where x is a dummy variable and is analogous to a fugacity in thermodynamics.

The sum defined in Eq. (2) converges only when $x\mu < 1$. At $x = 1/\mu = x_c$ one has a critical point for this ensemble of clusters. If $x < 1/\mu$ there are more small clusters than large ones. For $x > 1/\mu$ the reverse is true. At $x = 1/\mu = x_c$, clusters of all sizes exist in this ensemble. In terms of usual critical phenomena one can say that there is no characteristic cluster size.

The typical cluster size $\langle N \rangle$ is given by

$$\langle N \rangle = \frac{\sum N \Omega_N x^N}{\sum \Omega_N x^N}. \tag{3}$$

$\langle N \rangle$ diverges as we approach x_c from below. From Eq. (2) we obtain

$$Z(x) \sim (x - x_c)^{-\gamma}. \tag{4}$$

Here γ is the critical exponent that describes how cluster numbers decrease with cluster size.

To make the analogy with critical phenomena more complete, we need a second critical exponent that characterizes the correlation length of the system. Here we define the correlation length as the radius of a typical cluster. We can define a weight function $g_N(r)$ which gives the number of clusters that contain N sites and have a radius of gyration r:

$$r = \frac{1}{N}\sqrt{\sum_i |\vec{r}_i - \langle \vec{r} \rangle_{cm}|^2}. \tag{5}$$

The sum runs over all sites of the cluster and $\langle \vec{r} \rangle_{cm}$ is the center of mass of the cluster. Obviously

$$\Omega_N = \sum_r g_N(r). \tag{6}$$

The correlation length ξ is then given by the relation

$$\xi^2(x) = \frac{\sum_{N,r} r^2 g_N(r) x^N}{\sum_{N,r} g_N(r) x^N} \sim (x - x_c)^{-2\nu}. \tag{7}$$

Here ν is the second universal critical exponent that we were seeking.

In general we can define a variety of exponents, but we hope that as in critical phenomena, we can express them all in terms of these two, γ and ν.

The typical radius of gyration of a cluster of N sites is given by the relation

$$R_N^2 = \frac{\sum_r r^2 g_N(r)}{\sum_r g_N(r)}. \tag{8}$$

Using Eqs. (4), (6) and (7), we find

$$\sum_{N,r} r^2 g_N(r) x^N \sim (x - x_c)^{-\gamma - 2\nu}. \tag{9}$$

Hence

$$R_N^2 \sim \frac{N^{\gamma + 2\nu - 1} \mu^N}{N^{\gamma - 1} \mu^N} \sim N^{2\nu}. \tag{10}$$

On the other hand, we can define a "fractal dimensionality" of a cluster by

$$N \sim R_N^{d_f}. \tag{11}$$

Comparing (10) and (11) we get

$$d_f = 1/\nu. \tag{12}$$

Equation (13) is rather important as it relates the geometrical structure of the cluster with one of the critical exponents. This relation is valid for the general approach of writing a one-cluster problem in a generating function formalism as we have done. For percolation (in which we have more than one cluster) the situation is more complex.

We note as an aside that for lattice animals one can prove that[7]

$$\gamma = (2 - d)\nu. \tag{13}$$

Usually, however, the exponents γ and ν are unrelated, as in the self-avoiding walk that will be discussed later.

Richardson[12] has proved that for the Eden model $d_f = d$, i.e., clusters are compact, so $\nu = 1/d_f = 1/d$. But for the calculation of γ a peculiarity arises. In the Eden model at a given time ($t = N$) the probability of producing a cluster of size N is always unity, since at every time step we occupy a site and there is no reason (on an infinite lattice) for the growth to stop. So from (1) we find that $\gamma = 1$ always. This result is generic for most growth models (e.g., DLA).

In this context we also mention that an important open problem is to find a growth model that would generate lattice animals with the correct statistical weight. So far only a model that grows directed lattice animals exists, and that needs large amounts of informational input.[8]

THE DYNAMICS OF GROWTH

In general the dynamics of irreversible growth phenomena are restricted to special trajectories in phase space as opposed to usual thermodynamics where all paths are possible. The description of these processes is extremely difficult, so one tries to look at the growth process phenomenologically.

One such way is to look at growth as a surface problem. We define a function $P(r, N)$ that represents the probability that the next site is grown at time $t = N$ at a distance r from the original seed.[9] Thus $P(r, N)$ is a measure for the width of the growing surface. If we plot $P(r, N)$ against r for different N, the curves shift out with increasing N. The question is whether we can scale these curves (i.e., the curves collapse on a scaling function), thus relating the width of the growing surface with the number of sites in a cluster.

We start by assuming that such a scaling relation does exist. Then $P(r, N)$ must be of the form

$$P(r, N) \sim 1/R_N f(r/R_N). \tag{14}$$

Here R_N is the radius of a typical N-cluster and is included to normalize the probability, and $f(r/R_N)$ is a scaling function. Using $R_N \sim N^\nu$ we can rewrite P as

$$P(r, N) \sim N^{-\nu} f(r N^{-\nu}). \tag{15}$$

Then we have the following behavior for the typical radius of an N site cluster

$$\langle r \rangle_N = \int r P(r, N) dr \sim N^\nu, \tag{16}$$

and the mean width of the distribution of growth events is

$$\Delta_N = \left[\int (r - \langle r \rangle_N)^2 P(r, N) dr \right]^{1/2} \sim N^\nu. \tag{17}$$

This scaling relation is correct for the epidemics model introduced before.[10] We can see this if we plot $P(r, N) N^\nu$ vs $r N^{-\nu}$ (see Fig. 6). Note that all the data collapse onto one curve, thus demonstrating the validity of the scaling *ansatz* of Eq. (15).

However if we do the same for the Eden model we find that $\langle r \rangle_N \sim N^\nu$ and $\Delta_N \sim N^{\bar\nu}$ where $\bar\nu/\nu = 0.3 \pm 0.03$[11], i.e., $\nu \neq \bar\nu$. Thus a scaling law of the above form does not exist and it seems that $\bar\nu$ is another critical exponent which cannot be found from ν and γ. Thus the situation is more complicated for the Eden model.

Another question we can ask is if the number of growth sites 'G' scales with the total number of sites N of a cluster. We assume that

$$G \sim N^\delta. \tag{18}$$

One can also express the exponent 'δ' as a fractal dimension, namely the fractal dimension of the set of growth sites through the relation[13]

$$d_g = \delta \cdot d_f. \tag{19}$$

For the Eden model it can be shown[12] that

$$\delta = \frac{d-1}{d}. \qquad (20)$$

What can we say about δ for the epidemic model? Notice that the epidemic model is essentially the Eden model on a dilute lattice. This is because the sites that are blocked during the growth process with probability $(1-p)$, i.e., the immune sites, remain immune forever. They constitute the dilution of the lattice. Also as they are blocked independently from each other, and only once for all times, the epidemic model produces a percolation cluster at the value 'p.' Percolation involves starting with a lattice and occupying each site with probability p. The configuration that one obtains consists of many clusters of nearest-neighbor occupied sites. Percolation is a static model since no statement is made about the growth of the configuration. The epidemic model, on the other hand, produces only one cluster and is a growth model. However the one cluster it grows is a percolation cluster.

So we shall try to use the scaling ideas of percolation on the epidemic model. Let us first, as an exercise, calculate the exponent 'γ' of the epidemic model at P_c. If n_s denotes the cluster size distribution, i.e., the number of clusters consisting of 's' sites each, then we know from percolation that at $p = p_c$, $n_s \sim s^{-\tau}$ (power law decay) and at $p < p_c$ $n_s \propto e^{-as^x}$, i.e., larger clusters occur with exponentially decreasing probability. Close to p_c the scaling relation

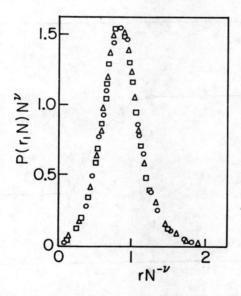

Fig. 6: Scaling plot of Eq. (16) for the model of epidemics.[10] $N = 800$: △, $N = 2000$: ○, $N = 4000$: .

$$n_s = s^{-\tau} f[(p - p_c)s^\sigma] \qquad (21)$$

holds. If in the epidemic model $p = p_c$ then the probability P_N that a cluster reaches a size N is

$$P_N = \frac{1}{p} \sum_{S=N}^{\infty} s n_s \sim \int_N^\infty s^{-\tau+1} ds \sim N^{2-\tau}. \qquad (22)$$

Using the relation (1) that defined γ, i.e., $P_N \sim N^{\gamma-1} \mu^N$, we get

$$\gamma = 3 - \tau. \qquad (23)$$

In two dimensions $\tau = 187/91$ for percolation, so $\gamma = 86/91$. Thus we know the statics well, but not the dynamics. Let us examine the dynamics of the epidemic model more closely.

In the conventional epidemic model we pick one of the growth sites randomly and occupy it with probability p or block it with probability $1-p$. The exponent δ associated with this model was evaluated numerically[10] to be $\delta = 0.40 \pm 0.015$. One can however think of several variants of this model. One example is to pick the growth sites with some probability distribution proportional to $\rho^{-\alpha}$, where ρ is the distance between the old and new growth site and α is a parameter of the model. This corresponds to doing a "Levy flight" instead

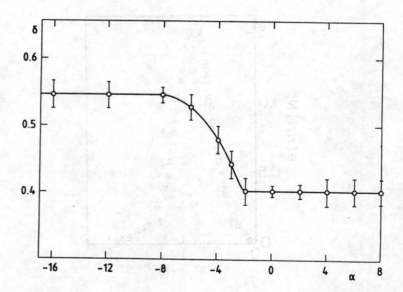

Fig. 7: Variation of δ with α.

of a random walk on the growth sites. In such a model δ changes continuously according to the value of α^{10} (see Fig. 7) without the static exponents γ and ν changing.

Another variant to the epidemic model is obtained by using the "first-in first-out" (FIFO) rule.[14] In this case we occupy the growth sites in the same order as they were created. The advantage of such a model is that we can analytically relate δ to the fractal structure of the cluster. One finds that

$$\delta = \frac{d_\ell - 1}{d_\ell}, \tag{24}$$

where d_ℓ is the "chemical dimension" of the cluster. To see Eq. (24) observe that because of the FIFO rule, first the nearest neighbors (NN) of the original seed are occupied, then the next NN to the just-occupied sites and so on. This process is equivalent to growing all the sites on chemical shells of increasing chemical distance centered around the original seed. One grows shell by shell. Consider all the growth sites that belong to a shell at a chemical distance ℓ from the seed. Then we can use the definition of the chemical dimension d_ℓ

$$N \sim \ell^{d_\ell}, \tag{25}$$

where ℓ is just the chemical length. Since G is the number of growth sites that lie on a chemical surface,

$$G \sim \ell^{d_\ell - 1} \sim N^{(d_\ell - 1)/d_\ell},$$

and since $G \sim N^\delta$, we have $\delta = (d_\ell - 1)/d_\ell$ as we wanted to show. This means that $\delta = 0.403 \pm 0.003$ using for d_ℓ the value of Ref. 14.

Another variant of the epidemic model is the FILO model in which the growth sites are occupied in the reverse order of their being occupied, i.e., the newest growth site is chosen next. For FILO one finds $\delta = 0.78 \pm 0.05$, i.e., twice as large as the δ for FIFO.[14]

If instead of being at p_c we are a certain distance $\epsilon = p - p_c$ away from the critical point, then instead of $G \sim N^\delta$ we need a more complicated relation[14]

$$G = N^\delta f[(p - p_c)N^\phi], \tag{26}$$

where ϕ is a crossover exponent.

For $p < p_c$ the interpretation of this behavior is quite simple. Figure 8 schematically shows the typical behavior of G as a function of N. For short time scales, the cluster does not realize that the system is not at p_c and the growth sites grow as $G \propto N^\delta$, i.e., exhibiting the behavior we would expect at p_c. For longer time scales, the system realizes that it is finite and the rate of increase of growth sites levels off. The decreasing tail goes to zero because we are dealing with finite systems and eventually no more growth sites are left to occupy.

In Fig. 8 G_{max} is the maximum number of growth sites that the system had at any time during its growth, while $\overline{G}$ is the mean number of growth sites that were available to the system averaged over the time of the growth process. If we plot G_{max} and $\overline{G}$ against $(p_c - p)/p_c$ (the normalized distance from the critical point) on a log-log plot, we get two parallel lines with slope δ/ϕ. This is consistent with the scaling relation Eq. (26) and gives us the numerical value $\phi = 0.5 \pm 0.1$.[14]

PHENOMENA OF STERIC HINDRANCE

We now turn to growth models that have only one growth site, namely walks. The cluster is defined as the path traced out by the walk. The advantage of this is that one can for these simple growth model more easily understand the effect that we shall deal with next: steric hindrance.

Random walk. The random walk is the simplest walk one can study. If q is the coordination number of the lattice and N the number of steps taken by the walk, i.e., the mass of the cluster, then

$$\Omega_N = q^N \quad \text{and} \quad \gamma = 1. \tag{27}$$

Also we know that

$$R_N^2 \sim N \Longrightarrow \nu = 1/2 \quad \text{or} \quad d_f = 1/\nu = 2, \tag{28}$$

i.e., a random walk in two dimensions will eventually cover the entire lattice or plane.

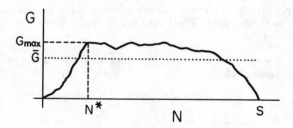

Fig. 8: Schematic plot of the number G of growth sites against the time N. G_{max} is the maximum value and $\overline{G}$ the mean value of this profile. S is the number of sites of the finished cluster.

The random walk, however, is not a realistic walk to deal with while dealing with the formation of polymers. For example, it does not take into account excluded volume effects. A more realistic walk is the self-avoiding walk (SAW). In the SAW one considers all the configurations that do not intersect themselves. All the walks have, by definition, equal weight. Thus we use the same formalism that we introduced for lattice animals. For two dimensions one knows that $\nu = 3/4$ and $\gamma = 43/32$.[15]

Kinetic growth walk (KGW). In the KGW,[16] instead of considering all the possible configurations as in the SAW, one *grows* walks with the self-avoiding condition. However we run into the previously discussed problem of clusters having different weights depending on their shape or their history. For example we can have clusters of the same N but with different weights as shown in Fig. 9a or clusters of the same N, the same shape, but with different histories and different weights as shown in Fig. 9b. As in the static equivalent, a KGW cannot retrace its path or crosslink. Thus several regions in the lattice may be closed to it because the path to that region is blocked by the trace of the same walk. This phenomenon is known as steric hindrance.

Steric hindrance, among other things, changes the growth process and can lead to the trapping of the growth walk. Trapping occurs when a growth walk finds that all its paths are blocked due to the traces of previous steps. This problem is especially acute in two dimensions, where a walk has a finite probability of returning to the region in which it started out. As one goes to higher dimensions this effect tends to become unimportant.

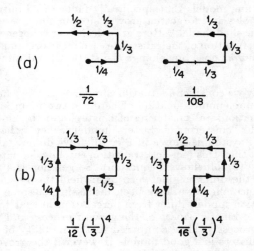

Fig. 9: Clusters of KGW. At each individual bond the probability occurrence is marked. The overall weight of a configuration, i.e., the product of the individual probabilities, is given below each configuration.

The probability of growing an N site cluster is identical to the probability P_N that a walk will not trap in N steps. Again the asymptotic behavior

$$P_N \sim \mu^N N^{\gamma-1}$$

is assumed. This yields $\gamma < 1$, as opposed to the Eden model, $\gamma = 1$. This difference of value of γ is solely due to trapping effects, i.e., the long range memory effect of the steric hindrance.

DLA. In a model like DLA, in principle all configurations are possible. However as a particle diffuses from outside the cluster, it has a greater probability of sticking to an external perimeter site than to an interior perimeter site. Thus the outer sites prevent particles from reaching sites in the interior. This effect is known as screening and is the steric hindrance effect of the arms of the cluster with respect to the diffusing particle.

Fjords, once formed in the interior of a DLA cluster, tend to stay—thus leading to a ramified cluster containing fjords of all sizes (see Fig. 2b). Clusters like DLA are "unfinished" clusters since one can never be sure that a diffusing particle will not by chance come into the interior at a later time. The growth sites in the interior are called forgotten growth sites because it is improbable that a diffusing particle will occupy them. One may then ask the question how to calculate best the fractal dimension of an unfinished cluster. Pragmatically one can simply define a waiting period and if a part of the cluster does not change over that period one considers it to be a fully-grown part and then calculates its properties like the fractal dimension.

One can also define models that specifically favor rapid outward growth, thus leaving many holes and forgotten growth sites in the interior. One such model is the epidemic shell model.[17] Here growth sites are chosen with a probability $P(r)$, which is a function of their distance r from the original seed particle: $P(r) \propto r^\alpha$. This *dynamics* does not change the static fractal dimension of the cluster.

Gelation. Finally we consider a situation similar to that we have described earlier, i.e., the phenomenon of gelation. There are two main kinetic models that mimic gel formation: (a) clustering of clusters and (b) kinetic gelation. We have described the "clustering of clusters" model earlier. Schematically the time evolution of the system is shown in Fig. 10. If we do not take steric hindrances into account then this (sol-gel) transition can be adequately described by mean-field theory (Smoluchowski equation). So gelation is a nice example demonstrating that the difference between mean field theory and reality lies in the steric hindrance effect we have been discussing at some length. The two mentioned gelation models differ from percolation in that they explicitly take into account the kinetic aspect of the growth process and the mobility. When the growth process is slow compared to the mobility of the clusters, then clustering of clusters is a good model. However if the growth process is extremely fast compared to the mobility of clusters, a kinetic gelation model is better.

In a kinetic gelation model[18] we start with a lattice. Each site is seeded with a monomer capable of forming a certain number of bonds. The growth process is started by randomly sprinkling a certain number of initiators on the lattice. Each of these initiators attaches itself to a monomer, breaking one of its double bonds and thus unsaturating it. Each unsaturated monomer thus has a free electron pair which then randomly criss-crosses the lattice, breaking double bonds and creating links between various monomers. Interlinked monomers are defined as clusters. Clusters grow and merge, finally creating in addition to several finite cluster, an infinite cluster (gel) that spans the entire lattice. In essence kinetic gelation is very similar to the mole's labyrinth model.

In this model of kinetic gelation, which realistically describes for instance the polymerization of polyacrylamide, steric hindrance effects of various kinds show up. Since the free electrons form long chains by some kind of random walks, the effects of trapping appear. Then clusters cannot crosslink with other clusters far apart but are forced to touch neighbors since they are not transparent. This makes up for the difference with the earlier mentioned mean-field Smoluchowski equations.

SUMMARY

In this lecture we have tried to give you a flavor of the state of the field at the moment. We have seen that there are several models of various degrees of tractability which have been proposed to describe different kinds of growth processes. We have tried to emphasize the link between growth processes and critical phenomena, briefly showing how we can define analogous statistical quantities for clusters. Yet the basic question of finding the source of and being about to predict the diversity of form from the simple growth rules is unanswered. We hope that you, the student, after hearing this set of lectures and those that follow will be motivated to try to apply some new innovative ideas in an attempt to fathom the mysterious relationship between growth and form.

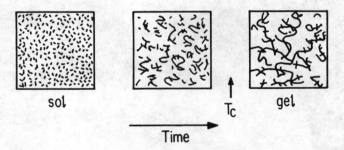

Fig. 10: Schematic representation of the sol-gel transition. In the beginning one has many small clusters. They start coagulating to larger clusters until at the sol-gel transition T_c an infinitely spanning cluster appears for the first time.

[1] M. Eden, Proc. Fourth Berkeley Symp. on Math. Stat. and Prob., ed F. Neyman, Vol. IV (Univ. of Calif. Press, Berkeley, 1961), p. 223.
[2] T. A. Witten and L. M. Sander, Phys. Rev. Lett. **47**, 1400 (1981).
[3] H. J. Herrmann, J. Phys. A **16**, L611 (1983).
[4] M. Kolb, R. Jullien and R. Botet, Phys. Rev. Lett. **51**, 1119 (1983).
[5] P. Meakin, Phys. Rev. Lett. **51**, 1123 (1983).
[6] D. Stauffer, Phys. Rep. **54**, 1 (1979).
[7] G. Parisi and N. Sourlas, Phys. Rev. Lett. **46**, 871 (1981).
[8] J. P. Nadal, B. Derrida and J. Vannimenus, Phys. Rev. B **30**, 376 (1984).
[9] M. Plischke and Z. Rácz, Phys. Rev. Lett. **53**, 415 (1984).
[10] A. Bunde, H. J. Herrmann, A. Margolina and H. E. Stanley, Phys. Rev. Lett. **55**, 653 (1985).
[11] R. Jullien and R. Botet, Phys. Rev. Lett. **54**, 2055 (1985).
[12] D. Richardson, Proc. Cambridge Phil. Soc. **74**, 515 (1973).
[13] F. Leyvraz and H. E. Stanley, Phys. Rev. Lett. **51**, 2048 (1983).
[14] H. J. Herrmann and H. E. Stanley, Z. Phys. B **60**, xxx (1985).
[15] B. Nienhuis, Phys. Rev. Lett. **49**, 1063 (1982).
[16] I. Majid, N. Jan, A. Coniglio and H. E. Stanley, Phys. Rev. Lett. **52**, 1257 (1984); J. W. Lyklema and K. Kremer, J. Phys. A **17**, L691 (1984).
[17] A. Bunde, H. J. Herrmann and H. E. Stanley, J. Phys. A **18**, L523 (1985).
[18] H. J. Herrmann, D. P. Landau and D. Stauffer, Phys. Rev. Lett. **49**, 412 (1982).

Antonio Coniglio, Naeem Jan and Mike Shlesinger

FORM: AN INTRODUCTION TO SELF-SIMILARITY AND FRACTAL BEHAVIOR

H. Eugene Stanley*

Center for Polymer Studies and Department of Physics
Boston University
Boston, Massachusetts 02215

We will organize this talk around three questions, questions that could be used to organize an introduction to almost any subject.

QUESTION ONE: WHAT IS THE PHENOMENON?

Exact Fractals

Consider an unusual two-dimensional object, the Sierpinski gasket (Fig. 1). We take a photograph of a triangle. On a large length scale L it appears to be solid but as we enlarge the negative (examining it on finer length scales), we detect the fact that it has holes that occur on all length scales. The density thus decreases when the length scale changes.

This behavior is certainly very much against our normal ideas of Euclidean geometry. Objects can be either solid (density one) or full of holes (density below one), but in both cases the density is *independent* of the length scale on which it is measured. If a truck full of donuts has mass M, then a truck whose cargo space is increased in edge from L to $2L$ will carry eight times the mass:

$$M(2L) = 8M(L) = 2^3 M(L). \tag{1a}$$

In general, for any positive number λ we can write

$$M(\lambda L) = \lambda^d M(L). \tag{1b}$$

---

* Based in part on notes taken by C. Amitrano and R. L. Blumberg Selinger.

Equation (1a) is a functional equation. The solution is obtained by setting $\lambda = 1/L$:

$$M \sim L^d. \tag{1c}$$

The density $\rho = M/L^d$ hence scales as

$$\rho \sim L^0. \tag{2}$$

The lower is the density, the smaller are the amplitudes that appear implicitly in (1c) and (2). No matter how low the density, the exponent in (1b) and (1c) is always the Euclidean dimension d.

The above results are so familiar that we do not need to use the formal language of functional equations to understand them. However it is exactly this language that *is* needed to describe fractal objects. For the Sierpinski gasket example of Fig. 1, the mass obeys the functional equation

$$M(2L) = 3M(L) = 2^{d_f} M(L)$$

with

$$d_f = \ln 3 / \ln 2 = \log_2 3. \tag{3}$$

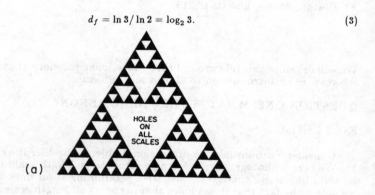

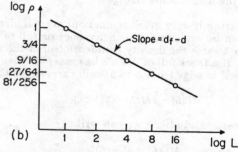

FIG. 1: The prototype "regular" or "exact" fractal, a Sierpinski gasket, shown here for a system of Euclidean dimension $d = 2$. On each iteration the density decreases, a generic feature of fractal objects. After Stanley (1984b).

Similarly, the density obeys

$$\rho(2L) = (3/4)\rho(L) = 2^{d_f - d}\rho(L), \qquad (4)$$

where $d - d_f$ is termed the co-dimension.

In general, fractal objects obey functional equations of the form

$$M(\lambda L) = \lambda^{d_f} M(L) \qquad [d_f < d]. \qquad (5a)$$

The solution of the general functional equation (5a) is

$$M(L) = L^{d_f} \qquad \textbf{[FRACTAL DIMENSION No. 1]}. \qquad (5b)$$

For this exact fractal there are no "correction to scaling terms"–the leading "scaling" term suffices for all values of L.

Statistical Fractals

The simplest example of a statistical fractal is an N-step random walk on, say, a square lattice (Fig. 2a). The walker, a Polya drunk, takes one step per time unit, so that $N = t$. Since there are 4 choices for the drunk for each next step, after a time t there are a total of 4^t distinct configurations. Most configurations are not fractals, yet certain average quantities obey functional equations identical to those obeyed by the Sierpinski gasket. To see this, let

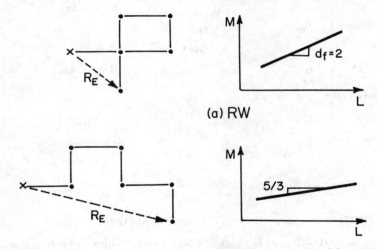

FIG. 2: Schematic illustration of (a) a random walk and (b) a self-avoiding walk, each of $M = t = 6$ steps. We show just *one* of the 4^6 possible walks, most of which have zero weight for the SAW case. Shown also are log-log plots of the relation between the characteristic length scale L (the mean end-to-end distance R_E) and M, namely the number of steps.

TABLE 1. "Rosetta Stone" connecting notation used here with that used by others. In the final column are relations among the 10 fractal dimensions; those relations that are in the conjecture stage are indicated by $\stackrel{?}{=}$.

	This lecture		Other Notation	Relations
Extrinsic Exponents				
(1) $N_f \sim L^x$	$x = d_f$		$x = D, \bar{d}$	(a) $d_f = y_h$
(2) $N_w \sim L^x$	$x = d_w$		$x = 2 + \theta$	(b) $d_w \stackrel{?}{=} \frac{3}{2} d_f$
				(c) $d_w \stackrel{?}{=} d_f + 1$
(3) $N_{min} \sim L^x$	$x = d_{min}$		$x = d_{min}$	(d) $d_{min} \stackrel{?}{=} d_f - d_{red}$
(4) $N_{resistors} \sim L^x$	$x = d_R$			(e) $d_R = d_w - d_f = d_w^{BB} - d_f^{BB}$
(5) $N_{hull} \sim L^x$	$x = d_h$			(f) $d_h \stackrel{?}{=} 1 + d_{red}$
(6) $N_{unscreened} \sim L^x$	$x = d_u$			(g) $d_u \stackrel{?}{=} (d_f - 1) + (d - d_f)/d_w$
(7) $N_{growth} \sim L^x$	$x = d_g$			(h) $d_g = 2d_f - d_w$
(8) $N_{backbone} \sim L^x$	$x = d_{BB}$		$x = d_{BB}$	
(9) $N_{red} \sim L^x$	$x = d_{red}$		$x = d_{red}$	(i) $d_{red} = y_T$
(10) $N_{elastic} \sim L^x$	$x = d_E$			(j) $d_E \stackrel{?}{=} d_{min}$
Intrinsic Exponents				
$N_w \sim (N_f)^x$	$x = d_s$		$\tilde{d}$	$d_s \equiv 2d_f/d_w$
$N_f \sim (N_{min})^x$	$x = d_\ell$		$\hat{d}$	$d_\ell \equiv d_f/d_{min}$

(a) Stanley (1977)
(b) Alexander and Orbach (1982)
(c) Aharony and Stauffer (1984)
(d) Havlin and Nossal (1984)
(e) Stanley and Coniglio (1984)
(f) Sapoval et al (1984)
(g) Coniglio and Stanley (1984)
(h) Stanley (1984a,b)
(i) Coniglio (1981)
(j) Herrmann and Stanley (1985)

us use for the length scale L the "range" of the walk–the Pythagorean distance from the origin after t steps, averaged over the ensemble of all t step walks. Then $L(t=1) = 1$ for the four 1-step walks, $L(2) = 2^{1/2}$ for the 16 2-step walks, $L(3) = 3^{1/2}$ for the 64 3-step walks, etc. In general,

$$L(t) = t^{1/2}. \qquad (6a)$$

The proof that (6a) holds for walks on any d-dimensional hypercubic lattice is an exercise for the student.

To make the formal correspondence stronger, we can imagine that after each step is placed a monomer, and the steps correspond to chemical bonds between monomers. Then the random walk traces out a "free-flight" polymer which ignores the Archimedes principle that two objects cannot occupy the same point of space. The total polymer mass M after t steps is proportional to t, so the mean Pythagorean distance from beginning to end of a polymer with N monomers is related to total polymer mass by

$$t \sim M(L) \sim L^2. \qquad (6b)$$

Clearly (6b) is identical in structure to (5b), suggesting that we may speak of a fractal dimension of a random walk. Hence, we write (Table 1).

$$M(L) \sim L^{d_w} \qquad \text{[FRACTAL DIMENSION No. 2]}, \qquad (7a)$$

with

$$d_w = 2 \qquad \text{[RANDOM WALK]}. \qquad (7b)$$

QUESTION TWO: WHY DO WE CARE?

"First Answer."

There are many answers to this question. The first answer is that *fractals occur in nature*; recognizing that an object is a fractal gives one predictive power if you know the fractal parameters characterizing it. The number of such objects is vast. Since even the one realization of fractals in nature that I have been involved with (Nittmann et al 1985; Daccord et al 1985a,b) is already treaded elsewhere at this school, I will resist the temptation to "pull out the family album" and show you pictures of every fractal that I know.

"Second Answer"

A second reason for studying fractals is the possibility of *gaining insight into extremely basic questions regarding the theory of critical points*. We illustrate this fact by citing one of the many examples where studying geometric phase transitions has given us insight into the nature of thermal phase transitions, *percolation*. There is compelling numerical evidence that percolation obeys Widom-Kadanoff scaling (Nakanishi and Stanley 1980,1981 and references therein). That is, the "Gibbs potential" $G(h,\epsilon) = \Sigma_s n(s,\epsilon)\exp(-sh)$ is a generalized homogeneous function, obeying a functional equation that for two variables is an obvious generalization of the one-variable homogeneity relation (5a):

$$G(\lambda^{y_h} h, \lambda^{y_T} \epsilon) = \lambda^d G(h, \epsilon). \tag{8a}$$

Here $n(s,\epsilon)$ is the number of s-site clusters per lattice site, and the two numbers y_h and y_T are called scaling powers. All critical exponents in percolation can be expressed in terms of these two scaling powers, so it would be nice if they had a simple geometric interpretation. The first paper introducing fractals into percolation noted in passing that

$$y_h = d_f, \tag{8b}$$

and wondered "what cluster property could be related to y_T" (Stanley 1977). Four years later it was found that

$$y_T = d_{red}, \tag{8c}$$

based on Monte Carlo calculations for d=2 (Pike and Stanley 1981) and exact arguments valid for all d (Coniglio 1981). Here d_{red} is the fractal dimension for the singly-connected bonds. Thus the Kadanoff scaling powers are simply given in terms of fractal dimensions, at least for percolation!

WHAT DO WE ACTUALLY DO?

A simplistic answer to this question is as follows. If you are an experimentalist, you try to measure the fractal dimension of things in nature. If you are a theorist, you try to calculate the fractal dimension of models chosen to describe experimental situations; if there is no agreement then you try another model.

In reality, this simplistic answer describes research in many fields. In the present subject, however, we can be usefully guided by the previous example of other scale-invariant phenomena, namely the study of critical phenomena. There the first task was the recognition that there were a denumerable number of scale-invariant quantities, so that a separate critical exponent could be assigned to each.

We shall see that "the" fractal dimension d_f is not sufficient to describe these phenomena. For example, percolation and DLA both have $d_f = 2.5$ when $d = 3$, yet any child can immediately see that these two fractals look completely different from each other. Hence there must be some other fractal parameter that differs. Later in this lecture we shall see that we must introduce a new fractal dimension, d_{min}, to describe the tortuosity of the fractal. We will see that $d_{min} = 1$ for DLA and $d_{min} > 1$ for percolation. In fact, by the time these lectures have ended, we shall have found it necessary to introduce a total of ten distinct fractal dimensions. Although this may sound overwhelming at the present time, we shall see that these arise in as natural a fashion as the ten or so distinct critical point exponents that most students know and use regularly. Just as the critical point exponents were found to be not all independent of one another, so also the ten fractal dimensions are not all independent quantities. Rather, they are related by simple relations not altogether unlike the scaling laws relating critical exponents.

Since one can introduce fractal dimensions that play the role of critical exponents, and since relations among the dimensions play the role of scaling laws, it is natural to ask "what about renormalization group?" It turns out that one can develop a very successful renormalization group for geometrical objects. No Hamiltonian appears, yet the essential idea of "renormalization" of critical parameters upon successive re-scalings still holds. In these lectures we will not have time to go into the renormalization group work in detail, so we refer the interested reader to a recent review (Stanley et al 1982).

Our first task is to clearly define the term fractal dimension. Suppose we plot on log-log paper against a characteristic length scale L some quantity q that can be interpreted as a "number of objects" (equivalently, a "mass"). If there is an asymptotic (large-L) region in which this plot becomes straight, then the slope is termed the fractal dimension d_q characterizing that quantity. The notation d_q, unlike other notations, has the advantage of being completely unambiguous: it is *always* the slope of a log-log plot of q against L (Table 1).

Let us illustrate the utility of our notation by deriving a simple expression for the spectral dimension. Suppose we wish to know the mean number of sites visited when a random walker stumbles around randomly on a fractal substrate. De Gennes (1976) has termed this problem the "ant in a labyrinth" but a "drunk in a gulag" might seem more descriptive of the actual picture

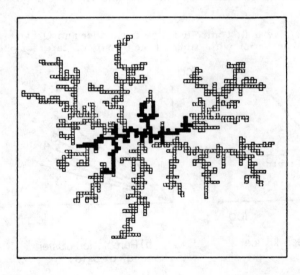

FIG. 3: A random walk of 2500 steps on a DLA fractal substrate with 1000 sites. The sites visited by this walk have been indicated by solid squares. This is a visualization of the de Gennes "ant in a labyrinth" problem. After Meakin and Stanley (1983).

[Fig. 3]. A random walker (drunken ant or drunken prisoner) has been parachuted onto a randomly-chosen site, which is then colored black. At each successive time step, the ant moves to a neighboring fractal site, and that becomes black also. At each time step t, the ant (*la fourmi*) calculates her range L, the rms displacement from the local origin where the parachute landed. Recall from (6b) that the fractal dimension d_w determines how t scales with L. We anticipate that d_w is considerably *larger* than 2 since many of the neighbors at each step are unavailable to the ant so she is obliged to return in the direction of her parachute point. Hence L increases much less fast with t than for an unconstrained RW (Fig. 4).

Now let us ask how the number of black visited sites (one "mass") scales with the time (a different "mass"). The ant fills compactly a region of the fractal, so the first, "black" mass scales with fractal dimension d_f. The second scales with fractal dimension d_w. Hence that the number of visited sites scales with time as the *ratio* of the two fractal dimensions,

$$<s> \sim t^{d_f/d_w}. \qquad (9a)$$

Note that one additional advantage of our proposed notation is that one sees by inspection that the critical exponent in (9a) is just the ratio of the fractal dimension of the two variables s and t that appear in (9a). In short, it is possible to "write down exponents" by inspection, and virtually impossible to make mistakes.

The quantity d_f/d_w was first introduced by Alexander and Orbach (1982) (henceforth AO) in connection with studying the density of states for phonon-

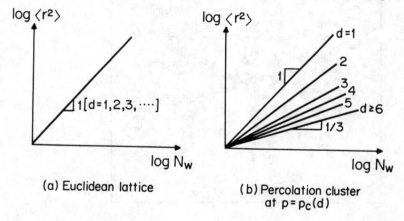

FIG. 4: Schematic illustration of the dependence of the mean square displacement on the number of steps of a random walk, plotted double logarithmically, for (a) a Euclidean lattice, and (b) a percolation cluster at the percolation threshold. For d above 6, the slope sticks at the $d = 6$ value of $1/3$—so that the fractal dimension of the walk sticks at $d_w = 6$.

like excitations on a fractal substrate. The quantity that appears is

$$d_s = 2d_f/d_w \quad \text{[INTRINSIC FRACTAL No. 1]}, \quad (9b)$$

and is called the fracton or spectral exponent. AO tabulated the quantities d_f, d_w and d_s for percolation fractals on a d-dimensional Euclidean lattice. They noted that while d_f and d_w change dramatically with d (below $d_+ = 6$), d_s does not. Hence they made the now-famous "AO conjecture" that for percolation,

$$d_s = 4/3 \quad [2 \le d \le 6]. \quad (10)$$

The phrase hyper-universal or super-universal was coined to express the possibility that an exponent could be independent not only of lattice details (as in ordinary universality) but also of d itself. Shortly after the AO conjecture was published, several studies were published that seemed to confirm it to higher accuracy (Pandey and Stauffer 1983; Havlin and Ben-Avraham 1983). Other studies tested whether the AO conjecture might hold for other random phenomena, such as random walks on DLA clusters (Meakin and Stanley 1983) or diffusive annihilation between random walkers who are constrained to move on a percolation fractal substrate (Meakin and Stanley 1984).

In the past few months, four independent groups have decided to put the AO conjecture to a very severe test. Using completely different computational methods all four groups concluded that d_s is about 2% larger than 4/3.

In a way, it is a pity the AO idea does not work, since if it did we would have a relation between two fractal dimensions, one of which describes "statics" (d_f) and the other of which describes "dynamics" (d_w). Accordingly, several workers are attempting to find another relation to replace (10).

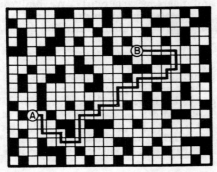

FIG. 5: Schematic illustration of the minimum path followed by a soldier in carrying an important message from General X (at position A) to General Y (at position B) in a battlefield that has been randomly infested by mines in sufficient concentration that the pair connectedness length ξ is larger than the Pythagorean distance L between A and B. This figure is adapted from Ritzenberg and Cohen (1984), where it was used to model the spread of electrical excitations in the heart tissue.

THE FRACTAL DIMENSION OF THE MINIMUM PATH

You are a soldier in a mine-infested battlefield. Your boss, General X, wants you to carry a message to his boss, General Y, who is 10 km. distant (Fig. 5). You do this, and note that it takes one hour. General Y asks you to carry the same message to his boss, General Z, who is 20 km. away. It is critical that the message be delivered within two hours. You tell the impatient General Y that it will, statistically, take *more* than two hours to travel twice the distance. You explain that when land mines are placed at random (by the presumably drunken enemy troops), they form clusters of all possible sizes. Hence the minimum path between two generals is a convoluted path since it must avoid these clusters. If the Pythagorean distance L is short, there is little probability of encountering a huge cluster. Since General X and General Z are separated by $2L$, it is possible that you will encounter *much* larger mine clusters and hence your path length will *more* than double. More precisely, you tell your boss that the minimum path ℓ obeys the functional equation

$$\ell(\lambda L) = \lambda^{d_{min}} \ell(L), \qquad (11a)$$

where $d_{min} \simeq 1.1, 1.3$ for $d = 2, 3$ percolation fractals (Herrmann et al 1984). The solution to (11a) is found in setting $\lambda = 1/L$,

$$\ell \sim L^{d_{min}} \qquad \text{[FRACTAL DIMENSION No.3]}. \qquad (11b)$$

Can one relate the exponent d_{min} to the other exponents? To do this, it is first important to develop some feel for this quantity. To this end, we have embarked on a program of calculating d_{min} for various fractal substrates, including percolation, lattice animals, DLA, and cluster-cluster aggregation (see Meakin et al 1984 and references therein). We find that in all cases except DLA (Fig. 6), d_{min} increases monotonically with d up to the critical dimension. At and above the upper marginal dimension d_+,

$$d_{min} = 2 \qquad [d \geq d_+]; \qquad (11c)$$

this means that the minimum path between two points in a fractal above the critical dimension has the *same statistics as a random walk* or Gaussian chain. Thus we can understand the critical dimension in a different fashion: all fractals have different values of d_f, d_w, etc. above d_+ but all have $d_{min} = 2$. As d is decreased, this continues to hold until at $d = d_+$, the shortest path between two points is "straighter" than a random walk, and $d_{min} < 2$. That $d_{min} = 1$ for DLA (Meakin et al 1984) is consistent with the idea that DLA has no upper critical dimension.

The utility of d_f in providing a quantitative characterization of a fractal form was discussed above–so also the utility of d_w in characterizing diffusion ("random walks") on a fractal substrate. What is the utility of d_{min}? Clearly d_{min} applies to physical phenomena that propagate *efficiently* from site to site, not re-visiting previous sites. In a rough sense, we can say that d_w describes random walk propagation where re-visiting fractal sites is possible while d_{min} describes a sort of self-avoiding walk where re-visits are not. Our "runner" finds the minimum path from General X to General Y by considering all possible self-avoiding walks from X to Y and choosing the shortest.

Physically, this is the same as imagining that every cluster site of the fractal had a tree on it. If the system is near the percolation threshold, then the trees form a self-similar substrate for length scales up to the connectedness length $\xi = (p - p_c)^{-\nu}$. Now at time $t = 1$ let us ignite the trees on the site occupied by General X. At $t = 2$, we ignite the trees on the neighboring sites of X, and so forth until after some time delay the site occupied by General Y is ignited. This time delay is of course simply the minimum path ℓ between X and Y, and we have described a very elaborate procedure of determining ℓ which consists of igniting successive "chemical shells" around General X.

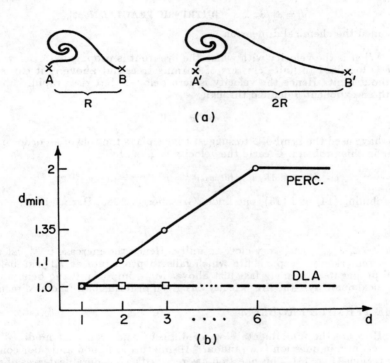

FIG. 6: (a) Schematic illustration of the significance of the fact that d_{min} is above 1: If the distance between points A and B is doubled, then the shortest path length increases by a factor $2^{d_{min}}$, which is larger than 2. (b) Numerical values of d_{min} for percolation (top curve) and DLA (bottom curve). The fact that d_{min} sticks at 2 above the upper marginal dimensions d_c means that the shortest path between two is a random walk. The fact that d_{min} does not seem to approach 2 for DLA means that there is no d_+ for DLA. This figure is based on data in Meakin et al (1984).

Having described the forest fire mechanism, we can now introduce a new intrinsic fractal dimension. Suppose we ask for the total mass of burning trees at time t. From the definition of the fractal dimension d_f, we have

$$M(t) \sim L^{d_f}, \qquad (12)$$

where $R = R(t)$ is the radius of gyration (or any length such as the caliper diameter, that scales with the radius of gyration). We asked for the dependence upon t of $M(t)$, where $t = \ell$ is the minimum path length between the newest shell of burning and the origin of the fire. Combining (11b) and (12), it follows that

$$M(\ell) \sim \ell^{d_\ell} \qquad [\ell \ll L^{d_{min}}]. \qquad (13a)$$

Here

$$d_\ell = d_f/d_{min} \qquad \text{[INTRINSIC FRACTAL No. 2]}, \qquad (13b)$$

is called the chemical dimension.

What is the velocity with which the fire front is propagated? Below p_c the fire is localized to finite clusters of burning trees but above p_c it can spread without limit. Hence the velocity is zero below p_c but rises rapidly above p_c with an exponent β defined through

$$v \sim (p - p_c)^\beta. \qquad (14)$$

We have used the symbol β to suggest that v plays the role of an order parameter in this problem. Clearly the velocity is given by

$$v = dR/dt = dR/d\ell = (d\ell/dR)^{-1} = R^{1-d_{min}} = (p - p_c)^{\nu(d_{min}-1)}. \qquad (15)$$

Combining (14) and (15), one finds (Grassberger 1985, Barma 1985)

$$\tilde{\beta} = \beta/\nu = d_{min} - 1. \qquad (16)$$

For $d = 2$, $d_{min} = 1.10$, very close to unity. Hence the increase in v just above p_c is remarkably steep: a fire which fails to propagate at all just below p_c will propagate extremely fast just above. More important, the behavior of a physical quantity—the fire velocity—is given in terms of a fractal dimension!

THE TERMITE PROBLEM

How are the fundamental laws of diffusion and transport modified when the medium in question is a random "AB-mixture" of good and poor conducting regions (Fig. 7)? This question has received a considerable degree of recent attention for two limiting cases: (i) The random resistor network (RRN)–or pure "ant" limit–for which B, the poor conducting species, has zero conductance, and (ii) The random superconducting network (RSN) or pure "termite" limit, for which A, the good conducting species, has infinite conductance.

The terms "ant" and "termite" arise from the fact that one can replace the conductivity problem with a diffusion problem using the Nernst-Einstein relation. For the RRN limit, no diffusion can occur on the component with zero conductance, so the constrained diffusion problem is rather like an "ant in a labyrinth" (de Gennes 1976). For the RSN limit, the diffusion can occur everywhere since both components conduct, but the fact that the good conductor species has zero resistance means that the diffusion is remarkably different in this region than elsewhere. Some years ago de Gennes (1980) invented the term "termite diffusion" to describe this subtle phenomenon. However to this date there has been no clear statement of exactly how to properly define or measure this phenomenon, in contrast to the "ant" limit where the diffusion is simply constrained to one component. There are many reasons for the current upsurge of interest in this problem.

(i) One reason is that there are many experimental systems that are random and inhomogeneous. For example, a rock is composed of tiny grains of different conductivities (to heat, to fluid flow, to electricity). To the extent that such inhomogeneous materials are also random, we may think of using a site-random description of this material: a "lattice-gas" description. One first coarse grains the material and then assigns to each cell one of two conductivities, σ_a and σ_b. Calculations based upon such a straightforward approach have been usefully compared with a wide range of experiments, from conductivities of thin films of lead depositions on an insulating substrate (roughly the RRN limit) to thin films of superconducting material vacuum deposited on a normal substrate (roughly the RSN limit). Moreover, ionic conductors mixed with a dispersed insulating phase represent random heterogeneous materials, where both limits seem to play an important role.

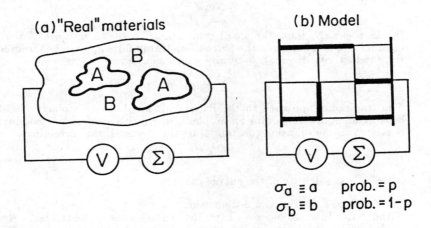

FIG. 7: (a) Schematic illustration of a random two-component composite material before coarse-graining. (b) Replacement by equivalent random network with two conductances $a = \sigma_a$ [probability p] and $b = \sigma_b$ [probability $1 - p$].

(ii) A second reason is related, perhaps, to the reason why the Ising model has always been of great interest: it is an extremely simple model that captures the essential physics of a realistic system in nature. The analog of the Ising model for random inhomogeneous materials is a mixture of sites (or bonds) randomly distributed on a lattice–see, e.g. , Fig.1b. The sites (or bonds) are assumed for simplicity to have only two possible values of the conductance,

$$\sigma = \begin{cases} \sigma_a & [\text{probability } p] \\ \sigma_b & [\text{probability } 1 - p] \end{cases}. \qquad (17)$$

By convention, we choose $\sigma_a > \sigma_b$, so that the ratio $h = \sigma_b/\sigma_a$ is always less than unity.

Conventionally, one wants to know the *macroscopic* magnetization of an Ising ferromagnet composed of elements (spins) whose *microscopic* property is a two-valued variable. Similarly, we *now* want to know the *macroscopic* conductivity which depends on all possible configurations of the *microscopic* elements (conductors) whose property is again a two-valued quantity (σ_a and σ_b). Just as the magnetization couples to a conjugate field H, the conductivity couples to a conjugate field h.

The two limiting cases mentioned above can now be discussed more precisely: (a) In the RRN limit, the large conductance is set to unity and the small conductance is set to zero. As the percolation threshold p_c is approached from above, the macroscopic conductivity approaches zero with a critical exponent μ,

$$\sum \sim (p - p_c)^\mu. \qquad (18a)$$

(b) In the RSN limit, the small conductance is set to unity, and the large conductance is infinite. As the percolation threshold is approached from below, the conductivity diverges to infinity with an exponent $-s$

$$\sum \sim (p_c - p)^{-s}. \qquad (18b)$$

The traditional approach to the RRN limit has been to replace Kirchhoff's laws by an equivalent diffusion problem, where the macroscopic conductivity is related to the diffusion constant D by the Nernst-Einstein relation,

$$\sum \sim nD, \qquad (19)$$

where n is the density of the charge carriers.

We place a walker on a d-dimensional lattice made of two kinds of bonds, A and B (for illustration: $d = 1$ here, the general-d case is discussed in Hong et al 1985). The walker carries two coins, weighted and unweighted, and a clock. Without loss of generality, let the origin be well inside a high-conductivity A region. At each tick of the clock, the walker tosses the unweighted coin and moves to the left or right depending on the outcome of the coin toss. When the walker comes to a site on the boundary between the A region and the B region, he tosses the other coin that is weighted with probability (Fig. 8)

$$P_a = f_a/(f_a + f_b) = 1/(1 + h), \qquad (20)$$

to stay in the A region, and a probability

$$P_b = f_b/(f_a + f_b) = h/(1+h), \tag{21}$$

to go outside into the B region. In the event that the walker steps outside the A region, then he must *slow* down by the ratio f_a/f_b. For example, if the conductivity of the B region is 10 times smaller than that of the A region, then f_b is 10 times smaller than f_a ($h = 0.1$) and the walker steps *only* after every 10 ticks of his clock.

Limiting cases of our random walk model are as follows:

(i) $h = 1$. There is no distinction between regions, no reflection on the boundaries ($P_a = P_b$), and no difference in walk speed on and off the A clusters.

(ii) $h \ll 1$. The walker now moves at one step per clock tick when he is on an A cluster, and is almost always reflected when he comes to the boundary. Extremely rarely he passes out of an A region and into a B region, whereupon he walks much, much slower—taking a new step only after his clock has made h^{-1} ticks. Statistically speaking, in a very large time $\gg h^{-1}$, the walker peforms $O(f_a)$ moves in the A region and $O(f_b)$ moves in the B region.

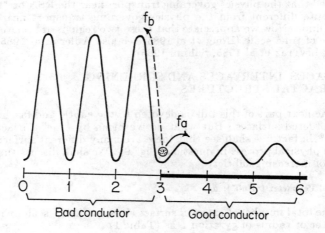

FIG. 8: Schematic illustration of our random walk model for a one-dimensional lattice, showing the presence of a bondary between a good-conductor cluster of conductors σ_a and a poor-conductor cluster of conductors σ_b. The corresponding jump frequencies f_a and f_b determine, through Eq. (21), the probability of a particle being reflected at the boundary. After Hong et al (1985).

Suppose we make a motion picture of the walker's motion. Then we see that the walker is reflected from the walls almost all of the time, and only very rarely—roughly once per h^{-1} trials—will come outside the cluster (see Hong et al 1985). When this does occur, his motion will slow down by a factor of h. If we watch this motion picture, perhaps we become impatient watching the walker in the B region and we speed up the motion picture projector by a factor of $1/h$ so that the walker is now taking *one* step per unit of time while in the B region. Then we are no longer impatient while the walker is in the B region. However, when he finally encounters an A cluster, he moves onto it with a high probability, $1/(1+h)$, and proceeds to move about the A cluster with a motion that is *also sped up by the same factor* $1/h$. Thus the orginal normally on an A cluster and *extremely slowly* on B clusters has suddenly been transformed into a "termite" who moves normally on B clusters and *extremely fast* on A clusters. *Indeed, the only difference between the two domains, "ant" (RRN domain) and "termite" (RSN domain), is the definition of the time scale.* This simple observation can be formalized in terms of a rigorous transformation (Hong et al 1985). That transformation in turn forms the basis of the scaling laws for the ant and termite limits of the general two-component random mixture.

Thus the two-component random mixture requires for its treatment the understanding of how to handle a diffusion process to which there are two time scales, not one. This problem has not been treated previously and is proving to be quite subtle in many respects. Until quite recently it was widely believed that the physics governing transport near the RSN or "termite" limit was quite different from the physics governing transport near the RRN or "ant" limit. Now we appreciate that these two regions are related by a simple change of time scale (Hong et al 1985; see also Adler et al 1985, Bunde et al 1985c, Leyvraz et al 1985, Sahimi 1985).

SURFACES, INTERFACES AND SCREENING OF FRACTAL STRUCTURES

The next part of this talk is devoted to the subtle and fascinating subject of disordered surfaces. But what do we mean by "the" surface of a fractal object? In fact, we shall see that there are many different surfaces, depending on the physical process in question (Fig. 9). We shall discuss these roughly in order of increasing subtlety.

External Perimeter ("Hull"): d_h

The total number of *external* surface sites, or "hull," scales with the caliper diameter or radius of gyration L as (Table 1)

$$N_{hull} \sim L^{d_{hull}} \quad \text{[\textbf{FRACTAL DIMENSION No. 4}]}. \quad (22)$$

For $d = 2$ percolation, d_{hull} appears to be about 1.74 ± 0.02 (Voss, this conference; Sapoval et al 1984), thus motivating the conjecture

$$d_{hull} = 1 + d_{red} = 7/4, \quad (23)$$

since $d_{red} = 1/\nu = 3/4$ exactly for $d = 2$.

Total Perimeter: d_f

We know that the total number of perimeter sites $N_{perimeter}$ scales in the same fashion as the total number of cluster sites,

$$N_{perimeter} \sim N_f \sim L^{d_f}, \tag{24}$$

for d-dimensional percolation (Kunz and Souillard 1978) where, for $d = 2$, $d_f = 91/48 = 1.896$. Hence the fact that $d_{hull} < d_f$ means that the ratio of *external* perimeter sites to *total* perimeter sites approaches zero at the percolation threshold. As the clusters get larger the internal perimeter sites ("lakefront" sites) completely swamp the external perimeter sites ("oceanfront" sites). For $d > 2$, internal perimeter sites are less commonly seen in finite computer simulations since it takes a lot of cluster sites to completely surround a "3-dimensional lake." An open question is the value of d_{hull} for $d > 2$: could it be that $d_{hull} = d_f$? Work is underway test this possibility.

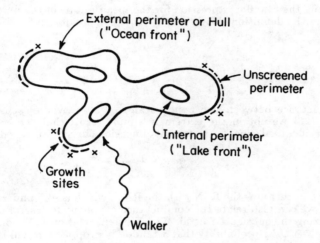

FIG. 9: Schematic illustration of four different fractal surfaces arising in the description of a percolation cluster. (a) The external "oceanfront" parameter or hull has a fractal dimension d_h. (b) The total perimeter has a fractal dimension d_f, equal to that of the total bulk mass of the cluster. Since $d_f > d_h$, it follows that the internal "lakefront" perimeter must have the same fractal dimension d_f of the total perimeter. (c) The unscreened perimeter where an incoming walker is more likely to hit has fractal dimension d_u (heavy solid lines). (d) The growth sites are those perimeter sites that form the living frontier of the cluster. These have fractal dimension d_g, but the nature of the G-site fractal depends on the actual mechanism of how the percolation cluster grows (see, e.g., Stanley et al 1984; Bunde et al 1985a,b; Herrmann and Stanley 1985).

Unscreened Perimeter: d_u

Coniglio and Stanley (1984) introduced the concept of the "unscreened" perimeter to describe that portion of the hull that is effective in termite motion:

$$N_{unscreened} \sim L^{d_u} \quad \text{[FRACTAL DIMENSION No. 5]}. \tag{25}$$

Moreover, they showed that the conductivity exponent for the RSN is simply related to d_u. Recalling the Nernst Einstein relation, we have

$$\sum \sim D \sim (R_{cluster})^2 \tau^{-1}. \tag{26a}$$

Here the jump frequency τ^{-1} scales as the fraction of cluster sites belonging to the unscreened perimeter,

$$\tau^{-1} \sim [N_{unscreened}/N_f]. \tag{26b}$$

Substituting the Stauffer expression for the mean radius of the finite clusters, $R_{cluster}$, and the definitions for d_u and d_f into (26), we obtain

$$\sum \sim L^{2-(d-d_f)} L^{d_u - d_f}. \tag{27}$$

Since $\sum \sim \epsilon^{-s}$, we have

$$\tilde{s} = s/\nu = d_u - (d-2). \tag{28}$$

The conductance between two points scales as $L^{\tilde{\zeta}_{RSN}}$ where $\tilde{\zeta}_{RSN} = \tilde{s} + (d-2)$. From (28) we obtain the extremely simple result that the conductance exponent is identical equal to the fractal dimension of the unscreened perimeter

$$\tilde{\zeta}_{RSN} = d_u. \tag{29}$$

Let us contrast now the RSN with the RRN, which is well understood. For the RRN, we can also relate the conductance exponent to fractal dimensions characterizing the substrate fractal. The Einstein relation (19) holds, but we must set $n = P_\infty$, the probability that a randomly dropped ant will land on the incipient infinite cluster. Now P_∞ scales as the co-dimension $(d - d_f)$, and D scales as L^2/time. Hence (Gefen et al 1983; Ben-Avraham and Havlin 1982)

$$\sum \sim P_\infty D \sim L^{d_f - d} L^{2 - d_w}. \tag{30}$$

Recalling that $\sum \sim \epsilon^\mu$ for the RRN limit, we have

$$\tilde{\mu} = \mu/\nu = (d-2) + (d_w - d_f). \tag{31}$$

The conductance between two point scales as $L^{\tilde{\zeta}_{RRN}}$, where $\tilde{\zeta}_{RRN} = \tilde{\mu} - (d-2)$. Hence for the RRN problem, (29) is replaced by

$$\tilde{\zeta}_{RRN} = d_w - d_f. \tag{32}$$

It is convenient to think of the resistance between two points as the "mass" of 1-ohm resistors that we would place in series between the two points in order to have the same resistance (Fig. 10). In this way, we can interpret $\tilde{\zeta}_{RRN}$ as a proper fractal dimension for some specific fractal object (the set of resistors)

$$N_R \sim L^{d_R} \qquad \text{[FRACTAL DIMENSION No. 6]}. \tag{33}$$

From (29) and (32) it then follows that

$$d_R = \begin{cases} -d_u & \text{[RSN]} \\ d_w - d_f & \text{[RRN]}. \end{cases} \tag{34}$$

From (34) we see that both the RRN and the RSN have resistance exponents d_R that are simply expressed in terms of fractal properties of the substrate. The expressions are completely different, of course, since the mechanism of transport is completely different (Fig. 11). From Fig. 11a, we see that the transport is determined by a "fisherman's net" structure, with a mesh size given by the connectedness length ξ. The strands of the net are made of singly-connected "red" bonds and multiply-connected "blue" bonds, the statistics of which will be described shortly. From Fig. 11b, we see that for the RSN just below p_c transport from one bus bar to the other is determined by the motion of charge carriers from one cluster to another—more precisely *out of* the unscreened perimeter of one cluster and *into* the unscreened perimeter of the next.

Thus the clusters in the RSN limit play the role of the nodes in the RRN limit. As one moves close to p_c in the RRN problem, the critical bonds are the singly-connected "red" bonds (the hottest). As one moves close to p_c in the RSN limit, the critical bonds are those bonds on the lattice which—if occupied—would connect two clusters. I have always called these pink (since they are "incipient" red bonds: once occupied they will become red); recently Stauffer suggested the term "anti-red" because in every sense they are the

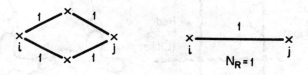

FIG. 10: One example of how a set of bonds (unit resistors) between sites i and j is replaced by an equivalent number (in this case $N_R = 1$) of unit resistors placed in series. Thus the resistance between i and j is equal to N_R, which in turn scales with the Pythagorean distance L between i and j with an exponent d_R (see Eq. (33)).

complement of the red bonds. Thus the red bonds control the physics of the RRN problem just above p_c, while the anti-red bonds control the physics of the RSN problem just below p_c. Coniglio (unpublished) has proved that the anti-red bonds have the same fractal dimension as the red bonds,

$$d_{anti-red} = 1/\nu. \tag{35}$$

Thus there is a certain symmetry between the RRN and RSN limits, which in some way should follow directly from the homogeneity theorems mentioned above. Work on this important topic is underway, and perhaps at this meeting some of you can help make progress along these lines.

Can we evaluate the fractal dimensions d_u and d_w appearing in (18) in terms of the fractal dimension d_f of the underlying substrate? Some progress along these lines has been made using arguments that require for their validity certain assumptions. In this section we will review a mean-field type argument (Coniglio and Stanley 1984) that

$$d_u = (d_f - 1) + (d - d_f)/d_w \quad \text{[CONIGLIO - STANLEY]}. \tag{36}$$

To this end, we must devise a method of probing the surface of a fractal object. The method we chose (Meakin et al 1985a) was to release random walkers, one at a time. When the random walker touched perimeter site i, a counter on site i was incremented by one unit (N_i becomes $N_i + 1$). After typically a million walkers have been released, statistics were done. Our analysis is based on the idea that only a relatively small fraction of the total perimeter will have a large probability of being contacted. Hence to analyze the

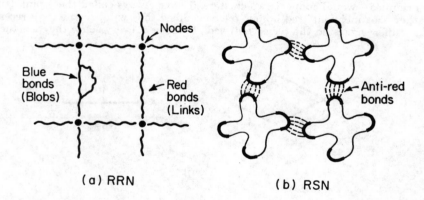

FIG. 11: Schematic illustration of the essential features of cluster structure in describing the conductivity of a general two-component random mixture in the limit of (a) the random resistor network, and (b) the random superconducting network. Adapted from Coniglio and Stanley (1984).

distribution function N_i ($i = 1, 2, \ldots, P$—where P is the total number of perimeter sites), we formed the moments μ_j defined through

$$[\mu_j]^j = \frac{N_T^{j+1}}{\sum_i N_i^{j+1}} = \frac{1}{\sum_i P_i^{j+1}} \sim [N_f]^{j\gamma_j}. \tag{37}$$

Here

$$N_T \equiv \sum_i N_i, \tag{38}$$

is the total number of incoming walkers, and

$$P_i = N_i/N_T, \tag{39}$$

is the probability that a given incoming walker will hit site i. The P_i are normalized to unity by virtue of (38).

First we calculated the γ_j for $j = 1-3$, and found that the Coniglio-Stanley mean field relation (36) was satisfied to within the accuracy of the calculations. We did notice a systematic dependence on j, so to test the possibility that the γ_j depend on j we extended the moment calculation to $j = 8$. The hinted dependence from $j = 1 - 3$ became much clearer (Fig. 12) and so we conclude that there is not a single exponent but rather an entire hierarchy of exponents (Meakin et al 1985a,b). This result has been confirmed by Halsey et al (1985).

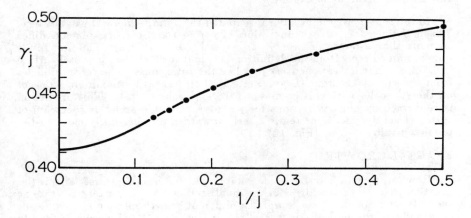

FIG. 12: The exponent γ_j characterizing the behavior of the jth moment of P_i. Here P_i, the basic quantity in surface growth, is the probability that perimeter site i is the next site at which the cluster will grow. In order to see if this apparent hierarchy or "spectrum" of exponents tends in the $j = \infty$ limit toward the expected limit, $1 - 1/d_f$, we have plotted these exponents against $1/j$. From Meakin et al (1985b).

Why is the Coniglio-Stanley relation wrong? Presumably because it smears out the interface or "active zone" of the fractal into a band, and then assumes that there is an equal probability of capture for all surface sites within this band. In reality, there is a continuous gradation in "temperature," with the outermost tips being immensely hotter and the deepest invaginations being extremely cold. This situation is reminiscent of that found by De Arcangelis et al (1985) for $N(V)$, the distribution of the number of bonds in the backbone across which the voltage drop is V. Here also there is a continuous gradation in temperature from the red bonds (the "hottest" in the sense that the full voltage drop of the entire cluster falls across each red bond) to the very cold bonds arising from the very long loops comprising the blobs.

This discovery of an infinite hierarchy of critical exponents–both in the voltage distribution of the percolation backbone and in DLA–is striking because normally one assumes that two exponents will suffice to describe a critical object. For example, we noted above that y_h $(= d_f)$ and y_T $(= d_{red})$ were sufficient to describe percolation. However when we "do something" to the fractal, such as put a battery across it or bombard it with random walkers, we introduce a new measure. Instead of each fractal site having weight 1, each site has a weight that depends on what we are doing to the fractal (e.g., each site has a voltage attached to it, or with each site we associate the number of hits on that site). Several groups (Meakin, Stanley, Coniglio and Witten unpublished; Turkevich and Scher 1985; Halsey et al 1985) are currently seeking to understand the meaning of this new measure and what we can learn from this infinite hierarchy of exponents.

In retrospect, we might have anticipated this infinite hierarchy in advance. This is because for two extreme values of j, $j = -1$ and $j = \infty$, exponents differ by more than a factor of two: $\gamma(-1) = 1$ and $\gamma(\infty) = 1 - 1/d_f$. The first result follows immediately from the definition (41) and the fact that the total surface in DLA scales with exponent d_f just like the total mass. The second follows from the recent theorem (Leyvraz 1985) that P_{max} (the maximum value of all the P_i) scales with cluster mass to the exponent $1 - 1/d_f$. This prediction is confirmed by our calculations for $j = 1 - 8$. Our hierarchy or spectrum of surface fractal dimensions tends clearly toward a number fairly close to the predicted value $1 - 1/d_f$ (Fig. 12).

FRACTAL GROWTH

How can we characterize the fashion in which a fractal grows? This is the question that we shall address in this final section. It is important to state at the outset an obvious fact: completely different growth mechanisms can lead, eventually, to the same static fractal object (see, e.g., the discussion in Bunde et al 1985a). For now, let us consider one of the simplest kinetic mechanisms of growth, the ant. Instead of dropping the ant onto a pre-formed fractal structure as we did before, we could instead drop the ant onto a Euclidean lattice but give her a set of rules with which she could form the fractal as she moves. This means that the ant would need the four-sided coin that all walkers have (taking a square lattice for now), but she would also need some dynamical mechanism of generating the ultimate static fractal. For percolation fractals, all she needs is a weighted coin so that each site can be blocked with probability $1-p_c$. The precise mechanics for the ant motion are described in Bunde et al (1985a).

After some time has evolved, the ant is moving around in a rather interesting region of space that is characterized by 3 sorts of sites (the terminology in brackets suggests an epidemic interpretation of this entire growth problem):

(i) cluster sites already visited ["sick"]
(ii) sites already tested and blocked by the second coin ["immune"]
(iii) neighboring sites that have not been tested ["growth"].

Thus from the reference frame of the ant, only the growth sites are special in the sense that only these enlarge her territory.

Why are the growth sites interesting? Subjectively, they represent the "open frontier" (Rammal and Toulouse 1983) of the growing fractal. Objectively, this disconnected set of sites has a well-defined fractal dimension d_g in the sense that the number or "mass" of growth sites obeys a scaling relation of the form

$$N_g \sim L^{d_g} \quad \text{[FRACTAL DIMENSION No. 7]}. \quad (40a)$$

Here, as always, L represents the cluster diameter or radius of gyration.

The number of growth sites has been calculated as a function of cluster mass N_f. We predict for the intrinsic exponent

$$N_g \sim (N_f)^x \quad \text{with} \quad x = d_g/d_f. \quad (40b)$$

Stanley et al (1984) obtained the first estimates of x, $x = 0.49$ [$d = 2$ percolation].

We now evaluate d_g for the two popular conjectured relations between d_w and d_f. Above, we discussed the Alexander-Orbach [AO] conjecture $d_w = (3/2)d_f$, and we mentioned the Aharony-Stauffer [AS] conjecture $d_w = 1 + d_f$. Thus

$$d_g = \begin{cases} d_f/2 = 91/96 = 0.9479 & \text{[AO]} \\ d_f - 1 = 43/48 = 0.8958 & \text{[AS]}. \end{cases} \quad (41a)$$

For the intrinsic exponent x, we then predict

$$x = \begin{cases} 1/2 = 0.5000 & \text{[AO]} \\ 1 - 1/d_f = 0.4725 & \text{[AS]}. \end{cases} \quad (41b)$$

Thus we see that the calculated values of x fall in between the AO and AS predictions.

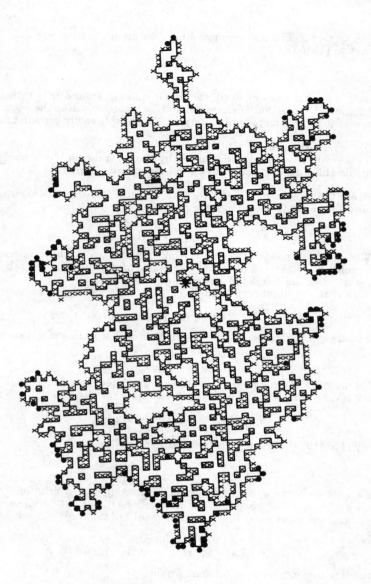

FIG. 13: Typical percolation cluster being grown by the "butterfly" mechanism for the case that each perimeter site has equal weight. This model is equivalent to the Eden model on a "diluted" lattice on which a fraction $1 - p_c$ of the sites have randomly been removed or "poisoned." After Bunde et al (1985a).

Leyvraz and Stanley (1983) considered the conditions under which the AO conjecture might hold, focussing on the need for complete statistical independence of the increments in N_g. They noted that this statistical independence certainly occurs for the Cayley tree, since it is impossible to have correlations on a loopless fractal. In this fashion, they understand the numerical result that AO holds for the Cayley tree $(d_w = 6, d_f = 4)$. In the asymptotic limit of a truly huge cluster—say the size of Corsica—they imagined that correlations in growth sites would begin to vanish even though Corsica has loops. To the extent that these correlations eventually drop off, we can expect that in the asymptotic limit the increments in Ng will be statistically independent and hence the distribution will be normal, with $x = d_g/d_f$ exactly 1/2 [AO].

Recently, the subject of growth sites has arisen in various contexts. One concerns a family of epidemic growth models, all of which eventually grow static percolation clusters (Bunde et al 1985a,b; Herrmann and Stanley 1985). Depending on the growth rules, the dynamic critical exponent dg can differ—even though the static critical exponent d_f is the same for all rules. We considered already one such epidemic model above. A second model is a variation of the above model in which the "ant" visit *only* growth sites, never re-visiting cluster sites. Clearly this requires the ant to make long-range jumps. In the simplest version of this model, the ant randomly chooses her next move from among all the existing growth sites, weighting them equally. This model can be simulated very rapidly (at least 100 times faster than the "walking" ant). Had we set $p = 1$, we would have obtained an Eden cluster, with fractal dimension $d_f = d$. When a fraction $1 - p$ of the growth sites are poisoned, our flying ant acts rather like a butterfly, choosing carefully to land only on a non-poisoned site. A typical cluster at $p = p_c$ is shown in Fig. 13.

FRACTAL GOEMETRY OF THE CRITICAL PATH: "VOLATILE FRACTALS"

How does oil flow through randomly porous material just at that point where it can "break through" and reach the surface? How does electricity flow through a random resistor network where the fraction of intact resistors is close to the percolation threshold p_c? How does one describe the shape of the incipient infinite cluster that appears near p_c when one considers the polyfunctional condensation of monomers? There are but three of a host of questions that one can ask that are of some practical interest and require for their answer a clear specification of cluster structure.

It has long been recognized (Kirkpatrick 1978, Shlifer et al 1979) that the dangling ends, shown as lines in Fig. 14, do not contribute to transport. Hence people began to focus on the problem obtained by decapitating all the dangling ends. The resulting structure is termed the backbone (Fig. 15). Is the backbone self-similar up to length scales L of the order of the connectedness length? If so, what is its fractal dimension d_{BB}? Many workers have verified that the backbone is indeed a fractal object, with

$$N_{BB} \sim L^{d_{BB}} \quad \text{[FRACTAL DIMENSION No. 8]} \tag{42}$$

for length scales L in the range $1 \ll L \ll \xi$. The fractal dimension d_{BB} sticks at the value 2 for $d > d_c = 6$, and then decreases "slowly" as d decreases below 6.

By $d = 2$, d_{BB} has decreased to about 1.65 (different workers obtain different values, since d_{BB} is quite difficult to estimate accurately). One of the problems is associated with obtaining a fast algorithm suitable for very large systems. It appears that Herrmann's burning algorithm (Herrmann et al 1984) is quite efficient and makes possible studying two-dimensional backbones in a box of size up to $(600)^2$ and $d = 3$ backbones up to $(60)^3$.

Some time ago, it was noted (Stanley 1977) that there are two kinds of bonds in the backbone:

(i) singly-connected ("red") bonds, which have the property that if they are cut, the backbone ceases to conduct, and

(ii) multiply-connected ("blue") bonds, which have the property that they can be cut without interrupting the flow.

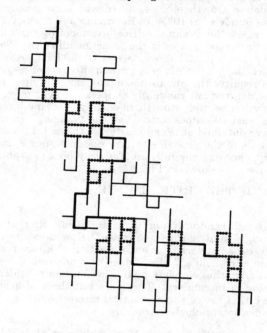

FIG. 14: A large machine-generated finite cluster just below the percolation threshold of the square lattice bond problem (Stanley 1977). The bonds of this cluster are of three sorts: dangling ends (light), singly-connected "red" bonds (heavy), and multiply-connected "blue" bonds (dotted). The red and blue bonds form the backbone, which (a) carries stress if the two ends of the cluster are pulled, (b) carries current if a battery is applied across the cluster, (c) carries fluid if the bottom of the cluster is oil and the top is my car [assuming the cluster represents a model of a randomly-porous material], (d) propagates spin order, if we view this as a large cluster of Ising spins just below p_c (see, e.g., Stanley et al 1976). This figure is taken from Stanley (1977).

The rationale for the terms red and blue is that the red bonds carry all the current from one bus bar to the other, and so are "hotter" than the blue bonds which "share" the current with the other blue bonds. In 1977, no one believed that the red bonds were a critical quantity. At the annual Canadian Physics Association meeting in November 1977, I posed the "red-blue" problem and an *undergraduate* member of the audience, Rob Pike, came up afterward to announce that he could solve the problem. In the ensuing 3 years, Pike succeeded in developing a computer algorithm that partitions bonds of a cluster into three separate colors: red, blue, and "yellow" (the dangling ends).

One of the worries that clouded the Pike project was the possibility that as p approached p_c, the mean number of red bonds would approach zero as the dangling ends found each other, transforming everything in between into a large blue blob. Fortunately, Pike's data showed clearly that all three functions, N_{red}, N_{blue} and N_{yellow} diverge (Pike and Stanley 1981):

$$N_{red} \sim L^{d_{red}} \quad \text{[FRACTAL DIMENSION No. 9]} \quad (43a)$$
$$N_{blue} \sim L^{d_{BB}} \quad (43b)$$
$$N_{yellow} \sim L^{d_f}, \quad (43c)$$

where it is understood that $1 \ll L \ll \xi$. The same year, Coniglio (1981,1982) provided a rigorous argument that for all d,

$$d_{red} = 1/\nu = y_T, \quad (44)$$

where $\xi \sim |\epsilon|^{-\nu}$ with $\epsilon = (p_c - p)/p_c$ and y_T is the thermal-like scaling field in percolation. Since

$$y_h = d_f \quad (45)$$

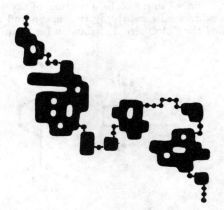

FIG. 15: Actual simulation of a backbone in site percolation just below p_c for a square lattice. The decomposition into blobs of all sizes from 1 to ∞ is apparent. After Herrmann and Stanley (1984).

(Stanley 1977), it follows that both scaling fields in percolation are equal to geometric properties of the incipient cluster! What about d_{BB}? To what is *this* fractal dimension related? This is a problem are working on just now.

ELASTIC BACKBONE

Suppose we regard each of the bonds in Fig. 14 as being a tiny spring (Fig. 16). Then if we pull on the two extremeties of the IIC, we find that only a subset of the backbone bonds, called the *elastic backbone*, carries stress. The fractal dimension of the elastic backbone is defined by

$$N_{Elastic} \sim L^{d_E} \qquad \text{[\textbf{FRACTAL DIMENSION No. 10}]}. \qquad (46)$$

Herrmann et al (1984) and Herrmann and Stanley (unpublished) have carried out extensive calculations of this quantity for $d = 2, 3$. It appears that within the 1 – 2% accuracy of the calculations,

$$d_E = d_{min}. \qquad (47)$$

Here d_{min} is the fractal dimension of the minimum path joining two points. This is not altogether unreasonable, since the elastic backbone is the union of all minimum paths—just as the backbone between two points is the union of all self-avoiding walks between the two points. If true, Eq. (47) means that the sort of loops which contribute to the elastic backbone do not occur on all length scales.

VERIFICATION OF RED-BLUE STRUCTURE OF BACKBONE: VOLATILE FRACTALS

Is there direct evidence that the red/blue (or, to use the term coined by Coniglio (1982), "links" and "blobs") picture of the backbone is valid? This is the question addressed recently by Herrmann and Stanley (1984). To this end, they studied the "blob size distribution function" for site—as opposed

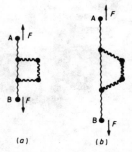

FIG. 16: Construction used in arguing that the elastic backbone is the union of all minimum paths from A to B. If each bond is a spring, then the resistance to elongation is provided by only one of the four bonds comprising the 4-bond "blob." After Herrmann et al (1984).

to bond—percolation (cf. Fig. 17). They found the probability Ps that a randomly-chosen site in the backbone belongs to an s-site blob scales, for fixed box size, as

$$P_s = sn_s \sim s^{-d/d_f} \sim s^{-\tau+1}. \tag{48}$$

Now suppose we double L. What happens to the backbone? As Fig. 5a suggests, the actual identity of the backbone changes, as tiny strands of red sites become part of a big blob. In general, the little blobs cascade into big blobs and the entire distribution shifts "downward" (Fig. 18). We say that the backbone is a volatile fractal, in that its identity depends on the box size.

Another volatile fractal is cluster-cluster aggregation, which was treated in the seminar by M. Kolb. Here the time plays the same role as the box edge L. If we double t, then little clusters get eaten up, and become part of bigger clusters. The detailed "blob size distribution scaling analysis" for volatile fractals (Fig. 17) is presented by Herrmann and Stanley (1984) for the backbone case, and by Vicsek and Family (1984) for the "cluster size distribution scaling analysis" cluster-cluster aggregation case. The formalisms, though developed independently, are completely identical under the transformation

$$L \longleftrightarrow t. \tag{49}$$

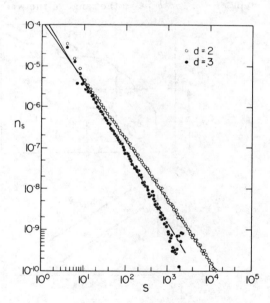

FIG. 17: Blob size distribution function for $d = 2$ and 3 for $L = 600$ ($d = 2$) and for $L = 60$ ($d = 3$). Note that the data are linear over many decades. After Herrmann and Stanley (1984).

From Fig. 15 it appears that the backbone is a pearl necklace, some of the pearls being swollen into a myriad of gargantuan shapes. Presumably, the backbone should be a string of pearls each of which is strung together by a drunken assembly line worker whose job is only to randomly choose a pearl from the pearl size distribution function P_s. To test this appealingly-simple possibility, we considered the statistical distribution of strings of "red sites" (1-site blobs). Specifically, we made a histogram of the number of strings with m sites against m. We found that this histogram decayed exponentially with m—one of the few examples in fractals where one must use semi-log as opposed to log-log paper! The parameters entering this exponential decay are calculated in terms of the total number of red sites and the total number of blobs.

We conclude by making connection with the problem of anomalous diffusion and transport mentioned above. There we emphasized the fundamental importance of the Nernst-Einstein relation connecting a transport quantity, the *macroscopic* conductivity, to the diffusion constant which measures *microscopic* motion. Applied to the full cluster, we obtained the relation $d_R = d_w - d_f$ of Eq. (34). What happens if we chop off all the dangling ends? Clearly the conductivity cannot change, so that d_R is unaffected. However d_f is changed to d_{BB} and d_w is changed to d_w^{BB}, which is defined as the exponent connecting the number of steps *on the backbone* to the range of the walker:

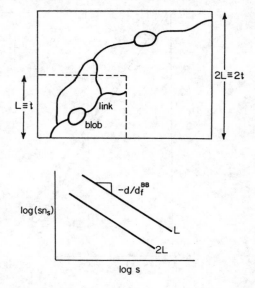

FIG. 18: (a) Schematic illustration of how small blobs become part of larger blobs when the box size L is doubled. (b) Schematic illustration of how the cluster size distribution $n_s(L)$ is uniformly depressed when L increases.

$N_w^{BB} \sim L^{d_w^{BB}}$. Hence the relation

$$d_R = d_w - d_f \qquad (50a)$$

becomes (Stanley and Coniglio 1984)

$$d_R = d_w^{BB} - d_f^{BB}, \qquad (50b)$$

Combining these two equations, we have the "invariance relation"

$$d_w - d_f = d_w^{BB} - d_f^{BB}, \qquad (50c)$$

which we sometimes name after my spouse, since she was mildly amused that this simple-looking equation was the only tangible result of a week's research in Naples with Antonio Coniglio. With this optimistic outlook on the future, perhaps we should bring these lectures to a close.

ACKNOWLEDGEMENTS

First and foremost, I wish to thank my collaborators in this research. These include A. Bunde [49], A. Coniglio [39], G. Daccord [33], Z. Djordjevic [38], F. Family [98], S. Havlin [972], H. J. Herrmann [57], D. C. Hong [82], N. Jan [809], F. Leyvraz [41], I. Majid [91], A. Margolina [7], P. Meakin [44], J. Nittmann [43], T. Vicsek [36], and T. Witten [1]. I also wish to thank many without whose helpful remarks the present understanding would not have occurred: A. Aharony [972], S. Alexander [972], Y. Gefen [972], P-G de Gennes [33], E. Guyon [33], B. Mandelbrot [33], I. Procaccia [972], and D. Stauffer [49]. Last and not least, I wish to thank my graduate students and colleagues, without whose insight I would not have arrived at my present appreciation for fractal materials. I also with to apologize that time and space considerations do not permit as coherent and complete a description of the subject of fractal materials as I had desired. Since earlier work is reviewed elsewhere (Stanley 1981, 1982a,b,c, 1983, 1984a,b, 1985; Stanley and Coniglio 1983), I've chosen to focus on ideas of recent months, oblivious to the adage that bringing a lecture "up to date" at time $-\tau$ renders it "out of date" at time $+\tau$.

LITERATURE CITED

Adler J, Aharony A and Stauffer D 1985 J Phys A **18** L129
Aharony A and Stauffer D 1984 Phys Rev Lett **52** 2368
Alexander S 1983 Ann Israel Phys Soc **5** 149
Alexander S and Orbach R 1982 J de Physique **43** L625
Barma M 1985 J Phys A **18** L277
Ben-Avraham D and Havlin S 1982 J Phys A **15** L691
Bunde A, Herrmann HJ, Margolina A and Stanley HE 1985a Phys Rev Lett **55** 653
Bunde A, Herrmann HJ and Stanley HE 1985b J Phys A **18** L523
Bunde A, Coniglio A, Hong DC and Stanley HE 1985c J Phys A **18** L137

Coniglio A 1981 Phys Rev Lett **46** 250
Coniglio A 1982 J Phys A **15** 3829
Coniglio A and Stanley HE 1984 Phys Rev Lett **52** 1068
Daccord G, Nittmann J and Stanley HE 1985a Phys Rev Lett (submitted)
Daccord G, Nittmann J and Stanley HE 1985b *Proc Les Houches Conference on Finely Divided Matter* eds M Daoud and N Boccara (Springer Verlag, Heidelberg)
deArcangelis L, Redner S and Coniglio A 1985 Phys Rev B **31** 4725
de Gennes PG 1976 La Recherche **7** 919
de Gennes PG 1980 J Phys (Paris) Colloq **41** C3-C17
Family F and Vicsek T 1985 J Phys A **18** L75
Feng S and Sen PN 1984 Phys Rev Lett **52** 216
Gefen Y, Aharony A and Alexander S 1983 Phys Rev Lett **50** 77
Grassberger P 1985 J Phys A **18** L215
Halsey TC, Meakin P and Procaccia I 1985 Phys Rev Lett (submitted)
Havlin S 1984 Phys Rev Lett **53** 1705
Havlin S and Ben-Avraham D 1983 J Phys A **16** L483
Havlin S and Nossal R 1984 J Phys A **17** L427
Herrmann HJ, Hong D and Stanley HE 1984 J Phys A **17** L261
Herrmann HJ and Stanley HE 1984 Phys Rev Lett **53** 1121
Herrmann HJ and Stanley HE 1985 Z Phys **60** xxx
Hong DC, Stanley HE, Coniglio A and Bunde A 1985 Phys Rev B **32** xxx
Kertész J 1983 J Phys A **16** L471
Kirkpatrick S 1978 AIP Conf Proc **40** 99
Kremer K and Lyklema JW 1984 J Phys A **17** L691
Kunz H and Souillard B 1978 J Stat Phys **19** 77
Leyvraz F 1985 J Phys A **18** xxx
Leyvraz F, Adler J, Aharony A, Bunde A, Coniglio A, Hong DC, Stanley HE and Stauffer D 1985 preprint for J Phys A Letters
Leyvraz F and Stanley HE 1983 Phys Rev Lett **51** 2048
Majid I, Jan N, Coniglio A and Stanley HE 1984 Phys Rev Lett **52** 1257
Margolina A, Nakanishi H, Stauffer D and Stanley HE 1984 J Phys A **17** 1683
Meakin P 1985 in *On Growth and Form—Proc. 1985 Cargèse Nato ASI* eds HE Stanley and N Ostrowsky (Martinus Nijhoff, Dordrecht, 1985)
Meakin P, Majid I, Havlin S and Stanley HE 1984 J Phys A **17** L975
Meakin P and Stanley HE 1983 Phys Rev Lett **51** 1457
Meakin P and Stanley HE 1984 J Phys A **17** L173
Meakin P, Stanley HE, Coniglio A and Witten TA 1985a Phys Rev A **32** 2364
Meakin P, Stanley HE, Coniglio A and Witten TA 1985b preprint
Nakanishi H and Stanley HE 1980 Phys Rev B **22** 2466
Nakanishi H and Stanley HE 1981 J Phys A **14** 693

Nittmann J, Daccord G and Stanley HE 1985 Nature **314** 141
Pandey RB and Stauffer D 1983 Phys Rev Lett **51** 527
Pike R and Stanley HE 1981 J Phys A **14** L169
Rammal R and Toulouse G 1983 J de Physique **44** L13
Ritzenberg AL and Cohen RJ 1984 Phys Rev B **30** 4038
Sahimi M 1985 J Phys A **18** 1543
Sapoval B, Rosso M and Gouyet JF 1985 J Physique Lett **46** L149
Shlifer G, Klein W, Reynolds PJ and Stanley HE 1979 J Phys A **12** L169
Stanley HE 1977 J Phys A **10** L211
Stanley HE 1981 in *Int Conf on Disordered Systems and Localization* eds C Castellani, C DiCastro and L Peliti (Springer Verlag, Heidelberg)
Stanley HE 1982a in *Proc NATO Advanced Study Institute on Structural Elements in Statistical Mechanics and Particle Physics* eds K Fredenhagen and J Honerkamp (Plenum Press, New York)
Stanley HE 1982b in *Physics as Natural Philosophy: Festschrift in Honor of Laszlo Tisza* eds A Shimony and and H Feshbach (MIT Press, Cambridge)
Stanley HE 1982c Prog Physics (Beijing) **30** 95 [in Chinese]
Stanley HE 1983 J Phys Soc Japan Suppl **52** 151
Stanley HE 1984a in *Kinetics of Aggregation and Gelation* eds F Family and D Landau (North Holland, Amsterdam)
Stanley HE 1984b J Stat Phys **36** 843
Stanley HE 1985 in *Encyclopedia on Polymer Science* (Wiley, New York)
Stanley HE and Coniglio A 1983 in *Percolation Structures and Processes* eds G Deutscher, R Zallen and J Adler (Adam Hilger, Bristol)
Stanley HE and Coniglio A 1984 Phys Rev B **29** 522
Stanley HE, Birgeneau RJ, Reynolds PJ and Nicoll JF 1976 J Phys C **9** L553
Stanley HE, Majid I, Margolina A and Bunde A 1984 Phys Rev Lett **53** 1706
Stanley HE, Reynolds PJ, Redner S and Family F 1982 in *Real-Space Renormalization* eds TW Burkhardt and JMJ van Leeuwen (Springer Verlag, Heidelberg)
Turkevich LA and Scher H 1985 Phys Rev Let **55** 1026
Vicsek T and Family F 1984 Phys Rev Lett **52** 1669

SCALE-INVARIANT DIFFUSIVE GROWTH

Thomas A. Witten III

Exxon Corporate Research Laboratory
Annandale NJ 08801, USA

INTRODUCTION

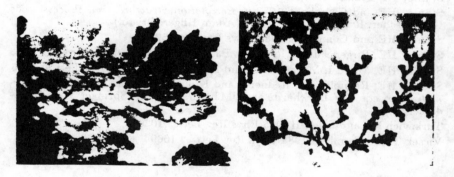

These figures (courtesy of Mac Lindsay and Paul Chaikin of the University of Pennsylvania) show an optical micrograph of zinc metal leaves electrodeposited on a piece of filter paper previously soaked in a zinc sulfide solution[1]. The deposition of the zinc is strongly irreversible, so that a zinc atom once condensed onto the deposit probably does not move appreciably afterward. As in the copper electrodeposition to be discussed by Ball, the rate-limiting step in the growth is probably the diffusion of the atoms from the medium to the deposit. The resulting patterns are fascinating both because of their scale-invariant appearance and because of their resemblance to living things.

In these lectures, I want to discuss the essential features that make such structures form. First I want to distinguish this type of structure from others considered up to now in the school. It is not obviously like a percolation cluster or a random animal, because it results from a growth process in a uniform medium. Neither is it like the clustering of clusters treated by Max Kolb and Paul Meakin. There growth occurred in large steps, in which each growth event (the fusion of two clusters) made a substantial difference in its size. Here growth occurs in indefinitely small steps; each zinc atom added to the deposit adds almost nothing to the structure. The growth is continuous.

In these talks I want to survey diffusive growth processes that models, to illustrate how scale invariance arises, and show generally how such models can be described mathematically. But neither of these models is suitable for treating the zinc leaves of Figure 1. For this, we must describe growth from a diffusing field. Accordingly, I'll review different types of growth from a diffusing field, for simple, nonfractal structures. I want to describe how matter may absorb a diffusing field, note the special features of fractal absorbers, and mention some limits on the rate of absorption.

With this preparation we can approach stochastic growth from a diffusing field. Computer simulations of this kind of growth show features very reminiscent of the zinc leaves. And the simulations appear to produce scale-invariant, fractal objects. I'll try to survey some of these properties, though new ones are rapidly being discovered. I'll illustrate how these growth models may be defined as a statistical mechanical ensemble and list the various approximations which have been used to make the models tractable.

STOCHASTIC GROWTH

The simplest "stochastic growth" model I can think of is a random walk. Here is a true random fractal whose scaling properties are completely understood! I'd like to take a minute to review how these scaling properties emerge, because I think there are important parallels with fractal objects generally. We imagine our walk as starting from the origin of a large (hyper) cubic lattice and proceeding by random, nearest-neighbor steps. All the scaling properties can be expressed in terms of the probability $u(r,t)$ that the walker is at the lattice point r at time step t. We may compute u at one time step from its values on the previous time step. If the walker visits r at time t, it must have been at a nearest neighbor of r at $t-1$. To reach r, the walker must both be next door at $t-1$, and must step in the proper direction on the tth step. If there are z nearest neighbors, the correct step occurs with probability $1/z$. Combining the probabilities,

$$u(r,t) = 1/z \sum_{\text{neighbors}} u(\text{neighbor}, t-1).$$

This equation must contain all the scaling and self-similarity of the walk, but this scaling is not particularly apparent. Indeed, this equation cannot have simple scaling properties, since it exactly describes short walks, which need not scale, as well as long ones, which do scale. We can hope to discover scaling by considering the limit of long times and large distances. Thus, we re-express the equation in terms of a small time step τ and lattice spacing "a":

$$\tau \frac{\partial u}{\partial t} + \frac{1}{2}\tau^2 \frac{\partial^2 u}{\partial t^2} + \cdots = \frac{a^2}{z}\nabla^2 u + O(a^4 \nabla^4 u).$$

To maintain a smooth functional behavior at the macroscopic scale, we must take τ and a to zero so that a^2/τ is a constant. This constant is of course the diffusion constant ς:

$$\partial u/\partial t = a^2/(z\tau)\nabla^2 u = \varsigma \nabla^2 u. \tag{1}$$

To relate the macroscopic distance R the walker travels to the elapsed time t, we are forced to use this constant. From dimensional analysis we have to get $R^2 \sim t$, and we are thus led to the conclusion that the path of the walker is a fractal with $D = 2$. [To verify that the walk satisfies strictly the definition of a fractal requires further checks.]

The general method followed for this random walk problem was to describe the system on a scale indefinitely larger than its elementary pieces. Then to obtain a finite description of the microscopic behavior (finite R in finite t) the microscopic units have to go together to zero in certain power relationships. This is the philosophy followed by renormalized field theory in predicting the fractal properties of phase transitions, self-avoiding walks, random animals and percolation clusters. These "well understood" fractals can all be treated in the framework of equilibrium statistical mechanics. Even though this framework may not be useful for treating our growth models, there must still exist some "renormalized" description which is independent of microscopic parameters like τ and a.

The random walk, though it can be viewed as a growth model, generates an equilibrium ensemble; viz. the ensemble of all distinct random walks. The feature of growth made no difference to the scaling. We now turn to another solvable growth model where the growth feature is crucial. This is the "penetrable Eden model."[1] In this model as in the random walk one again adds particles one by one to a particle at the origin of a lattice to form a connected cluster. The next particle is chosen at random, a nearest neighbor site is then chosen at random and the new particle is put there. The particles freely interpenetrate so that many may occupy a given site.

The scaling properties may again be found by considering a space dependent probability $G(r, t+1)$—the probability that the $t + 1^{st}$ particle is at the lattice site r. Any particle t_1 preceding the $t + 1^{st}$ may be the one picked to gain a neighbor. The probability that any one is picked is $1/t$. The probability that this one is next to r is $G(\text{neighbor}, t_1)$. Finally, the probability that the site r will be chosen out of the z possibilities is $1/z$. Combining these, we find

$$G(r, t+1) = 1/z \; t^{-1} \sum_{t_1=1}^{t} \sum_{\text{neighbors}} G(\text{neighbor}, t_1).$$

This can be readily be expressed as a differential equation by multiplying through by t and then subtracting the resulting equation at two adjacent time steps. One finds

$$\partial G/\partial(\log t) = a^2/z \nabla^2 G.$$

Evidently, the G function obeys the random walk equation, but with t replaced by $\log(t)$. Now instead of $t \sim R^2$ we have $t \sim \exp(R^2)$; the mass grows faster than any power of the radius. This is not a fractal object in any usual sense; but it is a useful point of departure for studying many growth models. This model is closely related to the impenetrable Eden model on a Cayley tree, to be discussed by other speakers.

DIFFUSION, ABSORPTION, AND GROWTH

Droplet Growth: Langer[2] has made an authoritative review of classical forms of growth by diffusion. The simplest problem involving growth from a diffusing field is the growth of a spherical droplet, treated and by Mullins and Sekerka[3]. We may imagine that a droplet of some chemical species is condensing from a supersaturated solution. The condensing particles move by diffusion in the solution; each executes a random walk. The concentration or probability density $u(r,t)$ of these walkers is given by Eq. 1. Particles join the droplet faster than they leave it; so that the droplet acts as a partial absorber of the diffusing field. The absorption probability per time step is related to an inverse length—the logarithmic derivative of u at the drop surface [For more careful treatment of the surface boundary condition see Ref. 2.] Eventually the droplet, of radius S, grows much larger than this microscopic length; then it is practically a perfect absorber. Suppose the absorption is turned on at time 0. What happens to the diffusing field if the growth of the droplet radius is negligible? This can be deduced by solving Eq. 1 given the absorbing boundary condition and the initial uniform u field. The behavior can be deduced without equations by considering the individual random walks. Imagine the trails of all the particles since time 0 if there had been no absorption. Since the particles were initially uniformly distributed, they still are. The effect of the absorption is to remove all the trails which intersect the droplet. What remains is no longer uniform. Particles now much farther than $(\varsigma t)^{1/2}$ from the droplet surface have trails too short to have touched the surface; thus the concentration is almost unperturbed beyond this distance. After a very long time the size R of the trails is much larger than any given distance r to the center of the droplet. The number of particles absorbed in a time step is the number of trails that first touch the droplet on that time step. A trail which does intersect the droplet has of the order of $(S/a)^2$ contacts. (Here we use the fractal property of random walks.) The total number of contacts is just the average density of walkers times the volume times the time: $S^d u(\infty) t/\tau$. The ratio of the total number of contacts to the number of contacts per trail is the number of first contacts, i. e. the number M of particles absorbed

$$M \simeq S^{d-2} a^2 u(\infty) t/\tau. \tag{2}$$

From this we see that the flux dM/dt attains a steady-state limiting value. The reasoning applies[4] to more general absorbers and trails, as discussed below. The reasoning must be modified in two dimensions; where a random walk has the same dimension as space. Then the typical number of contacts between a trail and the droplet increases indefinitely, as $\log(t)$. Accordingly, the flux declines as $\log(t)$.

The absorption by the aggregate is associated with a depletion of the concentration $u(r)$. For points r far from the aggregate ($r \gg S$), the amount of depletion relative to $u(\infty)$ is the probability that a trail through r intersects the droplet. This is the density of trail segments of that trail at the droplet—$a^2 r^{2-d}$—times its volume S^d. Of course this may also be deduced by solving Eq. 1 in the steady state limit. This is simply Laplace's equation, and the local concentration $u(r)$ may be thought of as an electric potential fixed at 0 on the droplet and equal to $u(\infty)$ at infinity.

As the droplet absorbs matter, it grows: $dS^d/dt = dM/dt \sim S^{d-2} u(\infty)$. Thus $S^2 \sim t u(\infty)$. The distance the radius grows is proportional to the distance a typical walker travels. But one may always decrease the proportionality factor as much as desired by decreasing the average concentration $u(\infty)$, thereby attaining the steady state limit. On the other hand, if the concentration is higher and the growth more rapid, the growth is closer to the initial regime, where a typical absorbed particle has only traveled a distance of order S.

Parabolic Growth: The droplet growth velocity slows continually with time; but there is another growth geometry that attains a constant velocity. This is the parabolic tip geometry first worked out by Ivantsov.[5] Here the droplet is replaced by a paraboloidal absorbing interface. For this parabolic shape, the scaling of mass with time is superficially that of a fractal: If the height is h, the mass scales as $h^{d-(d-1)/2}$, since each lateral dimension varies as $h^{1/2}$. The diffusing field u around such an interface can be readily solved analytically, since Eq. 1 is separable in parabolic co-ordinates. The interface grows outward in proportion to the local flux of the diffusing field onto it. Remarkably, this flux is of just the right form to maintain the parabolic shape; the interface simply translates along its axis of symmetry. Thus the problem at a later time is identical to the problem at an earlier time, and the growth velocity v must be constant in time. Thus it is convenient to view this system in a frame of reference moving with the interface. From this frame, the random walkers have an imposed drift. For distances r larger than the "diffusion length" $R \equiv \varsigma/v$ they have the fractal dimension of a straight line rather than that of a random walk. If we look at the problem on length scales much larger than this diffusion length the absorbed particles are essentially a uniform rain falling on the interface. The interface has a length scale of its own, the radius of curvature S_0 of the tip. When S_0 is much bigger than the diffusion length, the diffusive character of the particles is only evident in a thin boundary layer of size $\sim R$ around the interface.

Both the droplet and the parabola are unstable against deformation of the interface.[3,6] Any small initial deformation grows with time until large distortions result. This instability is already present in the limit of a thin boundary layer mentioned above.[7,8] To see this, imagine the parabolic needle distorted by a gentle wave whose length S is longer than the boundary layer thickness R, and whose amplitude h is smaller than this layer. In this limit, a section of the surface may be considered nearly flat. Qualitatively the lines of constant u are somewhat squeezed together around the peaks of the wave and spread apart at the troughs. The local absorption rate is proportional to the spacing between these lines, i.e., to the gradient of the field. Evidently the peaks absorb faster than the troughs; this makes them grow higher.

The amplitude h thus grows exponentially.[2] For small wavevectors $q =$

$1/S(qR \ll 1)$ the growth rate $\omega(q)$ is expandable in $(qR)^2$. This rate must vanish as $q \to 0$, since a $q = 0$ wave amounts simply to a shift in the position of the flat surface. Thus for small qR, one finds $\omega(q) \sim (v/R)(qR)^2 = \varsigma S^{-2}$. (For the complementary case $qR \gg 1$ one attains the "quasi-stationary" limit, independent of ς and R: $\omega(q) \simeq vq \simeq v/S$.) [These results correct and revise those presented in the lecture.]

Two basic relaxation processes compete with this wrinkling instability, and drive the surface towards its lowest free energy state. Thus, e.g., in the condensation of a liquid droplet the driving force for relaxation is the surface energy: the surface area times the surface tension σ. To treat the relaxation mechanisms quantitatively, we consider the relaxation of a small bump of height h and width $S \gg h$. Its surface energy U, less that of the flat surface, is given by $U \simeq \sigma S^{d-1}(h/S)^2$.

An important relaxation mechanism is diffusion: atoms evaporate from the surface, diffuse through the solvent, and condense elsewhere on the surface. We may infer the relaxation rate by equating the rate of energy dissipation by the diffusing current density j to the rate $\dot{U}$ at which surface energy is being lost. We assume that the relaxation is very slow on the time scale of diffusion (quasi-stationary limit). Then the current density occupies a volume of order S^d, and the total current jS^{d-1} is of order of the rate of loss of particles from the bump, i.e. $\rho \dot{h} S^{d-1}$, where ρ is the (number) density of the droplet. The dissipation rate $kT/\varsigma \int d^d r j^2$ is of order $kT\dot{h}^2 S^d \rho/\varsigma$. The rate of loss of surface energy is of order $\dot{U} \simeq \sigma h \dot{h} S^{d-3}$. Equating this to the dissipation rate yields $\dot{h} \simeq \sigma \varsigma S^{-3}/(kT\rho)h$, for general dimension d. (For sufficiently small S, the relaxation time $h/\dot{h}$ becomes shorter than the time S^2/ς needed to diffuse a distance S. In that limit[2] $\dot{h}/h \sim S^{-2}$.) Comparing this result with the unstable growth rate ω derived above shows that the relaxation dominates the growth at sufficiently small scales S when the growth is quasistationary. In the complementary regime where growth is fast compared to diffusion, both growth and relaxation have the same scaling with S.

Another mechanism of relaxation is by viscous flow in the droplet. The relaxation rate may be estimated as above. The typical current density j is of the same order as in the diffusive case, but the dissipation is now proportional to the squared gradient of this current rather than to the squared current itself. The expression for the dissipation now has an extra factor of S^{-2}. As a result, the relaxation rate is now proportional to S^{-1} rather than S^{-3}: $\dot{h} \simeq \sigma S^{-1}/\eta h$, where η is the viscosity. Thus, the hydrodynamic mechanism is comparatively more effective than the diffusive one at relaxing long-wavelength modes.

Detailed treatment of the instabilities sketched above[7,8] may be performed in the boundary layer regime $R \lesssim S$. This treatment has proven successful and promising as an explanation of the striking speartip shapes seen in dendritic solidification. These shapes have been carefully investigated experimentally.[9,10] The dendrites are observed to grow at nearly uniform velocity; the regular pattern of sidebranches lies within a roughly paraboloidal envelope. The average density within this envelope does not seem to become indefinitely small as one changes parameters of the system. Thus, the interior does not seem to be approaching a fractal structure.

Needle Growth: Up to now we have discussed the growth of a regular, smooth object through the absorption of a diffusing field. To some extent, we can also characterize absorption by irregular structures as well. An instructive case is a needle of length S. In two dimensions the steady state problem is a classic problem in electrostatics. To make it completely definite, we enclose the needle in a giant circle of radius L. We expect the absorption rate to be singular near the tips of the needle. To examine the nature of this singularity we make a needle tip by conformally transforming an infinite line along the x axis of the complex z plane, using the transformation $z' = z^2$. One finds that the flux ∇u onto the needle at distance r from the tip varies as $r^{-1/2}$. Even this simple object has a power-law distribution of absorption rates; we expect the distribution for a fractal to be at least this complicated. Lee Turkevitch will have more to say about this.

Though the details of absorption on a fractal are complicated, the global scaling of the absorption is simple. A fractal with dimension D and radius S absorbs mass at qualitatively the same rate as a solid sphere of that radius. We may use the argument of Eq. 2 above. The number of intersections at time t is $S^D u(\infty)$. The number of intersections per trail is S^{D+2-d}. As long as this latter is much greater than 1, the D power drops out, and the absorption scales like that of a solid object. If the number of intersections does not increase with S, i.e., $D < d - 2$, then a trail which makes contact typically does so of order once. Then the absorption is proportional to the number of particles in the fractal. All of the fractal has relatively equal access to the diffusing field, and there is negligible screening; the fractal is "transparent" to the walkers. The "brownian trail" methods described above may be used to treat many types of interactions with arbitrary fractals.[11]

STOCHASTIC DIFFUSIVE GROWTH

We have dealt in the last section with deterministic aspects of diffusive growth; viz. growth or absorption of a continuum diffusing field. In this section we want to consider cases where diffusive growth produces random fractal structures. First I'll catalogue the different types of random diffusive growth models along with various measured properties. Then I'll describe how these models may be treated formally, to obtain some understanding and some constraints on their behavior. Finally I'll sketch various attempts to explain the fractal dimension of diffusive growth.

Descriptive: Paul Meakin has described in his lecture the basic diffusion-limited aggregation model, the so-called Witten-Sander model.[12] As he noted, the density correlations of this model appear to have a scale-invariant form. This is most readily seen by examining the pair correlation function $C(r)$ defined as $\langle \rho(0)\rho(r)\rangle/\langle\rho\rangle$, where the average $\langle ...\rangle$ is taken over all positions and orientations of a large aggregate. In a fractal object of dimension D, $C(r) \sim r^{D-d}$. The DLA model has correlations of this form, with $D \simeq 1.7$ in two dimensions. We shall discuss below some recent modifications to this picture.

The fractal behavior of this model can be obtained by slightly perturbing the solvable penetrable Eden model described above. We call this variant "penetrable DLA."[13] In the penetrable Eden model each particle was added at random next to existing particles. In penetrable DLA a random walker is launched at a great distance from the existing cluster as in ordinary DLA. Whenever the walker arrives next to a cluster particle, it may remain there and join the cluster. The new feature is that this sticking event occurs with only a finite probability p. The probability that the walker is absorbed upon visiting a site is the number of particles on adjacent sites, times p. If the walker is not absorbed, it continues to walk through the aggregate. Evidently, if p is small enough, the walker may traverse any finite cluster many times before being absorbed. Then the growth probability at a site is simply that of the penetrable Eden model. After about $p \log p$ particles have grown, a walker touching the cluster has a probability of order unity of being absorbed. From then on the interior is strongly screened, and the growth takes on the fractal properties of DLA. The mass scales as the radius to the 1.7 power. The pair correlation function $C(r)$ has the same power-law behavior[13] as in ordinary DLA, even for very small sticking probability p. This shows that DLA behavior does not require an explicit limitation of multiple occupancy. Unlike the self-avoiding walk or random animals or percolation, one may freely allow multiple occupancy without changing the scaling properties.

Wilkenson[14] has devised another variant which shows that discrete increments of growth are not necessary to obtain DLA behavior. In his model, a deposit grows from a deterministic diffusing field through the bonds of a random lattice network. These bonds have different widths, so that a given amount of material deposited results in varying amounts of growth. The bonds vary randomly from site to site over a limited range of widths. When the deterministic growth equations are integrated numerically for a given random lattice, a pattern with the appearance and the scaling properties of DLA emerges. The resemblence to DLA is reduced when the variability of the bonds is decreased.

Wilkenson's model is actually expressed in the language of fluid flow. The mathematical correspondence between diffusive growth and Darcy flow of a fluid boundary was pointed out by Patterson.[15] Darcy flow occurs when there is some solid structure in contact with the fluid which absorbs its momentum. Flow in a gel, a porous rock, or between parallel plates is of this type. In such situations the hydrodynamic equations are replaced by a simple Ohm's-law relationship between the fluid velocity v and the gradient of the pressure. When a fluid of negligible viscosity displaces one of viscosity η in such a medium, the interface has a local velocity equal to that of the adjacent fluid, i.e., ∇p. If the fluid is incompressible, $\nabla^2 p = 0$. Further, the pressure is constant along the interface, since it is uniform throughout the displacing fluid (in the absense of surface tension). Thus the growth equations are just those used above to describe our condensing droplet.

The variants discussed above show that allowing penetrability or reducing the randomness affect the resulting growth only on a local length scale. We now discuss some modifications which dominate at large scales. If the growth occurs from a gas of random walkers of density $u(\infty)$ rather than from a single walker, then the rate of growth is not negligibly small relative to the motion of the walkers. As in the case of droplet growth or the parabolic tip, the nonzero growth rate introduces a length scale R beyond which the fractal correlations of the random walkers break down. One may show[16] that this breakdown occurs when the average density of the aggregate decreases to that of the initial gas of walkers. At this point the aggregate has a radius R given by $R^{D-d} \simeq u(\infty)$. This is what one observes in simulations. Clusters much larger than the length scale R have local spatial correlations like DLA, but are globally of uniform density. This behavior gives a practical limit on the range of DLA scaling behavior. In experiments, such as the one Robin Ball will describe, there is always a nonzero density of random walkers. There must be a corresponding limit on the range of fractal scaling behavior.

One obvious variant which must affect the global structure is to grow DLA from an extended seed such as a line or a plane. In Meakin's[17] studies of these geometries, he found that the average density $\langle \rho \rangle$ of the deposit scaled with its thickness h in a predictable way: $\langle \rho \rangle \sim h^{D-d}$. This scaling is what one would predict assuming that the local correlations $C_h(r)$ for points at distance h from the seed are unaffected by the seed for $r \lesssim h$, and that $C_h(h) \sim \langle \rho \rangle_h$. This gives indirect evidence that the local spatial correlations are unaffected by the shape of the seed.

Another property also may be predicted by assuming that the spatial correlations $C(r)$ of DLA are those of a normal fractal. When DLA is grown from an extended substrate, the deposit consists of a set of disconnected trees with a broad distribution of masses s. Racz and Vicsek[18] showed that the distribution of these masses may be related to the fractal dimension D. Their treatment can easily be extended to a general fractal substrate of dimension D_1; a linear substrate has $D_1 = 1$; a planar substrate has $D_1 = 2$. If a very thick DLA deposit grows from this substrate, trees with a broad range of masses s and heights h are generated. The Racz-Viscek reasoning uses two basic assumptions. First, an individual tree has the same height-mass scaling as in an entire DLA cluster: $s \sim h^D$. Second, the distance between two neighboring trees of height h or more is of the order of h. To motivate this, we consider a surface at a distance h from the substrate. A tree of height greater than h is expected to have an intersection with this surface of size $\sim h$. Growth of another tree within distance h is thus inhibited, but growth for distances much greater than h is unaffected.

Given these assumptions, one may infer the probability $P(h)$ that an arbitrary substrate site has a tree of height at least h. If a given site does have such a tree, we draw a sphere of radius h around that substrate site. By our second assumption above, the growth from substrate sites within this sphere is inhibited; the trees within it have height of order h or less. Thus, out of h^{D_1} sites, only about one has height of order h or greater: $P(h) \sim h^{-D_1}$. Using our first assumption, the probability $P(s)$ that the mass of a tree exceeds s goes as $s^{-D_1/D}$. The probability $n(s)$ that the mass is exactly s is $n(s) = dP/ds \sim s^{-D_1/D-1}$. Meakin[19] has found distributions $n(s)$ consistent with this prediction.

Several striking properties of the basic two-dimensional DLA have been recently discovered. These show that DLA has a more regular structure than better-understood random fractals like percolation clusters. An important structural property related to transport on a DLA cluster is the length of the path connecting arbitrary points at distance r well within a large cluster. As Stanley and collaborators discovered, the length of the connecting path appears to scale linearly with r for DLA in various dimensions d: $d_{min} = 1$. As Stanley noted in his talk, all the better-understood fractals have $d_{min} > 1$.

A DLA cluster has unusual regularity in its overall shape. This may be measured for example by measuring the moment $\langle r_x^2 \rangle = S_x^2$ in various directions x and then comparing S_{max} to S_{min}. For an ordinary fractal such as a random walk, there is an anisotropy of order unity: S_{max}/S_{min} approaches some definite value greater than 1. But in DLA[20] the ratio appears to approach unity. In off-lattice simulations, the shape becomes more and more circular. In on-lattice simulations, the regular shape which emerges reflects the shape of the lattice, as we discuss below.

Other recent results concern the local homogeneity and isotropy of DLA. The well-understood fractal structures can be shown to be locally homogeneous and isotropic.[21] This property concerns correlations $\langle \rho(r_0)\rho(r_1)\ldots\rho(r_n)\rangle$ for the r_i confined within a neighborhood of radius R at distance T from the center of an aggregate of radius S. A definition of local homogeneity is that such correlation functions become independent of the position of the neighborhood (thus independent of T) in the limit $R \ll T$ and $R \ll S$. "Finite-size" corrections to this limit are of the order $(R/T)^{power}$ or $(R/S)^{power}$. Local anisotropy means that the correlations above are invariant under an overall rotation of the vectors r_i, up to similar finite size corrections.

Meakin[22] and Kolb[22] have investigated the isotropy question numerically by comparing two-point correlations in the direction toward the seed and in the perpendicular direction. The data suggest that the correlations fall off with a steeper power in the perpendicular direction. It will be important to clarify this effect, since this picture violates the simple fractal ideas used to understand several properties discussed above. This picture seems to suggest, for example, that the width of DLA trees grown from a surface becomes much narrower than their height.

In complement to the anisotropic correlations noted above is the lattice anisotropy effect[23]. If a DLA cluster is grown on a lattice, the broken rotational symmetry of the lattice is reflected in the overall envelope in which the aggregate lies. Meakin and Turkevich discuss this effect in their contributions.

Other information about the outer boundary concerns measurements of the "active zone"—the region where particles are currently being absorbed. As Meakin has mentioned, the next particle to be added to a large cluster has a probability distribution of distances from the center with a spread of order R. This must be true for any fractal structure with open spaces of order R connecting it with the exterior. But the most active sites need not be so broadly distributed. To see this we consider an "asterisk" aggregate made of, say, six radial needles. There the active zone defined by Racz and Pliscke[24] has a width of order R. But if each site is weighted with, say, the cube of its absorption probability, then the width of the distribution becomes of order

1 asymptotically. The width of the active zone depends on the definition of activity used.

Rigorous results

We have noted the similarity between DLA growth by the adsorption of random walkers and other forms of growth controlled by diffusing fields. In this section we make this connection explicit by deriving exact growth equations for DLA in which the random walkers are replaced by a diffusing field[1,25,13]. These growth equations completely specify the ensemble of clusters by giving the statistical weight of any cluster.

A particular aggregate may be specified by giving the density ρ for all lattice sites x. For ordinary DLA with perfect sticking, ρ is zero or one; for the penetrable DLA introduced above, ρ may span a large range of (non-negative) values. To characterize the ensemble of aggregates we must somehow give the relative statistical weights of various aggregates $\{\rho(x)\}$. For this it suffices to give the probability $P_{\{\rho\}}(y)$ that for a particular aggregate $\{\rho(x)\}$, the next particle is added at a given point y. With this, one can compute the probabilities of all $N+1$–particle aggregates given the probabilities of all N-particle aggregates, and hence determine the absolute probability of a given aggregate.

The probabilities $P_{\{\rho\}}(y)$ can be readily computed, because any cluster with a given density profile (no matter how it grew) has the same probability to gain its next particle at y. In computing this probability we are considering the ensemble of all possible ways in which the next walker can be adsorbed onto the given aggregate; i.e., the ensemble of all random walks around that cluster.

The probability that growth occurs at y on a particular time step t is the probability that the walker is at y then, times the conditional probability that a walker there is absorbed. The probability that the walker is present is $u(y,t)$. This probability may be found explicitly by solving a diffusion equation like Eq. 1. The absorption effect occurs between steps of the random walk. If the walker is now at y, it terminates with a probability $Q(y) \equiv p \sum_{\text{neighbors}} \rho(\text{neighbor})$. (If this number should exceed 1, Q is understood[13] to be 1.) Given $u(y,t)$ the probability that the walker survives the random absorption is $u'(y,t) = (1 - Q(y)) u(y,t)$. This $u'(y, t-1)$. Thus

$$u'(y,t) = (1 - Q(y)) u(y,t)$$
$$= (1 - Q(y)) \sum_{\text{neighbors}} u'(\text{neighbor}, t-1).$$

We may rearrange and express this equation in terms of the lattice Laplacian "∇^2" $u'(y) \equiv \sum_{\text{neighbors}} u'(\text{neighbor}) - z u'(y)$:

$$\Delta u'/\Delta t = (1 - Q(y)) \nabla^2 u' - z Q(y) u'.$$

From u' we may determine the growth probability $P_{\{\rho\}}(y)$. The probability of growth at time t is just the probability of absorption $u(y,t)Q(y) =$

$Q(y)/(1-Q(y))u'(y,t)$. The total probability of growth is a sum over t: $Q(y)/(1-Q(y))\sum_t u'(y,t)$; this is $P_{\{\rho\}}(y)$. Performing the $\sum_t$ on the u' equation yields

$$0 = (1-Q(y))\,{}^{\omega}\nabla^{2\omega}u(y) - zQ(y)u(y),$$

where
$$Q(y) \equiv p \sum_{\text{neighbors}} \rho(\text{neighbor}) \equiv p(z\rho(y) + {}^{\omega}\nabla^{2\omega}\rho).$$

Then, the growth probability is given by

$$P_{\{\rho\}}(y) = {}^{\omega}\nabla^{2}u(y)/z.$$

We note that the diffusion equation of this problem is in the steady state limit—the time derivitive has disappeared. This is because there is strictly no growth of the aggregate in the time before the next walker is adsorbed.[25] The limit of perfect absorption is attained when the sticking parameter $p=1$. In the limit $p \to 1$, $u(y)$ goes to zero for all sites adjacent to occupied sites; the $u(y)$ equation may be thought of as a Laplace's equation with the boundary condition that $u=0$ adjacent to the aggregate.

We have seen above that DLA scaling properties occur even when $p \ll 1$; then both ρ and u vary smoothly in space, and the growth can be expressed in the quasi-continuum language derived here. The language suggests further simplifications, which will be discussed below.

One advantage of this formulation is that it provides a natural way to describe lattice anisotropy. The lattice enters the equations through the use of the lattice "∇^2" instead of the true differential ∇^2. The lattice effects are thought not to arise from the diffusing field "∇^2"u, but rather from the confinement of cluster particles to a lattice—i.e., "∇^2"ρ. The finite difference "∇^2" may be expanded in true derivitives ∇^2, ∇^3, etc. For sticking confined to an oriented triangle of sites, discussed by Meakin at this meeting, there is a nonzero ∇^3 correction. For a square lattice, the lowest-order correction is ∇^4. For a triangular, six-coordinated lattice, the lowest correction is of order ∇^6. Since the three cases have qualitatively different anisotropies, one might expect each to have qualitatively distinct effects on the growth.

A further rigorous result about DLA is a bound[4] on the rate of growth, and hence on the fractal dimension. It is obtained by considering a large DLA cluster growing in a finite but very small concentration $u(\infty)$ of walkers. The outermost particle in the cluster has a distance $S(t)$ from the seed. This S may increase only if a walker is absorbed on the outermost particle. But the rate of absorption onto such a particle is limited. It can be no greater than the rate of absorption onto the initial seed. The total rate of absorption $\dot{M}$ is also constrained. The aggregate absorbs the diffusers like a perfect spherical absorber whose radius is of order S. As shown above, $\dot{M} \simeq u(\infty)S^{d-2}$. Now $dS/dM = \dot{S}/\dot{M}$. Since $\dot{S}$ is bounded above, we infer $dS/dM < 0(S^{2-d})$. If M is to grow as S^D, then D may not be less than $d-1$. Thus D must increase steadily as the dimension of space d increases. This property distinguishes

DLA from more tractable models, in which D becomes simple to calculate and independent of d in high dimensions.

Approximations to DLA

Since DLA was introduced, a number of heuristic models and approximations have been introduced to account for its fractal structure. I list here a representative sampling of these, to serve as in introduction for further reading. The models may be grouped into four classes. First, early attempts were based on the scaling properties of equilibrium structures like random walks. Standard energy-entropy balance arguments[26] for polymer conformations were modified heuristically to take account of the diffusive growth condition[27,28].

A second class of models may be called "shape hypotheses." They start from the premise that the essential feature needed to explain DLA scaling is some qualitative information about the form of the structure. The earliest models of this type[29,30] constructed regular, hierarchical structures, which resembled DLA simply in being branched structures. These structures had D values qualitatively like those measured for DLA. Pietronero[31] introduced a "branch competition" model with the additional feature of random branching. The recent work of Turkevitch and Scher[22] and of Ball et al[22] likens the distribution of adsorption probabilities of DLA to that of a right-angle wedge. It accounts explicitly for the diffusive feature of the DLA growth. Halsey, Meakin and Proccacia[22] extend this idea to address the question of how the hypothetical shape maintains itself. These theories will be discussed later in the school.

There has been a persistent hope that DLA, like other random fractals, should become simple to treat in high dimensional space. One model of the high-dimensional limit is a Cayley tree. And DLA becomes much simpler in that limit; it becomes equivalent to an Eden model, since there is no screening. The growth from one site is not reduced by absorption by more exposed sites. Many exact properties[32,33] can be treated in this limit. A third class of model exploits the Cayley tree limit to learn about DLA. The behavior of this Cayley-tree DLA seems at odds with numerical evidence about high-dimensioned DLA.[34] The numerical data suggests that $d - D \to 1$ in high dimensions. For the Cayley tree, $d - D$ seems to be 0 up to logarithms. Parisi and Zhang[34] have extended this picture by including screening effects approximately for less than infinite dimensions.

A final approach is based on the similarity of various diffusive growth processes, with widely varying amounts of explicit randomness. This approach could be characterized as perturbing in the randomness. To this class belong studies of the instability of a smooth surface under diffusive growth.[2,7,8] Complementary to this work are the continuous density models motivated by the DLA growth equations presented above. One may take the probability $P_{\{\rho\}}(y)$ as the actual growth rate $\partial \rho / \partial t(y)$. This yields a mean-field[35] model, and gives a scaling of density with size consistent with $d - D = 1$. Perturbing this model with randomness[36,13] appears as a promising avenue for studying DLA.

CONCLUSION

Our understanding of DLA is in a state of flux. The recently suggested anisotropy properties described above have become obvious only with the largest scale simulations now practical. These show that our intuitions about what features are important for accounting for DLA are not to be trusted. They cast doubt on previously accepted laws based on the notion that DLA has only one macroscopic length scale. These studies raise the suspicion that the asymptotic behavior of large DLA may be quite different than what has been observed to now. And they suggest that the scaling of DLA may be qualitatively richer than that of structures we now understand.

[1] T. A. Witten and L. M. Sander, Phys. Rev. B **27**, 5686 (1983).
[2] J. S. Langer, Rev. Mod. Phys. **52**, 1 (1980).
[3] W. W. Mullins and R. F. Sekerka, J. Appl. Phys **34**, 323 (1963); ibid. **35**, 444 (1964).
[4] R. C. Ball and T. A. Witten, Phys. Rev. A rapid commun. **29**, 2966 (1984).
[5] G. P. Ivantsov, Dokl. Acad. Nauk. SSSR **58**, 567 (1947).
[6] J. S. Langer and H. Mueller-Krumbhaar, Acta Metall. **2**, 1081, 1689 (1978); H. Mueller-Krumbhaar and J. S. Langer, Acta Metal. **2**, 1697 (1978).
[7] E. Ben-Jacob, N. Goldenfeld, J. S. Langer, and G. Shoen, Phys. Rev. Lett. **51**, 1930 (1983); Phys Rev. A **29**, 330 (1984).
[8] R. C. Brower, D. A. Kessler, N. Koplik and H. Levine Phys. Rev. A **29**, 1335 (1984).
[9] P. W. Voorhees and M. E. Glicksman, Acta Metall. **32**, 2001, 2013 (1984).
[10] R. Trivedi and K. Somboonsuk, Scripta Metallurgica **18**, 1283 (1984).
[11] T. Witten in *Physics of Finely Divided Media*, Proceedings of Les Houches Workshop 3/85, M. Daoud, ed. (Springer, to be published).
[12] T. A. Witten and L. M. Sander, Phys. Rev. Lett. **47**, 1400 (1981).
[13] T. A. Witten, Y. Kantor, and R. C. Ball, to be published.
[14] D. Wilkenson in Les Houches Proceedings, cf. Ref. 11 above.
[15] L. Paterson, Phys. Rev. Lett. **52**, 1621 (1984).
[16] T. A. Witten, Jr. and Paul Meakin, Phys. Rev. B **28**, 5632 (1983).
[17] P. Meakin, Phys. Rev. A **27**, 2616 (1983).
[18] Z. Racz and T. Vicsek, Phys. Rev. Lett. **51**, 2382 (1983)
[19] P. Meakin, Phys. Rev. B **30**, 4207 (1984).
[20] P. Garik, Phys. Rev. A **32** (1985), in press.
[21] E. Brézin, J. Phys. (Paris) **43**, 15 (1982).
[22] See this author's contribution to this book.
[23] R. M. Brady and R. C. Ball, CECAM Workshop, Orsay France, Sept. 1984, unpublished; see also Meakin's contribution to this book.
[24] M. Plischke and Z. Racz, Phys. Rev. Lett. **53**, 415 (1984).
[25] L. Niemeyer, L. Pietronero and H. J. Wiesmann, Phys. Rev. Lett. **52**, 1033 (1984).

[26] J. Isaacson and T. C. Lubensky, J. Phys. (Paris) Lett. **41**, L469 (1980).
[27] M. Tokuyama and K. Kawasaki, Physics Letters A **100**, 337 (1984).
[28] H. G. E. Hentschel and J. M. Deutch, Phys. Rev. A **29**, 1609 (1984).
[29] T. Vicsek, J. Phys. A **16**, L647 (1983).
[30] S. Alexander, private communication (1982).
[31] L. Pietronero and H. J. Wiesmann, J. Statistical Phys. **36**, 909 (1984).
[32] R. M. Bradley and P. N. Strenski, Phys. Rev. B **31**, 4319 (1985).
[33] G. Parisi and Y.-C. Zhang, Phys. Rev. Lett. **53**, 1791 (1984).
[34] G. Parisi and Y.-C. Zhang, to be published.
[35] R. Ball, M. Nauenberg, and T. A. Witten Phys. Rev. A **29**, 2017 (1984).
[36] M. Nauenberg and L. Sander, Physica A (Netherlands) **123A**, 360 (1984).

Hans Herrmann

DLA IN THE REAL WORLD

R. C. Ball

Cavendish Laboratory
Madingly Road
Cambridge CB3-0HE ENGLAND

INTRODUCTION

This talk will introduce you to some experimental contexts where the diffusion limited aggregation (DLA) model of Witten and Sander has been found to be relevant. I must stress that while many of the comments are my own, the original research is that of the authors referred to.

The crucial features of the original DLA model are that the cluster growth is limited by the rate of diffusion of incoming particles, and that they stick rigidly and irreversibly on contact with it. The examples discussed below do not all share these physical mechanisms, but involve also electrical conduction, fluid flow, dielectric breakdown and heat conduction limiting the growth rate; furthermore the 'deposit' is a fluid in some cases. What the examples do all share is a common underlying mathematical form, which is briefly outlined for DLA below.

Consider the diffusion equation for the probability distribution of an incoming particle, equivalent to the *average* concentration $u(\underline{r},t)$ of such mobile particles if we consider many diffusing at once:

$$\frac{\partial u}{\partial t} = D\nabla^2 u, \tag{1}$$

where D is the diffusion coefficient. Because the particles stick irreversibly on the cluster and never escape, the cluster presents a perfect sink and we have the boundary condition that $u(\underline{r},t) = 0$ around the cluster surface. Far away, where particles are supplied isotropically, we have $\underline{u}(r,t) = u_\infty$ at large radius $r = R_\infty$ where for $d > 2$ dimensions one can also let $R_\infty \to \infty$.

If the cluster is made of solid density σ then we have the *average* rate of advance of its interface v_n given in terms of the normal derivative of diffusant concentration by

$$\sigma v_n = D \frac{\partial u}{\partial n}, \qquad (2)$$

in order to conserve material.

These equations define the average motion of the system at each time step (and are exactly the Ivantsov equations for diffusion limited solidification with no surface tension). Because they are unstable with respect to the growth of roughness of the interface, it is an important defect that they omit both a short scale cut-off and noise. The short scale cut-off is very naturally provided with DLA on a lattice and it is easy to generalize the equations to take this into account, but the noise is more crucial. Because the deposition is quantized, the interface advance is subject to shot noise, and the perturbations induced by this are amplified indefinitely with time into the beautiful dendritic fingers of DLA.

Finally equation (1) contains a large scale cut-off, the diffusion length ξ, below which scale the diffusion field u is 'quasi-stationary,' i.e., Laplace's equation

$$\nabla^2 u = 0 \qquad (3)$$

is a good approximation. The DLA model in which only one particle enters at a time is the extreme implementation of this limit. In general we have

$$\xi \simeq \sqrt{Dt} \gg R$$

for the model to apply to a growth out to radius R in time t.

Summarizing mathematically, it suffices for the interface advance to be given by

$$v_n = \text{(constants)} \ \frac{\partial \phi}{\partial n} + \text{(shot noise)}$$

where $\phi(r,t)$ is a scalar field obeying $\nabla^2 \phi = 0$, $\phi = 0$ on the interface, ϕ isotropic far away. Short distance effects must limit the growth of local surface roughness at a fixed scale, but note that the equations do not demand a solid deposit so long as its interface motion is appropriately constrained.

DIFFUSION LIMITED ELECTRODEPOSITION OF COPPER *See Brady and Ball, Nature* **309**, *225 (1984)*.

Here it was the time dependent kinetics of the deposition process that was measured. The deposit was solid copper, its growth limited by diffusion of Cu^{2+} ions.

Copper sulphate solution was electrolyzed under conditions such that at the cathode only the deposition of copper contributed to the current. Thus there could be no net current of sulphate ions: either (as normally occurs) the electrical contribution to the flux of sulphate ions is balanced by convection of sulphate-rich solution, or else—as in this case—the sulphate ions (and an added excess of sodium sulphate) screen the electric fields from the solution. This screened regime was only achieved by inhibiting convection, using a small amount (1% w/w) of high polymer to viscosify the solution. The result is that only ionic diffusion contributes to the flux of copper ions towards the cathode, which acts as a perfect sink.

The size of the cathode was dominated by the deposited copper, so that the current I (measured electrically) was limited by the overall radius of the deposite R. It is easily shown that

$$I = \beta R,$$

so that the current was a measure of radius; on the other hand by Faraday's law the mass deposited is given by

$$M = 1/\alpha \int I dt = Q/\alpha,$$

where α and β are known coefficients. Thus in principle the current I and its time integral, the charge Q, give the radius-mass relation for the growth. In practice, to avoid constants of integration, the quantity

$$(I\frac{dI}{dt})^{-1} \propto \frac{dM}{dR^3},$$

giving a spherically averaged density of deposition, was plotted logarithmically against $I \propto R$ (see Fig. 1). Comparing to $dM/dR^3 \propto R^{d_f - 3}$ gave an inferred fractal dimension $d_f = 2.43 \pm 0.03$, in agreement with simulations of DLA in three dimensions (e.g., P. Meakin, Phys. Rev. A **27**, 604 (1985)).

This fractal behavior was measured over the range 20 to 300 μm, but the amplitude of the power law scaling was consistent with fractal behavior down to the observed scale of metallic grains. Thus it would appear that the grain was the effective unit of deposit and provided a scale with shot noise: either a grain was seeded and grew, or it did not. Electron microscopy revealed the deposits to be solid and faceted up to a scale of about 40nm, and dendritic thereafter, so the fractal behavior probably extended over four decades of length—far in excess of any DLA simulation to date.

ELECTRODEPOSITION OF ZINC METAL LEAVES *See Matsushita, Sano, Hayakawa, Honjo and Sawada, PRL* **53**, *286 (1984)*.

The leaves were grown in a two-dimensional geometry and their two-point correlation function $C_2(r)$ obtained by image analysis. The metallic zinc deposit grew in a thin interfacial layer between a shallow layer of zinc sulphate solution, whose electrical conductance probably limited the growth, and a (deep) covering of n-butyl acetate (see Fig. 2).

The preference for deposition on the interface is a specific effect, and has the great benefit that the very thin deposits obtained grow radially quite rapidly. The electrolyte was at concentration 2 Molar and 4 mm deep, so that convection should have been effective in balancing the flux of sulphate ions. Then the net (two-dimensional) electrical current density (of zinc ions) would be given by

$$\underline{J} = \sigma \underline{E},$$

where σ is a solution conductivity multiplied by the depth of the fluid. To avoid space charge build-up we have

$$0 = \nabla \cdot \underline{J} = \sigma \nabla^2 \phi,$$

where ϕ is the electrostatic potential playing here the mathematical role of concentration in DLA. Because the metallic deposit is essentially at equipotential, ϕ has the appropriate boundary conditions, and the advance of the leaf is proportional to its normal derivative if constant thickness of the leaves is assumed. The latter was not measured.

The depth of electrolyte provides a natural lower cut-off to the two dimensional behavior. Below this scale one has three-dimensional transport supplying a deposit in two dimensions, which for DLA must be compact by the causality bound of Ball and Witten (Phys. Rev. A **29**, 2966 (1984)). The Japanese authors did not quantify the cut-off in their paper, but it appears consistent with the above.

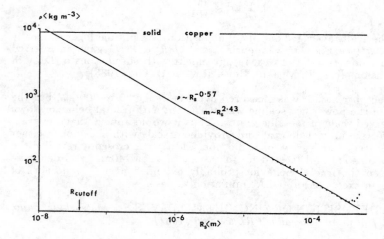

FIG. 1: Logarithmic plot of $\rho = dM/dR^3$ vs radius R_s for a diffusion limited copper electrodeposit, both quantities inferred by electrical measurement (see text). The observed scale of granular structure is indicated by R_{cutoff} and agrees well with the intercept of solid copper density. Reproduced from Brady and Ball, Nature **309** (1984).

For low enough applied voltages, $C(r) \sim r^{d_f-2}$ was observed over a radius range of about 30, with $d_f = 1.66 \pm 0.03$ in agreement with simulations of DLA in two dimensions (e.g., Witten and Sander, Phys. Rev. B **27**, 5686 (1983)). For higher voltages $V > V_c = 8$ volts, higher values of d_f were observed which the authors tried to associate with ballistic aggregation.

FRACTAL GROWTH OF VISCOUS FINGERS *See Nittman, Daccord and Stanley, Nature* **314**, *141 (1985)*.

The fingers of one fluid penetrating into another were observed in a two-dimensional geometry (Hele-Shaw cell) between two fixed glass plates. An example is reproduced on Fig. 3. Because the flow rate must be zero at the plates there is, at scales beyond the plate separation, a Poiseuille drag on the fluid proportional to the two-dimensional flow rate. Thus flow is proportional to pressure gradient (Darcy's law),

$$V = -\alpha \nabla p.$$

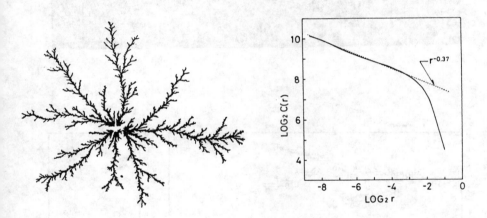

FIG. 2: (a) Ohmically-limited growth of zinc metal leaf from a thin (4 mm) layer of electrolyte. (b) Two-body correlation function indicating a fractal dimension of 1.63 for this image. Reproduced from Matsushita et al, Phys. Rev. Lett. **53**, 286 (1984).

For an incompressible fluid obeying Darcy's law, the constraint $\nabla \cdot v = 0$ leads to $\nabla^2 p = 0$ consistent with analogy to the DLA model, as pointed out by Paterson (PRL **52**, 1621 (1984)). The boundary conditions at the interface are only consistent with DLA if the viscosity of the driving fluid is negligible compared to that of the fluid displaced. Then the pressure variation in the less viscid fluid is negligible and the interface is at near constant pressure. A further complication is the pressure drop across a curved interface due to surface tension (capillary) forces. This would give a short scale cut-off length λ_c which varies as $v^{-1/2}$, which is *not* constant as in the DLA model.

The beauty of their experiment is that Nittman et al used water (dyed) and a viscous high polymer solution as their two fluids, giving a large viscosity ratio but negligible interfacial tension. As a result the thickness of the cell should have intervened as a short scale cut-off of the fingering in two dimensions, but this was not quantitatively assessed.

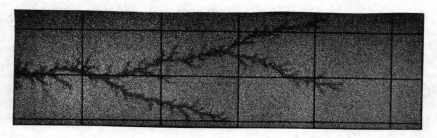

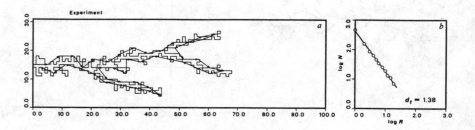

FIG. 3: Viscous fingers created by (dyed) water advancing into polysaccharide solution between two plates. (b) The interface measured with a ruler of length $R = 5$ on a digitized image. (c) Logarithmic plot of the number of ruler lengths N vs R whose slope gives $-d_f$. Reproduced from Nittmann et al, Nature **314** (1985).

A complication to interpreting the experiment is that the polymer solution was significantly shear thinning, i.e., less viscous at high flow rates. This alleviated otherwise experimentally excessive pressure drops but also means that the Paterson argument does not strictly apply.

The fractal scaling of the interface was measured rather than that of the bulk of the fingers (see Fig. 3). For DLA (which is all interface) this would make no difference, and for the fluid fingers this obviated influence of fattening of basal fingers on the measured fractal dimension. Some fattening was observed, but not quantified. The resulting fractal dimension was $d_f = 1.40 \pm 0.04$ which is low for DLA in two dimensions.

The authors attributed the low fractal dimension to finite size effects and supported this with computer simulation data for DLA growing in a corresponding strip geometry. However, the use of absorbing boundary conditions on the side walls, instead of reflecting conditions (for no net flux corresponding to no net flow into the wall), will have exaggerated size effects in the simulation. The shear thinning effect of the polymer will have disproportionately favored the growth of the fastest growing fingers, leading to fewer fingers and so another cause of lowered fractal dimension to be taken into account.

Paterson discusses the analogous Darcy law system of fluids flowing through a porous medium, and hence the possible importance of such fingering in secondary oil recovery. There one displaces oil from porous rock by injecting (generally less viscous) aqueous media, and viscous fingering could in principle occur on vast scales. However the viscosity ratios are generally modest (< 10) and a significant proportion (30 – 50%) of the oil is believed to be displaced compared to the vanishing proportion that pure DLA would predict.

Further work with a Hele-Shaw cell has recently been performed by Ben-Jacob et al (submitted to Phys. Rev. Lett.), using air displacing (Newtonian) glycerin from a point entry in the center of the cell. This system is complicated by being controlled by its (flow dependent) capillary length. Its geometry is the same as that of simple 2d DLA cluster growth and so there is no immediately favored direction; thus they were able to explore the consequence of supplying a preferred direction by means of lines ruled on the plates. They found the interesting result that branching is hindered when the lines are povided.

FRACTAL STRUCTURE OF DIELECTRIC BREAKDOWN PATTERNS
See Niemeyer, Pietronero and Wiesmann, PRL 52, 1033 (1984).

The forked patterns of electrical discharge from a central point electrode towards an outer circular one were observed in compressed SF_6 over a glass plate (see Fig. 4). This presents a mixture of two- and three-dimensional problems, because although the breakdown was confined to the two-dimensional geometry, the electrostatic fields controlling it were not.

The electrostatic potential around a discharge obeys the correct boundary conditions for DLA in the approximation that the ionized region is a good enough conductor to be considered at equipotential. However, it is not entirely clear quantitatively how the advance of the spark is related to the electric field at its tip. The authors considered a class of models with

$$v_{\text{tip}} \propto E_{\text{tip}}^n$$

for various different possible n. Only the simplest assumption of $n = 1$ leads to DLA, and simulations showed that the fractal dimension of such models decreases continuously with increasing n. It is important to note however that these simulations were performed with the diffusion field as well as the deposit confined in two dimensions.

Digital image analysis of the breakdown patterns gave a fractal dimension of about $d_f = 1.7$ (consistent with two-dimensional DLA) where, as with the viscous fingers, apparent spark width was not allowed to bias the measurement *except* that in this case sparks below a certain brightness would have been lost, as noted in the paper.

The experimental d_f was held up as indicating that the system was governed by $n = 1$, equivalent to 2d DLA. However the three-dimensional freedom of the electric field is a serious complication, because of the causality bound that $d_f \geq 2$ for DLA grown from 3 dimensions—even if the growth itself is restricted to a plane. Thus the present author is led to conclude that *if* dielectric breakdown advance is governed by a power of the field, then $n > 1$ distinctly.

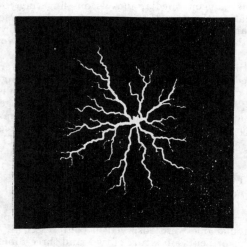

FIG. 4: Time integrated photograph of a surface leader discharge on a 2 mm glass plate on SF_6 gas at 3 atmospheres. Reproduced from Niemeyer et al, Phys. Rev. Lett. **52**, 1033 (1984).

This is further supported by consideration of dielectric breakdown fully in three dimensions, for which $n = 1$ would imply $d_f = 2.4$. Whilst it is hoped that it will not be possible to demonstrate this at Cargese, fork lightning has a fractal dimension apparently less than two.

DENDRITIC SOLIDIFICATION PATTERNS *See Langer, Rev. Mod. Phys.* **52**, *1 (1980) and refs. below.*

Solidification fronts between solid and supercooled liquid may be limited in their advance by diffusion of compositional mismatch between liquid and solid (as in alloys) or by conduction of latent heat. Generally both the capillary length from surface tension and the diffusion length from the overall advance rate are important constraints on the behavior. There have been no explicit claims that the growths are fractal, and attention has focussed on understanding the beautiful *regular*, side-branched dendrite structures which can be formed (see for example Fig. 5). Quantitative prediction of this non-linear behavior has been studied as an archetype of pattern selection problems (see Ben-Jacob et al, PRL **53**, 2110 (1984) for a recent reference).

A simplifying limit of the solidification front problem is the Boundary Layer Model (Ben-Jacob et al, PRL **51**, 1930 (1983)) in which the diffusion

FIG. 5: Snowflake adapted from a photograph by Nakaya. Reproduced from Langer, Rev. Mod. Phys. **52**, 1 (1980).

length is taken to be small compared to the local scale of the dendritic features. Interestingly it was shown that material anisotropy of the solid is a critical parameter for the occurrence of dendritic side branching. Too high anisotropy gave just needle-like growths. This corroborates with the Hele-Shaw cell observations that anisotropy hinders the branching of viscous fingers.

Very recently Ball, Brady, Rossi and Thompson (submitted to PRL) have observed a related phenomenon for anisotropic DLA in two dimensions. The vertical faces of the particles were deemed to be 'stickier' than the horizontal ones, and in agreement with a simple theory clusters were observed to grow indefinitely anisotropic in their profiles. Thus the clusters became needles, and also ceased to be fractal beyond a scale which increased as the sticking probabilities in the two different directions were made more equal.

Thus although the Boundary Layer Model, capillary length limited viscous fingers and DLA represent different limits for the controlling lengthscales of the deposition problem, all exhibit sensitivity to material anisotropy, losing their large scale dendritic behavior in favor of needle growth. In all cases this came as a surprise, and suggests that the general notion of growth direction might play an important role in a more formal understanding of the deposition of diffusing particles.

CONCLUDING REMARKS

Together these examples show that a 'diffusion' law of deposition comes in many disguises, but this alone does not suffice to drive DLA. Microscopic surface details are crucial and even anisotropy can destroy the fractal behavior. Nevertheless, the basic speculation of Witten and Sander (PRL **47**, 1400 (1981)) that DLA belongs to a universality class of real material behavior, has been experimentally substantiated in both two and three dimensions.

Tom Witten

PERCOLATION AND CLUSTER SIZE DISTRIBUTION

Dietrich Stauffer

Institute of Theoretical Physics
Cologne University
5000 Köln 41 WEST GERMANY

What am I doing here, except to look at the sea and the bathers? Apparently the organizers did not know that this summer school is on growth and forms and that percolation usually is not a growth process. Thus I will try to make the best of their error: I will use percolation theory to explain some of the concepts that are also used by the "growth people." Percolation is simpler than many growth processes, and thus much more is known about some percolation properties than about the corresponding properties in various growth processes.

Thus I will talk about what percolation is, how to simulate it on a computer, and how kinetic concepts can be introduced into it. This overview will constitute my first talk. In the second talk I will go more into the details of how the static cluster size distribution is described, and how one can calculate from it universal properties.

Let us mention where percolation concepts may be useful. Sputtered films of randomly mixed conducting and insulating materials seem to follow quite precisely the rules of percolation theory. Also in dilute ferromagnets at very low temperatures one has a regime where percolation concepts apply. Less clear are its usefulness for gelation of branched polymers or for fluid flow through porous media. A better overview over these various applications is given in the Introduction of Ref. 1. A textbook for students is also available on the theoretical aspects of percolation,[2] including computer simulations.

1. BASIC CONCEPTS
1.1 DEFINITION OF PERCOLATION

What is percolation? Let every site of a large lattice be randomly occupied with probability p, and be randomly empty with probability $1 - p$. A cluster then is a group of occupied neighboring sites, that means: Every site of a cluster is connected with every different site of the same cluster by at least one chain of sites such that neighboring sites on that chain are only one lattice constant apart. Occupied sites on different clusters are not connected by such chains. I thing Fig. 1 tells us better than these words, what is a cluster and what is percolation.

Figure 1 marks particularly the largest cluster in the system. If that largest cluster extends from one end of the system to the other we say that it percolates through the system. For small concentration p of occupied sites it is highly unlikely that such a large cluster exists. For p close to unity, however, the largest cluster will comprise most of the occupied sites and thus will touch all opposite sides of the system. In the limit of infinitely large systems, the largest cluster is infinitely large, too, if it percolates. There exists one unique threshold concentration p_c such that in infinitely large systems there is an infinitely large percolating cluster present for p larger than p_c but not for p smaller than p_c. Of particular interest is the border case $p = p_c$ where we call the largest cluster an "incipient infinite cluster." Its fractal properties will be discussed below.

After this definition of the problem you will perhaps agree with me that percolation is one of the simplest problems of statistical physics (at least, its definition). That's why the organizers selected it for me. In the following part I will try to explain how one can solve percolation problems, particularly if one has a good computer available. I will postulate several scaling laws. The more difficult task of explaining these scaling relations through a geometric interpretation was wisely left by the organizers to the next talk from the Camorra Napolitana.

1.2 COMPUTER SIMULATION

Computers cannot replace thinking but they may support it. In the ideal case the researcher finds out, after many hours of computer time, that the results suggest a certain exact solution, and then he finds a mathematical proof for that solution. Unfortunately, such ideal cases are rather rare, and thus I will not talk very much about exact solutions.

Another advantage of computers is that they can produce nice pictures of the objects you are dealing with, and that again may trigger your own thinking. However, in our field of reseach most properties we are looking for are averages over many clusters or other objects. In one single cluster one may be tempted very much to find just that property one wants to find, and other people may find other properties in the same picture. Thus when things become complicated you always have to average over many points by computer and not by eye; the computed averages may then be analyzed properly with a pocket calculator whereas beautiful high-resolution color graphics may not solve your problem. In general, the main problem of computational statistical physics lies in the interpretation data, not in their production.

For example, every computer is necessarily using a finite time and a finite memory to get any result, whereas physicists are often interested in the

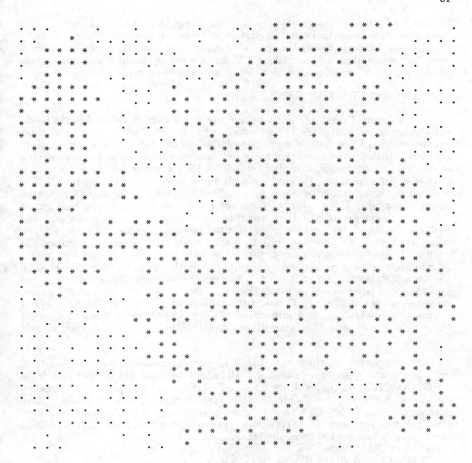

FIG. 1: Example of square lattice at its percolation threshold, $p = 0.59273$. The largest cluster contains 500 sites in this 35 x 35 lattice and is marked by $*$, the smaller clusters are maked by dots, whereas empty sites and isolated occupied sites are not shown. No periodic boundary conditions were used in the cluster analysis.

behavior of infinitely large systems and infinitely large times. Thus all your computer data need to be extrapolated to these asymptotic limits. Usually this extrapolation is the main source of error, and not the purely statistical error of your computer simulation. Thus in case of doubt trust your brain more than your computer, but let the computer support your brain.

1.3 RANDOM NUMBER GENERATION

The computer methods most people are using here is called Monte Carlo simulation because it uses random numbers. Since Cargese does not have roulette tables, I use instead a programmable pocket calculator to get random numbers r distributed homogeneously between zero and unity. Let r be such a random number; then I multiply r by 899 (or 997) and take only the fractional part of the product, i.e., I throw away all digits before the decimal point. The result is again a number between zero and unity. This process can be repeated again and again. If your initial value for r was not too bad, you will get in this way a sequence of numbers which are not correlated in any obvious way and are roughly homogeneously distributed between zero and unity. (If you want to cheat in this pseudo-roulette game, start with a simple number like $r = 0.2$.) If in any computer experiment you want to repeat the same sequence of random numbers, you merely have to start with the same "seed," i.e., with the same initial value for r. Then you will get the same results again, provided you made no programming error, the computer is working properly, electric power does not fail, etc. This reproducibility of random numbers is quite practical if you want to test improvements in your program: You can check if the results are exactly the same if your modifications mean that they should be.

Big computers do the same things as programmable pocket calculators, only faster and with more numbers. (I hope I will not get a libel suit from CDC, Cray, IBM or Digital Equipment now.) For IBM computers, where each integer is stored in 32 bits, you merely have to replace 899 by 65539. More precisely, in the FORTRAN computer language you get from a random integer IBM between zero and $2^{31} = 2147483648$ (always start with an odd integer IBM!) another such integer IBM via the statements
    IBM = IBM*65539
    IF(IBM.LT.0) IBM=IBM + 2147483647 + 1
(If you replace the last sum simply by 2147483648 the computer goes on strike; give him something to do! The product of two positive integers may become negative for computers since they may treat the first bit as the sign bit.) The multiplication in the first line will give in general a number greater than 2^{31}, and the computer throws away the leading bits. This throwing away of the leading part corresponds to taking the fractional part with the pocket calculator. If one insists in getting real numbers between zero and unity, and not integers up to 2^{31}, one may multiply each such integer with $1/2^{31}$, but quite often one saves computer time by working with the large integers directly.

It is well known that this random number generator has flaws. If you use three consecutive random numbers to calculate the positions of a point in a 20 x 20 x 20 cubic lattice, and if you want to fill up this lattice by selecting randomly a lattice point, filling it up if still empty, and repeating this process again and again, then you will find out that most points never get filled within the computer time theoretically required to fill all but a few points. One can avoid these difficulties by giving each lattice point not three coordinates i,j,k but only one which runs from zero to N, the total number of points. (This trick will also save computer time in many cases.) With less than 2^{30} different random numbers called in a project, this random number generator then is reasonably good and fast (particularly if the two lines above are written directly into the program instead of being called as a subroutine or function).

If you have a computer with 60 or 64 bits for a random number you get better results if you replace the multiplicator 65539 by a much larger number. The big 60-bit CDC computers use for their function RANF, with $0 < \text{RANF} < 1$, the number 20001207264271730565, where each digit here has to be interpreted as an octal, not as a decimal number. Again one can save computer time (C. Kalle, private communication 1983) if one puts the two lines into the program and works with large integers, instead of calling the function RANF and using real numbers up to unity. On a CDC Cyber 205 vector computer, a large number of RANF random numbers can be efficiently produced by Zabolitzky's tricks described in Ref. 3.

In most problems these 60- or 64-bit random numbers are random enough. If one has only 32 bits, or if one has problems[4] even with 64-bit RANF, one may use a Tausworthe shift generator for which an IBM assembler program was published.[5] Imagine you have already 250 different random integers $I(K)$ produced in some good, though perhaps slow, way (for example, by calling a random number RANF for every bit of the integer and setting that bit equal to unity if and only if RANF $< 1/2$). Then the computer gives new random integers through

$$I(250+K)=\text{XOR}(I(K),I(K+147)),\ K = 1,2,3...$$

where XOR stands for the bit-by-bit exclusive or. (XOR(left,right) is true if the Left governs and the Right is in opposition, or if the Right governs and the Left is in opposition. If both parties form a coalition, or if none of them really governs, XOR is false. In other words, XOR is true if and only if the truth values of its two arguments are different.) Again, Marsaglia[6] has criticized this XOR method and suggested several alternatives.

In summary, random number generation is an art, not a science. I recommend that you program in a simple and efficient way whatever runs on the computer. If the results are suspicious, compare them with those obtained from a different random number generator. And the rest should be left to the experts.

1.4 COMPUTER PRODUCTION OF PERCOLATION CLUSTERS

With the help of random numbers we now want to produce a picture similar to Fig. 1.

For this purpose we have to decide for a given lattice site if it is occupied or empty, provided the probability to be occupied is p. We achieve that aim by taking it first empty, and then making it occupied if a random number r is smaller than p. It is obvious that this procedure occupies that site with probability p; for a number r randomly distributed between zero and unity is with probability p in the interval between zero and p, and with probability $1 - p$ in the interval between p and unity. If we use random large integers like IBM above, we simply compare IBM with IP where IP, defined once at the beginning of the calculation, is p*2**31 for 32-bit computers. If we use the random number generator RANF for real numbers between zero and unity, then

$$I = \text{PPLUS1-RANF}$$

with PPLUS1 $= 1 + p$ defined one at the beginning, gives $I = 1$ for occupied sites (random number smaller than p) and zero otherwise. Since computers prefer calculating to thinking thus an arithmetic assignment may be faster than logical if-condition.

In this way we can go regularly, like a typewriter, through the whole lattice, and with a fresh random number for each site fill the whole lattice randomly as required for percolation. This method is the simplest way, can be taught easily to beginners, and has been used for lattices containing more than 10^{10} sites.[7] In courses on statistical physics, this method has repeatedly been used to introduce students to numerical analysis. Each pair of students is given 15 pennies to produce random bits, and a piece of paper with a 15 x 15 triangular lattice. Throwing these coins, the students determine line by line which sites are occupied and which sites are empty. Then the students mark the various clusters and count them as a function of size. Finally, the cluster numbers of the whole classroom are added up, and the results are drawn on a log-log plot. The negative slope of a straight line gives the effective value (typically, 1.5 is found) of the Fisher exponent τ ($= 187/91$; see below). The whole procedure can be finished in 90 minutes.

One sees already from such a simple classroom experiment that boundary effects are very important. Theoretically, the number of isolated sites in the triangular lattice at $p = p_c = 1/2$ should be (lattice size)/128 since that site must be occupied, the six neighbors must be empty. Actually, the observed numbers are higher since the boundaries cut larger clusters into smaller pieces. Often the use of periodic boundary conditions reduces these effects from the finite system size. Then the non-existing left neighbor of the left-most site in one line is identified with the existing right-most site of that line, with an analogous definition for top and bottom neighbors.[7] Nevertheless, boundary effects even in lattices containing 10^{10} sites are important at the threshold, and thus a different "growth" method[8-10] has been devised which, of course, is also more appropriate for this summer school.

For this growth method, one builds up one and only one cluster. First one starts with an occupied site. Then the shell of neighbors surrounding that site is investigated by selecting a random number for each site and deciding in this way whether it is occupied or empty. In a second step, one looks at the shell of neighbors of previously investigated occupied sites, then comes the third step, and so on. The number of such steps needed to reach a given occupied site is sometimes called the "chemical distance"[11] and counts how many nearest neighbor jumps from one occupied site to the next are needed to reach that site from the origin. It may also be identified with a time[9] and then measures how fast an epidemic[10] spreads through a population, provided this population is a rigid set of randomly occupied lattice sites. The growth process stops if all sites of the latest shell turn out to be empty. Then one simply counts the number of occupied sites in that finite cluster in order to know its "mass." By repeating this process again and again, one finds clusters of various sizes and shapes. The number of such clusters appearing in the Monte Carlo statistics is proportional to the probability that the origin belongs to such a cluster.

Thus this cluster growth method automatically gives the cluster size distribution without further analysis. It also avoids all boundary effects provided the cluster grown in this way did not touch anywhere the boundaries of the computer memory reserved for the growth. (The lattice filling method described above gives slightly wrong results even for small clusters since the boundaries may cut large clusters into smaller pieces.) A disadvantage is that above the threshold most growth attempts will lead to an infinite cluster and have to be terminated without giving information on the cluster statistics.

A modified cluster growth method got its name from "invasion percolation"[12] but has nothing to do with the French invading Corsica; the Schlumberger group is interested in fluids like oil or water invading the pores of sand(stone). This method can be described[12] as follows: Every site of a lattice is associated with a random number r between zero and unity. The origin is regarded as occupied. We then occupy that perimeter site with the smallest random number, determine the new perimeter, occupy the perimeter site with the lowest random number, etc. (The perimeter sites are all those sites which are not yet occupied but are neighbors to occupied sites.) There is no probability p in this growth mechanism, but the percolation threshold p_c corresponds to that value for the minimal random number which is chosen as very large clusters continue to grow.[12]

A fourth method[13] gives only properties of clusters of a given size, and not the probability to get such clusters in random percolation. It is similar to reptation in the simulation of linear polymers, but more complicated because of the branching of percolation clusters. It is suitable mainly to simulate those relatively large clusters which occur very seldomly in random percolation, like clusters of several hundred sites for probabilities p close to zero.

So far, the last two methods have seldomly been used, and we now return to the first method of filling up a lattice. How do we count clusters?

1.5 COMPUTER ANALYSIS OF PERCOLATION CLUSTERS

The total number of clusters containing s sites each, divided by the number of sites in the lattice, is called the cluster number n_s. The fraction of sites belonging to the largest cluster is called the percolation probability P. Finally, the second moment of the cluster size distribution, $\chi = \sum_s s^2 n_s$, is called the susceptibility and is a measure of the mean cluster size (we omit the largest cluster from this sum). A randomly selected site sits with probability $n_s s$ in a

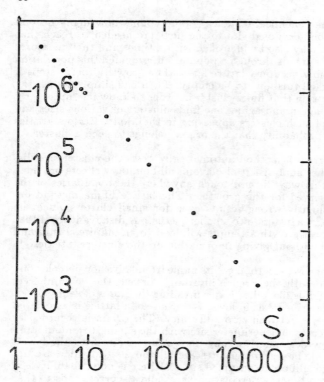

FIG. 2: Log-log plot of cluster numbers at p_c in a simple cubic lattice with 216 million sites. We show the number of clusters containing at least s sites each; thus a straight line through the data should have slope $1 - \tau = -1.2$ for large cluster size s.

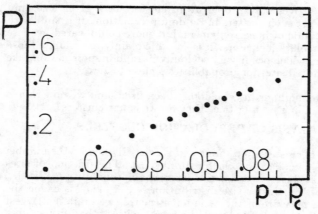

FIG. 3: Log-log plot of fraction P of sites belonging to the largest cluster versus distance from threshold.

cluster of "mass" s, and with probability P in the largest cluster. The average number of sites in a cluster to which a randomly selected site belongs is the susceptibility, at least below the threshold.

All these quantities thus can be calculated only if we know how to find out which sites belong to one cluster. This task is trivial in the cluster growth method where only one cluster is produced at a time. If instead we fill the whole lattice randomly at the beginning, the *a posteriori* analysis is not so trivial. The Hoshen-Kopelman algorithm[14] is widely used for this purpose since it allows to store only one line at a time for two-dimensional lattices, or one plane in three dimensions. A detailed description, together with a complete FORTRAN program and a sample output, is given in Appendix 3 of Ref. 2 and thus not repeated here. Rapaport[7] improved it by incorporating periodic boundary conditions, splitting big lattices into smaller subsections, and writing the central parts of the programs in assembler language. On an IBM 3081 computer, he needed only 3 microseconds per site for the filling and analysis of a 160000 x 160000 triangular lattice, the largest system ever simulated so far to my knowledge.

The largest 3d percolation samples are simple cubic 600 x 600 x 600 lattices simulated for the purpose of this school on a medium-size Masscomp 500, which needed nearly 5 hours for each run. Figure 2 gives the cluster numbers at the percolation threshold, Fig. 3 the percolation probability P (size of largest cluster), Fig. 4 the susceptibility (mean cluster size). We will interpret in the next section the scaling character of these results. Already now we can conclude qualitatively that the percolation probability goes to zero continuously, that the percolation threshold is about 0.312, that the susceptibility diverges whether one approaches this threshold from above or from below, and that the cluster numbers decay asymptotically with a simple power law.

Lots of people are also interested in fractal dimension. Thus Fig. 5 gives the size of the largest cluster at the threshold, as a function of system size. We see that this "incipient infinite cluster" has a mass increasing for large systems as L^D where L is the linear dimension of the lattice, and the fractal dimension D is close to 2.5, definitely lower than the Euclidean dimension 3.

1.6 ANTS AND OTHER KINETIC ASPECTS

These figures (1-5) give static properties only. We mentioned already that the cluster growth method can be interpreted[9,10] as a kinetic process; it leads, however, to the same static end results. In Corsica perhaps we should interpret it as a forest fire[15] rather than an epidemic,[10] and in fact this fiery interpretation again has led to the "burning" way to determine the so-called backbone.[16] The backbone is that part of the lattice which carries an electric current if occupied sites are conducting, empty sites insulating, and if a voltage is applied to the lattice.

Can we also look at kinetic processes happening after the cluster production process? We can. One of the best studied models is the "ant in the labyrinth," an insect species invented by the French genetic engineer de Gennes,[17] raised at the beginning[18] mainly in Marseille, France and Pomona,

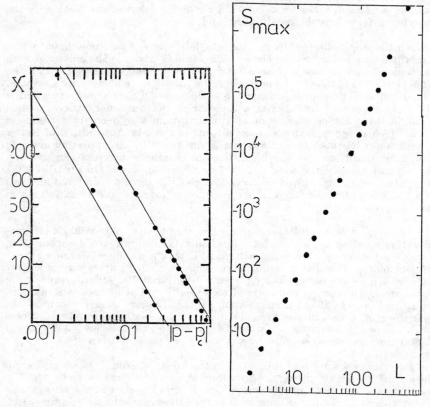

FIG.4: Log-log plot of susceptibility versus distance from threshold. The two parallel lines have a slope $-\gamma$; their distance indicates that the susceptibility below p_c is about 8 times larger than that above p_c

FIG.5: Log-log plot of largest cluster size versus linear dimension of lattice at the percolation threshold. The data follow a straight line asymptotically, with slope = D = fractal dimension near 2.5.

California, and now infecting many parts of the world. Very simply we construct a random lattice as usual in percolation theory, let an ant start on one of the occupied sites, and let it diffuse only on that cluster to which this occupied site belongs. Diffusion here means that at every time step the ant selects randomly one of the nearest-neighbor directions and moves to that nearest neighbor if and only if that neighbor site is occupied. Thus we have a random walk ("ant") performed on a random structure ("labyrinth").

Of particular interest is the behavior right at the percolation threshold. The incipient infinite cluster at p_c is neither really three-dimensional nor a finite object but is a fractal with a fractal dimension D larger than zero but smaller than the Euclidean dimension. Similarly the ant starting on the incipient infinite cluster is neither trapped in a finite cluster nor is it normally diffusing as on an infinite regular structure. Its motion can be described by some special fracton or spectral dimension.[19] Forgetting about these complicated words, we look at Fig. 6 which shows that at the three-dimensional percolation threshold the rms distance travelled by the ant increases asymptotically as (time)$^{0.2}$, if we average over all clusters as starting points. This exponent 0.2 interpolates between the exponent zero below the threshold (trapping in finite clusters) and the exponent $1/2$ obtained above p_c as for normal diffusion on regular lattices. Therefore one calls this effect "anomalous" diffusion.[20]

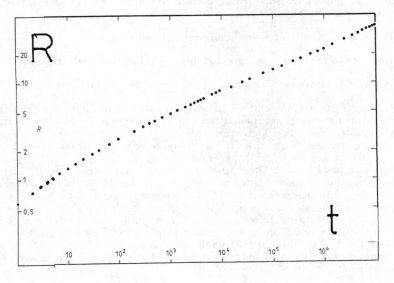

FIG. 6: Log-log of rms distance travelled by ants in a labyrinth at the percolation threshold. We show the average over many ants and many 256 x 256 x 256 lattices, with starting points on both finite and infinite clusters. Asymptotically the slope of the straight line through the data has a value near 2.0 and can be called the inverse walk dimension $1/d_w$. From Pandey et al (3).

Darwinistic evolution (also called publish-or-perish pressure) also leads to the breeding of other insect species besides these "ants." "Parasites" move on so-called lattice animals; "termites" run very fast on the clusters and very slowly off the clusters (besides they allowed me to write wrong scaling theories); and "butterflies" jump to more than nearest-neighbors only. Not surprisingly, butterflies are particularly beautiful since their anomalous diffusion exponents may depend on the fashion in which the probability to jump over a large distance decays with increasing distance.

What have these insects to do with growth? Let us look at the set of sites visited by an ant or other insect on its random walk through the random lattice. Initially, this set is just the origin, and in the course of time (time measured as the number of step attempts) it grows. If the ant started on a finite cluster, this set of visited sites will finally engulf the whole cluster. If it starts on the infinite cluster above the percolation threshold, it will grow without limits similar to diffusion in regular lattices. Right at the percolation threshold it grows, too, without limits, but slowly. Numerically, the number of visited sites increases roughly[19,21] but not exactly as $(\text{time})^{2/3}$.

It is not necessary to fill the whole lattice randomly before diffusion starts. Instead, one may grow the percolation cluster while the insect is diffusing on it and thus again avoid disturbing boundary effects. The ant then at every step has three possibilities:

It wants to move to a site already visited before: allowed.

It wants to move to a site previously determined as blocked: prohibited.

It wants to move to a new site: select new random number.

In the third case, depending on the new random number, the previously undetermined site is either made occupied in which case the ant moves there; or it is made empty in which case we call it a blocked site and let the ant stay where it was. "Growth sites" are neighbors of visited sites which have not yet been tried by the ant. The set of visited sites can grow only if the ant moves onto a growth site. Much effort has been placed into finding out how the number of growth sites depends on the number of visited sites, and where the growth sites are located.[22] The laws governing growth sites may lead to much-sought relations between diffusivity and static percolation properties. From Einstein we learned already that diffusivity and mobility (electrical conductivity) are proportional to each other. Many questions still remain open. Thus at the end of this part I merely emphasize that these problems have been used repeatedly to introduce young students, beginning from age 16, to scientific programming.[23]

2. SCALING THEORY
2.1 MATHEMATICAL ASSUMPTIONS

Scaling theory can be interpreted in physical or geometrical terms, or simply postulated mathematically. Since the first approach is presented by Coniglio in this book, the last approach is used here.

Experience with phase transitions or other asymptotic aspects of physics has indicated that most functions $F(x,y)$ of two variables approach the simpler form

$$F(x,y) = x^A f(y/x^B), \qquad (1)$$

if both x and y approach zero. If one has variables approaching infinity instead, one identifies x or y with the reciprocal value of these variables and again has x and y going to zero. Equation (1) is not a mathematical theorem, and indeed with a combination of exponentials and logarithms you may construct a function which even with suitably defined x and y does not obey Eq. (1). However, most simpler functions have this property. For example, $F(x,y) = (x + \sin(x/y))/(y + xy)$ looks quite complicated, but if x and y go to zero with a fixed ratio x/y, then in the numerator the x becomes negligible compared with the sine, in the denominator the product is less important than the y, and we get asymptotically $F(x,y) = \sin(x/y)/y = (x/y)\sin(x/y)/x$. Thus A in Eq. (1) is -1, $B = 1$, and $f(z) = \sin(1/z)/z$ in this simple example.

Of course, such results are valid only asymptotically if x and y approach zero at constant y/x. Similarly, $\ln(1 + x)$ "equals" x only asymptotically for small x; otherwise correction terms are important, $\ln(1 + x) = x - x^2/2 + \ldots$ We call assumptions like Eq. (1) the form of the leading scaling term with the scaling function f, the scaling variable $z = y/x^B$, and the (critical) exponents A and B. If more than the leading term is discussed as in $\ln(1 + x) = x - x^2/2$, we call the second term the leading correction term. Present numerical analysis seldomly says something reliable about higher-order correction terms, like $x^2/3$ in our logarithmic example.

How do we analyse data in order to check the validity of the scaling assumption, Eq. (1), and to determine its parameters? I ignore here the problems arising from the leading corrections to scaling and am satisfied with exponents correct to within, say, 10%. If x and y are reasonably small such an analysis is possible; only for extremely good data can one make a reliable correction-to-scaling analysis and may obtain exponents with much better accuracy. A detailed and careful estimate of statistical errors for Monte Carlo data is not very helpful if one does not know the systematic errors arising from the dangerous correction terms. However, for this summer school we simplify all these problems by ignoring them.

If our values for x and y are sufficiently small to make the correction terms unimportant, we can test Eq. (1) by plotting the scaled quantity $G = F/x^A$ versus the scaled variable $z = y/x^B$. Independent of whether Eq. (1) is valid or not, we get one curve of $G(z)$ versus z for every different value of x, provided we have so many good data that we can make a reasonable plot. (One data pair does not give a good curve, of course.) If Eq. (1) is valid, however, these curves for different x all collapse into a single curve, the function $f(z)$. Thus scaling means that different functions $F(x,y)$ for different x are similar to each other: If F is scaled by some power of x, and y by some other power of x, then the resulting scaled quantity $G = F/x^A$ is a function of only one scaled variable $z = y/x^B$, instead of being a function of two independent variables x and y. This similarity can be compared with the known similarity of all circles: If each circle is reduced in its radius by a factor porportional to the square root of its area, then all these different circles look the same.

Often we are concerned with the behavior of $F(x,y)$ for special cases like $y = 0$. Then Eq. (1) predicts a simple power law: F varies as x^A for small x. Plotting $\log(F)$ versus $\log(x)$ we thus should get data following a straight line with a slope of A. These log-log plots, used already in Figs. 2-6, are a standard way to find the critical exponents.

Experience has shown that in two and three dimensions most critical phenomena at continous phase transitions, and most cluster growth properties, are described by such scaling assumptions. For example, near the ferromagnetic Curie point the quantity F might be the magnetization, x the magnetic field, and y the temperature difference to the Curie temperature. If one has three or more variables, more complicated scaling assumptions apply which we ignore here. Also exceptions are known; for percolation in six and more dimensions these scaling assumptions are problematic. Thus we now restrict ourselves to dimensions larger than unity and smaller than six, trust the scaling assumption of Eq. (1), and compare it with computer simulations.

2.2 SCALING OF CLUSTER NUMBERS

We had already defined n_s as the number (per lattice site) of clusters containing s sites each. It depends on two variables $x = 1/s$ and $y = p - p_c$. Thus not surprisingly we follow Eq. (1) and postulate

$$n_s = s^{-\tau} f((p - p_c)s^\sigma), \qquad (2)$$

for large clsuters near the percolation threshold. Here τ and σ are exponents to be determined numerically or by exact solutions. (More precisely one needs correction terms to the scaling function f as well as to its argument; but we ignore these corrections here.) Equation (2) is a generalization of the Fisher droplet model[24] where

$$n_s = q_0 s^{-\tau} \exp(-const(p - p_c)s^\sigma) \qquad (3)$$

above and at p_c; Eq. (3) is not exact but a useful example for scaling and universality.

For two dimensions, recent high-precision data[7] are still fully compatible with Eq. (2), and thus again we restrict ourselves to three dimensions here. In the simulations used in Figs. 2-5 also the cluster numbers were determined; from Fig. 2, with n_s varying as $s^{-\tau}$ for large s right at the percolation threshold $p = p_c$, we estimate $\tau = 2.2$. Multiplying n_s with s^τ we should get the same function if we plot the product as function of $z = (p - p_c)s^\sigma$. From Figs. 3 and 4 we can estimate $\sigma = 0.45$, and this value is now used in Fig. 7. We see that large clusters follow roughly the scaling equation (2) as desired. For small clusters, the quantity $1/s$ is not yet small enough to apply scaling ideas with sufficient precision, and thus small clusters are ignored in our Fig. 7. The scaling function $f(z)$ is seen to have one maximum at negative argument; it decays asymmetrically to zero in its wings.

According to our present knowledge, this scaling law for the cluster numbers is valid for all dimensions between unity and six.

2.3 CONSEQUENCES OF CLUSTER NUMBER SCALING

If we believe Eq. (2), what results does it imply for the quantities plotted in our earlier figures, like the susceptibility χ (mean cluster size) or the percolation probability P? This question can be answered by a simple exercise of integration: If only large clusters are important, then in the sum for the susceptibility we can replace the sum by an integral running from zero to infinity.

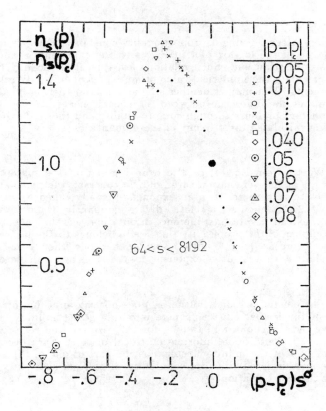

FIG. 7: Cluster numbers, normalized by their threshold values, versus the scaling variable z. Data for different p are seen to follow the same curve.

(We always ignore the largest cluster in this sum.) With the scaling variable $z = (p - p_c)s^\sigma$, the differentiation rule $dz/z = \sigma ds/s$, and p above the threshold we get

$$\chi = \sum_s s^2 n_s ds = \int_0^\infty s^2 n_s ds = \int_0^\infty s^{3-\tau} f(z) ds/s$$
$$= (1/\sigma)(p - p_c)^{-\gamma} \int_0^\infty z^{\gamma-1} f(z) dz,$$

with the critical exponent

$$\gamma = (3 - \tau)/\sigma, \qquad (4)$$

and an integral giving a proportionality constant. Thus we have related the slope $-\gamma$ of the log-log plot in Fig. 4 to the exponents σ and τ of the cluster size distribution. Such relations are typical for scaling laws: One does not calculate the numerical value of an exponent but only relates it to other exponents. You may remember similar tricks in thermodynamics where one calculates the difference between the specific heats at constant volume and at constant pressure through compressibility and thermal expansion, without having calculated directly any of these quantities.

We see from Eq. (4) or also from Fig. 4 that the susceptibility diverges strongly at the percolation threshold. In contrast, the percolation probability P goes to zero there, with an exponent (usually called β) somewhat smaller than 1/2. To calculate a related diverging quantity we look at the derivative dP/dp instead. (Note the difference in notation: p is the probability that a site is occupied, P the probability that a site belongs to the largest cluster.) If P goes to zero as $(p - p_c)^\beta$ with β smaller than one, then dP/dp diverges at the threshold with a negative exponent $\beta - 1$. Let us now calculate this exponent.

The cluster size distribution n_s does not say anything directly about the size of the largest cluster. There is one largest ("infinite") cluster present above the threshold, and thus n_s for this size s is the reciprocal lattice size. Thus we use additional information to calculate P: Each occupied site which is not part of the infinite cluster must be part of one of the smaller clusters, since even isolated sites are counted as clusters of size unity. Therefore we have the exact relation

$$p = P + \sum_s n_s s, \qquad (5)$$

since $n_s s$ is the probability that a site belongs to a cluster with s sites. Below the percolation threshold in very large systems, P is zero and thus p equals the sum in Eq.(5). Above the percolation threshold we differentiate Eq. (5) with respect to p and get (with $f' = df/dz$)

$$dP/dp = 1 - \sum_s sd(n_s)/dp$$
$$= 1 - \sum_s s^{1-\tau} df/dp$$
$$= 1 - \sum_s s^{\sigma+1-\tau} f'(z)$$
$$= 1 - \int_0^\infty s^{\sigma+1-\tau} f'(z) ds$$
$$= 1 - (1/\sigma)(p - p_c)^{\beta-1} \int_0^\infty z^{(2-\tau)/\sigma} f'(z) dz,$$

and thus $\beta - 1 = \tfrac{\tau-\sigma-2}{\sigma}$ or

$$\beta = (\tau - 2)/\sigma. \tag{6}$$

If instead of this power law for P we want to see the scaling behavior of the total number of clusters, we can use similar tricks but now have to calculate its third derivative with respect to p. Only that derivative diverges at the threshold. The total number of clusters as well as its first and second p-derivative remain finite at $p = p_c$. Such finite leading terms, added to a function diverging in some higher derivative and vanishing at the critical point, are often called "background" contributions. It is quite difficult to determine such weak singularities by computer simulations.

Finally, we look at the correlation length ξ which is the typical cluster radius

$$\xi^2 = \sum_s R_s^2 s^2 n_s / \sum_s s^2 n_s, \tag{7a}$$

or the rms distance between two randomly selected sites in one cluster. Here R_s is the radius of gyration of a cluster with s sites. Here ξ diverges with the exponent ν at the percolation threshold. "Hyperscaling" gives

$$\nu d = \gamma + 2\beta \tag{7b}$$

in d dimensions. The exponent ν is related to the fractal dimension D, defined here through

$$s \propto R_s^D, \tag{8}$$

Integrations similar to those leading to Eqs. (4) and (6) give

$$\nu = 1/(D\sigma), \tag{9}$$

which in turn can be expressed as

$$D/d = (\gamma + \beta)/(\gamma + 2\beta)$$
$$= (d - \beta/\nu)/d$$
$$= 1/(\tau - 1), \tag{10}$$

with the help of Eqs. (4), (6), (7) and (9).

In summary, we see that with two of the six exponents $\sigma, \tau, \beta, \gamma, \nu$ and D we can calculate the other four through the four scaling laws, Eqs. (4), (6), (7) and (9). This nice property is not restricted to percolation clusters abut applies to many phase transitions. The fractal dimension D alone does not determine all exponents. It describes aspects of the behavior right at the threshold, or for distances smaller than the correlation length. For example, Fig. 5 shows the size of the largest cluster at p_c as a function of the linear dimension L of the system. Since this "infinite" cluster has a radius of the order of L, we see from Eq. (8) that the size of the largest cluster is proportional to L^D. Figure 5 thus gives a fractal dimension D near 2.5 for the simple cubic lattice.

As mentioned above, such scaling laws do not give the numerical value of any critical exponent alone; they merely relate these exponents to each other in the sense of a phenomenological theory. More microscopic theories, like the computer simulations of Figs. 3, 4 and 5, the exact numerical investigation of small and intermediate clusters, or purely mathematical arguments, are needed to calculate the actual values of these exponents. Such calculations are also necessary to fine p_c.

Presently, $p_c = 1/2$ is known exactly for the triangular lattice, and estimated to be about 0.6962, 0.59273, and 0.3116 for the honeycomb, square, and simple cubic lattice. The two-dimensional exponents are widely believed to be exact: $\beta = 5/36$, $\gamma = 43/18$, $\nu = 4/3$, $\tau = 187/91$, $\sigma = 36/91$, $D = 91/48$ whereas in three dimensions the corresponding estimates are roughly 0.44, 1.76, 0.88, 2.20, 0.45 and 2.5. Reference 25 collects some recent papers with numerical estimates for these quantities in two to five dimensions.

As an exercise let us now make a wrong calculation "proving" $\tau = 2$ and thus, via Eq. (6), $\beta = 0$ and $D = d$, in contradiction with everything we know about percolation. For p approaching the threshold from below, $\Sigma_s n_s s$ approaches neither infinity nor zero but a constant, i.e., p_c, as Eq. (5) shows. Thus its critical exponent is zero:

$$(p_c - p)^0 \propto \sum n_s s$$
$$= \int_0^\infty n_s s \, ds$$
$$= \int_0^\infty s^{1-\tau} f(z) ds$$
$$= (1/\sigma)(p_c - p)^{(\tau-2)/\sigma} \int_0^\infty |z|^{-1-\beta} f(z) d|z|$$
$$\propto (p_c - p)^\beta,$$

from Eq. (6). Thus we see that β must be zero. Our error lies in the assumption that the integral over z exists. This is not true, for $f(z)$ approaches a constant for $z = 0$, as Fig. 2 showed us already: the cluster numbers n_s for a given s have a finite value $\propto s^{-\tau}$ at $p = p_c$, i.e., at $z = 0$. Thus for small z the integrand varies as $z^{-1-(\tau-2)/\sigma} = z^{-1-\beta}$, which is about $z^{-1.4}$ in three dimensions.

Obviously, the integral over $z^{-1.4}$ diverges if the lower boundary is zero, and the proportionality factor above does not exist.

The meaning of this diverging integral is that we were not allowed to replace the sum by an integral. At $p = p_c$, $\Sigma n_s s$ varies as a zeta function instead of an integral, except that $n_s \propto s^{-\tau}$ is valid only for large s, see Fig. 2. In other words, the sum over s in Eq. (5) is dominated by small clusters for which neither our scaling theory nor the approximation of sums by integrals is valid. (To derive Eq. (6) we therefore looked at the derivative of this sum, which diverges due to large clusters; thus scaling and integration apply.)

This discussion required that $f(z)$ is finite for $z = 0$. If, on the other hand $f(z)$ vanishes with a sufficiently large power of z for $z \to 0$, then the situation is different: Now the integral may be finite, the main contribution comes from large clusters, the sum may be replaced by an integral, and $\tau = 2$ if mass conservation as in Eq. (6) is valid. Such effects can happen for certian growth models,[27] though not for percolation in more than one dimension.

To mention a related fallacious argument, the reader might be inclined to believe that $\Sigma R_s^2 n_s s / \Sigma n_s s$ varies as $\xi^2 \propto (p_c - p)^{-2\nu}$, since ξ is the typical cluster radius; note the "minor" difference to definition (7a). Again the sum in the denominator is p and thus remains finite, and only in the evaluation of the numerator can we replace the sum by an integral and apply the cluster number scaling. Then we find this ratio to diverge as $(p_c - p)^{\beta - 2\nu}$ only.

Let us finally investigate another fallacy: Kunz and Souillard[26] have shown that above the threshold $\log(n_s)$ varies for large s as $-s^{2/3}$ in three dimensions. Scaling applies for large s only. Thus slightly above the percolation threshold we have

$$\log(s^\tau n_s) \propto -(p - p_c)^{2/(3\sigma)} s^{2/3}, \tag{11}$$

from Eq. (2). If p approaches p_c from above for a fixed s, we thus have a singular behavior of $n_s(p)$, i.e., some higher derivatives diverge. Such singularity is quite surprising since for every finite s the cluster numbers are finite polynomials[1] in p and thus no higher derivatives are allowed to diverge. This time the argument is wrong because of the order in which limits are taken: the Kunz-Souillard requirement applies to the limit of large s at fixed p and thus $z \to \infty$; the requirement for nonsingular behavior at the threshold deals with the limit of p going to p_c at fixed s and thus $z \to 0$. Thus Eq. (11) is compatible with both requirements if we restrict Eq. (11) to be valid for large $z = (p - p_c)s^\sigma$. Then the Kunz-Souillard theorem describes the decay of the scaling function $f(z)$ for large z, but not its behavior for intermediate or small z: no contradiction occurs. As a mathematical example, one may take

$$\log(f(z)) = -(const + z)^{2/(3\sigma)}$$

to find above and at p_c a behavior compatible with the scaling theory of Eq. (2), the Kunz-Souillard theorem related to Eq. (11), and the above requirements for smooth behavior at $p = p_c$. However, this *Ansatz* fails below the threshold where asymptotically $\log(n_s)$ varies as $-s$, another requirement for the scaling function $f(z)$ at large negative arguments z.

2.4 UNIVERSALITY

In the preceding section, we mentioned numerical values for the percolation threshold which were different for honeycomb, square and triangular lattices. On the other hand, we mentioned only one set of critical exponents for all two-dimensional lattices. The organizers had not requested me to shorten my manuscript and so lack of space cannot be the reason for this simplification. The real reason is called universality and is simply a generalization of the law of corresponding states.

You may remember that in the discussion of the van der Waals equation for fluids all material-dependent quantities drop out if the pressure is normalized by the critical pressure, the temperature by the critical temperature, and the density by the critical density. In particular, the critical exponents for specific heat, density difference and compressibility are the same for all fluids in the van der Waals theory. Modern scaling theory tells us that the van der Waals equation is wrong for three-dimensional fluids with short range interaction, but some remnants of it remain: the critical exponents are the same for all fluids in three dimensions, and a different set of exponents applies to all two-dimensional fluids. In a similar fashion, all short-range random percolation models seem to have, according to our present knowledge, the same critical exponents if the dimensionality d is the same.

That universality principle does not assert that *all* critical properties are independent of the lattice structure in, say, two dimensions. For example, the percolation probability $P = B(p - p_c)^\beta$ has the same exponent β but not the same amplitude B and not the same threshold p in all two-dimensional lattices. (If the law of corresponding states would be valid, even B or B/p would be universal, but numerical evidence shows that this is not the case.)

Nevertheless we can also look at amplitude universality.[28] The Fisher droplet model contains two lattice-dependent parameters, the prefactor q_0 and the constant in the exponential. The exponents σ and τ are required to be universal, the two parameters are not. In a similar fashion, we may write the basic scaling assumption, Eq. (2), as

$$n_s = q_0 s^{-\tau} U(\text{const}(p - p_c)s^\sigma), \tag{12}$$

where again q_0 and the constant as well as p_c depend on the lattice structure whereas the exponents σ and τ as well as the scaling fucntion U are universal, i.e., depend only on the dimensionality d. Such universality assumptions with two material-dependent scale factors are common also for thermal phase transitions.

If we believe Eq. (12), and there are good reasons to believe it,[28] we only have to look at such combinations of critical quantities where the prefactor and the proportionality constant in the argument of U cancel out. The easiest quantity to look at seems to be the susceptibility ratio $C_>/C_<$, defined through $\chi = C_>(p - p_c)^{-\gamma}$ above and $\chi = C_<(p_c - p)^{-\gamma}$ below p_c. Thus this amplitude ratio is the ratio of the susceptibility above and below the threshold, at equal small distances from the threshold. If we repeat the derivation of Eq. (4) with the universality equation (12) we see that the two lattice-dependent factors cancel out of this ratio; only the ratio remains of two integrals involving the universal scaling function U, Eq. (12), at positive and at negative arguments of the scaling variable $Z = \text{const}(p - p_c)s^\sigma$. This universal ratio has played an important role in the comparison of kinetic gelation with percolation (see Landau's talk); it is 0.1 in three and 0.005 in two dimensions.

SUMMARY

I hope I have been able to convey the impression that the scaling theory of percolation clusters is an easy entry into scaling for phase transitions or for growth models. It has been used repeatedly to teach young scientists what actual research can be about, how to use computers, how to analyse numerical data etc. It does not require complicated prerequisites like relativity, Maxwell equations, quantum mechanics, or temperature. Its application to the real world[1] has not been the aim of this review.

I thank H. J. Herrmann for writing up what I was supposed to say, and him, J. Kertész and N. Jan for helpful comments on the manuscript.

[1] G. Deutscher, R. Zallen and J. Adler, eds, *Percolation Structures and Processes*, Ann. Israel Phys. Soc. 5 (Adam Hilger, Bristol, 1983).
[2] D. Stauffer, *Introduction to Percolation Theory* (Taylor and Francis, London, 1985).
[3] R. B. Pandey, D. Stauffer, A. Margolina and J. G. Zabolitzky, J. Statist. Phys. 34, 427 (1984).
[4] C. Kalle and S. Wansleben, Comput. Phys. Comm. 33, 343 (1984).
[5] S. Kirkpatrick and E. Stoll, J. Comput. Phys. 40, 517 (1981).
[6] G. Marsaglia, "Random Number Generation," in *Encyclopedia of Computer Science*.
[7] D. C. Rapaport, J. Phys. A 18, L175 (1985) and to be published.
[8] P. L. Leath, Phys. Rev. B 14, 5064 (1976).
[9] Z. Alexandrowitz, Phys. Letters A 80, 284 (1980).
[10] P. Grassberger, J. Phys. A 18, L215 (1985) and to be published.
[11] S. Havlin, B. Trus, G. H. Weiss and D. Ben-Avraham, J. Phys. A 18, L247 (1985).
[12] D. Wilkinson and M. Barsony, J. Phys. A 17, L129 (1984).
[13] D. Stauffer, Phys. Rev. Lett. 41, 1333 (1978).
[14] J. Hoshen and R. Kopelman, Phys. Rev. B 14, 3428 (1976).
[15] G. MacKay and N. Jan, J. Phys. A 17, L757 (1984); E. Guyon, Percolation et matiere en grains, sect. 4.5, Academie des Sciences, Institut de France, séance du 22 juin 1981.
[16] H. J. Herrmann, D. C. Hong and H. E. Stanley, J. Phys. A 17, L261 (1984).
[17] P. G. de Gennes, La Recherche 7, 919 (1976).
[18] C. Mitescu and R. Roussenq, Chapt. 4 in Ref. 1.
[19] S. Alexander and R. Orbach, J. Physique (Paris) 43, L625 (1982).
[20] Y. Gefen, A. Aharony and S. Alexander, Phys. Rev. Lett. 50, 77 (1983).

[21] R. Rammal, J. C. Angles d'Auriac and A. Benoit, Phys. Rev. B **30**, 4087 (1984).
[22] H. E. Stanley, I. Majid, A. Margolina and A. Bunde, Phys. Rev. Lett. **53**, 1706 (1984).
[23] D. Stauffer, Eur. J. Phys. **5**, 6 (1984).
[24] M. E. Fisher, Physics **3**, 255 (1967).
[25] P. Grassberger, Ref. 10, and three preprints to be published in J. Phys. A by J. Kertész, by M. Sahimi, and by N. Jan, D. C. Hong and H. E. Stanley.
[26] H. Kunz and B. Souillard, J. Stat. Phys. **19**, 77 (1978).
[27] T. Vicsek and F. Family, Phys. Rev. Lett. **52**, 1669 (1984).
[28] A. Aharony, Phys. Rev. B **22**, 400 (1980).

Leo Kadanoff

SCALING PROPERTIES OF THE PROBABILITY DISTRIBUTION FOR GROWTH SITES

Antonio Coniglio

Istituto di Fisica Teorica
Mostra D'Oltremare, Pad. 19, 80125 Napoli, Italy

Center for Polymer Studies and Department of Physics
Boston University, Boston, Massachusetts 02215 USA

A growth model can be characterized by the set of probabilities $\{p_i\}_{i \in \Gamma}$ that each site at a given time on the external perimeter Γ becomes part of the aggregate. All quantities of interest both static and dynamic can be expressed in terms of the p_i. Equations for the set of p_i are given for DLA and other growth models using the electrostatic analogy of the dielectric breakdown model. Due to this electrostatic analogy the scaling properties of the probability distribution are related to those of the voltage distribution in a random resistor and random superconducting network at the percolation threshold. An infinite set of exponents is necessary to fully characterize the moments of the distribution which are related to the surface structure of the aggregate.

INTRODUCTION

What are the relevant parameters to fully describe the essential properties of a growth process? The answer to this question is extremely important in order to be able to understand the complexity and the richness of this exciting field of growth phenomena. It is clear for example that an aggregate cannot be fully characterized by its fractal dimensionality. DLA and percolating clusters in three dimensions have roughly the same fractal dimensionality, yet they have a completely different structure.

A possible way to fully characterize a growth model is by assigning at each time step the growth sites probability distribution (GSPD) $\{p_i\}_{i \in \Gamma}$, where p_i is the probability that site i becomes part of the aggregate. In most cases the growth occurs on the perimeter Γ of the aggregate (Fig. 1). t is the number of steps to grow the aggregate and usually coincides with the number of particles in the aggregate.

The outline of this lecture is as follows. We first show how the probability of each configuration can be expressed in terms of the GSPD and how the static and dynamic properties can be calculated in a growth model. The scaling properties of the voltage distribution in a random resistor network and a random superconducting network is then reviewed, and connected to the scaling properties of the GSPD. Finally, a method is presented to actually calculate the GSPD.

STATIC AND DYNAMIC CONFIGURATIONS

The probability $p(c)$ that a given cluster configuration of N particles occurs in a growth model is given by

$$p(c) = \sum p_{1,...N},$$

where $p_{1,...N}$ is the probability for a given sequence of N steps and is given by the product of probabilities of adding a particle at each step. The sum is over all sequences of N steps leading to the configuration c (see, e.g., H. Herrmann, this volume). Since the aggregate evolves in the course of time, one might ask how does one calculate static properties such as the fractal dimension of the aggregate.

To answer this question consider an imaginary box of length L. Let an aggregate grow from a seed at the center of the box regardless of the presence of the box. After a large number of time steps t the aggregate will start to grow outside the box. Consider a cluster configuration c inside the box. The probability $p(c, L, t)$ that such a configuration is realized at time t is given by

$$p(c, L, t) = \sum p_{1,...t}, \tag{1}$$

where $p_{1,...t}$ is the probability that the given sequence made of t steps leads to the configuration c in the box. The equilibrium distribution is given by

$$p(c, L) = \lim_{t \to \infty} p(c, L, t).$$

In this way we can calculate any static quantity in the box L and then take the thermodynamic limit $L \to \infty$.

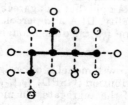

FIG. 1: Example of an aggregate made of 6 particles. The growth sites are represented by empty circles. For any given growth site there is a probability p_i of becoming part of the aggregate.

As an example, the mass or number of particles in the box $M(L,t)$ at time t is given by

$$M(L,t) = \sum_c p(c,L,t)M(c), \qquad (2)$$

where $M(c)$ is the mass in the configuration c. The mass in the box at equilibrium $M_{eq}(L)$ is given by

$$M_{eq}(L) = \lim_{t \to \infty} M(t,L),$$

from which the fractal dimensionality d_f can be calculated

$$d_f = \lim_{L \to \infty} \frac{\ln M_{eq}(L)}{\ln L}. \qquad (3)$$

This approach has been used in Ref. 1 to give very good estimates of the fractal dimensionality of DLA.

To find the dynamic behavior, using a scaling approach,[1] we write

$$M(L,t) = M(L,\infty) f\left(\frac{t}{\tau}\right). \qquad (4)$$

Here $M(L,\infty) \sim L^{d_f}$, $\tau \sim L^z$, where τ is a characteristic time above which the mass inside the box does not change anymore. We have $f(x) \sim const$ for $x \gg 1$ and $f(x) \sim x^{d_f/z}$ for $x \ll 1$, since the mass must be independent of L for large enough L. Moreover for $x \ll 1$ we also have $M(L,t) = t$ since all the mass is inside the box. Thus $z = d_f$.

VOLTAGE DISTRIBUTION IN RANDOM RESISTOR NETWORK AT THE PERCOLATION THRESHOLD

From the GSPD one can obtain not only the static and dynamic quantities of interest but also much more detailed information on the capability of each perimeter site to grow and therefore a better knowledge of the surface structure. In DLA for example the "hottest" sites which are more likely to grow are those at the tips of the cluster, for which the growth probability assumes the highest value, while the very "cold" sites deep inside the fjords are characterized by a very small value of the growth probability.

What are the scaling properties of the probability distribution? To provide an answer, note that in the dielectric breakdown model[2] the GSPD is related to the voltage distribution on the surface of the aggregate. An insight of what the properties of the GSPD are can be gained by looking at the voltage distribution in a random resistor network at the percolation threshold p_c, which has been recently studied.[3]

Consider a d-dimensional hypercubic lattice of size L. Suppose that each bond has a probability p of being active and $1-p$ of being non-active. From percolation theory (see D. Stauffer, this volume) we know that in the limit of infinite system there exists a percolation threshold p_c, above which an infinite cluster of active bonds is present. Right at p_c the bonds in the spanning configurations, as noted by Stanley,[4] may be partitioned in dangling bonds that do not contribute to the electrical resistance and the remaining backbone bonds. The backbone bonds may also be divided into links ("red" bonds), which are singly connected, and the remaining multiply-connected bonds ("blobs"). Whether the blobs could be neglected, as in the Skal-Shklovskii-de Gennes model,[5] or the links, as in the Sierpinski gasket model,[6] is a non-trivial question. It is now well established on the basis of exact results that both links and blobs are critical. The number of links L_1 diverges as $L_1 \sim L^{1/\nu}$ in any dimension when ν is the connectedness length exponent.[7] As d approaches 6, the blobs become less and less important, until they become irrelevant for $d \geq 6$. Strong numerical evidence confirms the validity of this picture.[8]

In order to characterize further the structure of the backbone bonds we associate a unit electrical resistance to each bond and apply a unit voltage at the opposite boundary of the cell connected by the percolating cluster (Fig. 2). Each bond is characterized by a voltage drop V equal to the current I flowing through it. The maximum voltage drop $V_{max} = I_{max}$ occurs in the links.

In this way each bond can be labelled by a number

$$\alpha = \frac{V}{V_{max}} = \frac{I}{I_{max}}$$

between 0 and 1, or in a more picturesque way by a corresponding color code. The higher the value of α, the hotter the bonds. The hottest bonds are the links or red bonds for which $\alpha = 1$. The coldest are those that carry zero current. If $N(\alpha)$ is the number of bonds corresponding to the value α we can construct the moments $M(k)$ and their relative exponents $\tilde{\varsigma}(k)$

$$M(k) = \sum_\alpha \alpha^k N(\alpha) \sim L^{\tilde{\varsigma}(k)}. \tag{5}$$

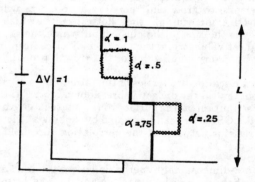

FIG. 2: A backbone configuration made of links and blobs at p_c. The bonds with the same voltage drop $\alpha = V/V_{max}$ are indicated in the same way.

$M(0)$ is the number of bonds in the backbone, therefore $\tilde{\varsigma}(0) = d_B$, the fractal dimension of the backbone; $M(2)$ is the resistance and $\tilde{\varsigma}(2) = \tilde{\varsigma}_R$ the resistivity exponent; $M(4)$ is related to the magnitude of the noise[9] and $M(\infty)$ coincides with the number of links, consequently $\tilde{\varsigma}(\infty) = 1/\nu$.

Are these exponents all independent of each other? We know in critical phenomena that the moments of the order parameter all scale with the "gap" exponent. Similarly in percolation the moments of the cluster size distribution at p_c scale as $\Sigma s^k N(s) \sim L^{k d_f}$ where d_f is the fractal dimensionality of the critical cluster and $N(s)$ is the number of clusters of s-sites.

The presence of one single independent exponent is due to the fact that there is one dominant critical cluster of size $s^* \sim L^{d_f}$ which is responsible for the critical behavior. In the voltage distribution problem, the exponents $\varsigma(k) = \nu \tilde{\varsigma}(k)$ have been calculated[3] analytically on a hierarchical model (Fig. 3), which describes the properties of the percolating backbone. The result for $d = 2$ is

$$\varsigma(k) = 1 + \frac{\ln(1 + 2^{-k})}{\ln 2}, \tag{6}$$

which reproduces extremely well the data from computer simulation for 2-dimensional random percolation. From (6) it follows that there is an infinite set of independent exponents describing the moments of the voltage distribution. Physically this means that there is no characteristic value of the voltage which dominates. All values of the voltage are equally important. This may also explain the failure of many attempts trying to find relations among various critical exponents related to the geometrical properties of the percolating cluster.

A generalization of the hierarchical model to higher values of d leads to the following prediction for $\varsigma(k)$

$$\varsigma(k) = 1 + \frac{\ln(1 + \lambda 2^{-k})}{\ln 2/\lambda}, \tag{7}$$

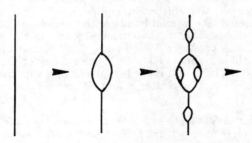

FIG. 3: Hierarchical lattice for the case $\lambda = 1$ corresponding to 2 dimensions. Starting with a single bond, it is replaced by the unit cell shown at the next level. This procedure is then repeated indefinitely (from Ref. 3).

with $\lambda = \frac{1}{4}(6-d)$ for $2 \leq d \leq 6$ and $\lambda = 0$ for $d = 1$ and $d \geq 6$. The expression (7) is in very good agreement with the existing numerical data in random percolation.

For d above the upper critical dimensionality $d_c = 6$ all exponents coincide with the exponents relative to the links $\varsigma(k) = 1$. This result has the physical explanation that above $d = 6$ the blobs are irrelevant and only the links are important. Consequently the moments of the voltage distribution are dominated by only one voltage drop $V = V_{max}$ across the links. On the other hand the maximum dispersion in $\varsigma(k)$ occurs at $d = 2$ where the blobs are relatively most important.

In conclusion, an infinite set of independent exponents is necessary to describe the moments of the voltage distribution. These are related to the structure of the percolating backbone. The more complex the structure the larger the dispersion in the exponents. This shows how this geometrical problem contains much more richness than ordinary critical phenomena, where the moments of the order parameter are dominated by the critical fluctuations of the order parameter, and are therefore expressed in terms of only one independent exponent.

So far we have considered the voltage distribution in the interior of the percolating cluster in the random resistor network. Similar results are obtained by studying the voltage distribution on the surface of the percolating cluster in a random superconducting network.

In this problem, superconducting bonds are present with probability p and normal bonds carrying a unit resistance with probability $1 - p$. For small values of p we have finite superconducting clusters in a background of normal resistors. As $p \to p_c$ the conductivity diverges. For a finite cell of length L at p_c the non-spanning configurations are characterized by very large clusters almost touching each other, each one attached to one of two opposite faces (Fig. 4).

Inside the superconducting clusters there are islands of normal resistors. If a unit voltage is applied between the opposite faces of the cell the bonds in the same superconducting cluster are equipotential. Therefore there is no current flowing through the normal bonds of the islands. Other bonds which do not carry current are those that connect two sites in the same superconducting cluster. These are the coldest bonds. The hottest bonds, which we call "bridges" are those normal bonds on the perimeter of the very large superconducting clusters, such that if one is replaced by a superconducting bond a percolating superconducting cluster is formed, joining the opposite faces of the cell.

Similarly to the backbone links it can be proved that the fractal dimensionality of the bridges is $1/\nu$, namely $N_B \sim L^{1/\nu}$ where N_B is the number of bridges.[10] The bridges are the hottest bonds corresponding to the highest value of the voltage drop $V_{max} = 1$. We can introduce the voltage distribution $N(V)$ for the normal resistor bonds with the moments $M'(k)$ given by

$$M'(k) = \Sigma n(V) V^k \sim L^{\tilde{\varsigma}(k)}. \tag{8}$$

The sum is over all the normal bonds. In $d = 2$ from duality argument follows[10,11] $\tilde{\zeta}'(k) = \tilde{\zeta}(k)$, and therefore the moments[8] are described by a set of independent exponents which in the limit $k \to \infty$ tend to $\tilde{\zeta}'(\infty) = 1/\nu$.

Finally, if in (8) the sum is restricted to the bonds on the surface of the critical cluster, another set of critical exponents is obtained, $\tilde{\zeta}''(k) \leq \tilde{\zeta}'(k)$ which in the limit $k \to \infty$ tend to $1/\nu$, i.e., the critical exponent related to the bridges.

SCALING PROPERTIES OF THE GROWTH SITES DISTRIBUTION PROBABILITY

What we have learned from the properties of the voltage distribution on the fractal surface of the percolating cluster can be applied to the GSDP in various growth models. We start with the dielectric breakdown model.[2] In this model the growth probability p_i at the perimeter site i is given by $p_i \sim (\phi_i)^\eta$ where ϕ_i is the potential at the perimeter site i obtained by solving the discrete Laplace equation with the boundary conditions $\phi = 0$ for each point of the aggregate and 1 outside a circle at large enough distance. When the parameter η equals 1 one recovers the DLA model (see T. Witten, this volume). Note that the GSDP is related to the voltage distribution on the fractal surface. Therefore we expect similar scaling properties as for the voltage distribution in the random resistor and random superconducting network, namely[1,12]

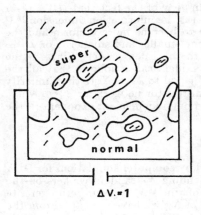
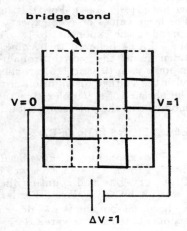

FIG. 4: Random superconducting network in the continuum (a) and discrete version (b). The bold lines are the superconducting bonds. The dotted lines are the normal resistors among which 4 are "bridge" bonds.

$$\sum_{i\in\Gamma}\left(\frac{p_i}{p_{max}}\right)^k \sim R^{y(k)}, \qquad (9)$$

where R is the radius of the aggregate and p_{max} is the highest probability of becoming part of the aggregate, relative to the "hottest" site. We expect all exponents $y(k)$ to be independent. As in the voltage distribution this would correspond to the fact that the GSPD is very broad and there is no characteristic value of the probability which dominates, giving rise to a "gap" exponent.

Strong arguments[13,14] suggest that $1/p_{max} \sim R^{d_f - 1}$. Therefore $y(1) = d_f - 1$ while $y(\infty)$ gives the exponent relative to the number of hottest sites. All the other exponents describe a weighted average of the surface sites. Large values of k give more weight to the "hot" sites at the tip, small values of k distribute the weight more uniformly to all the sites.

Another quantity which has received attention is[12,15,16]

$$\left[\frac{1}{\Sigma p_i^k}\right]^{1/(k-1)} \sim R^{x(k)}, \qquad (10)$$

where

$$x(k) = y(1) - \frac{y(k) - y(1)}{k-1}. \qquad (11)$$

Since the exponents $y(k)$ are expected to be all independent, the same is expected to be true for the exponent $x(k)$.[12] In view of the fact that $y(1) = d_f - 1$ we have $x(\infty) = d_f - 1$. For DLA in $2d$, $x(1) = 1$ is the information dimension.[16,17] For large values of d we expect the GSPD to become more uniform and the spread in the exponent $x(\infty) - x(1)$ to become smaller and smaller. For $d = \infty$ the DLA is believed to become identical to the Eden model. Since in the Eden model the growth probabilities p_i are identical, it follows that all the exponents in (9) and (10) are identical and equal to $d - 1$. Note the similarity with the voltage distribution in the random resistor network for $d \geq 6$, where the absence of a complex structure leads to one independent exponent in the moments of the voltage distribution.

EQUATIONS FOR GSPD

The exponents (11) have been measured by computer simulations for DLA in $d = 2$ for $k = 2\cdots 8$. This was done by sending N_T random walkers on the DLA aggregate and counting the number of times N_i a given perimeter site i was hit by the walkers. For large values of N_T we have $p_i \simeq \frac{N_i}{N_T}$. From the p_i the set of exponents was determined and indication of a set of independent exponents tending towards $x(\infty) = d_f - 1$ was found.

A different approach was used in Ref. 1 to actually calculate GSDP. Using the electrostatic analogy in the continuum version the density probability $p(x)$ at site x on the surface is given by (Fig. 5)

$$p(x) = -\vec{n}(x) \cdot \nabla \phi(x) = |E(x)|, \qquad (12)$$

where $\bar{n}(x)$ is the normal at the site x on the surface of the aggregate and $\phi(x)$ is the electrostatic potential satisfying the Laplacian equation with the condition that $\phi = const$ on the conductor and zero at infinity; the constant is chosen in such a way that the total charge on the conductor is equal to unity. The solution for $\phi(x)$ is given by

$$\phi(x) = \int_s \frac{\sigma(x')}{|x-x'|^{d-2}} ds', \qquad (13)$$

for $d > 2$ while for $d = 2$, $|x-x'|^{2-d}$ in the integrand is substituted with $\ln|x-x'|^{-1}$; $\sigma(x') = -\bar{n}(x') \cdot \nabla \phi(x')$ is the charge density. From (12)

$$p(x) = -\bar{n}(x) \cdot \nabla_x \int_s \frac{p(x')}{|x-x'|^{d-2}} ds'. \qquad (14)$$

This equation can be discretized and solved numerically for the $p(x)$, from which the moments have been calculated giving results consistent with the computer simulation results of Ref. 12. Equations for GSPD on lattice have also been given in Ref. 13.

In conclusion, the emphasis of this lecture has been on the distribution probability of growth sites. This distribution probability has scaling properties similar to the voltage distribution in random resistor networks for which an infinite set of exponents is necessary to fully describe the richness of the structure of the geometrical properties of the aggregate. For a more general theory of the GSPD see L. Kadanoff, this volume.

This lecture was based on work done in collaboration with C. Amitrano, L. deArcangelis, F. diLiberto, P. Meakin, S. Redner, H. E. Stanley and T. Witten. I also would like to thank C. Amitrano, F. diLiberto and L. Peliti for a careful reading of the manuscript.

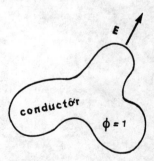

FIG. 5: Electrostatic analogy for various growth models. The density growth probability at the perimeter site is given by $p(x) \sim |\vec{E}(x)|^\eta$.

[1] C. Amitrano, A. Coniglio and F. diLiberto, to be published.
[2] L. Niemeyer, L. Pietronero and H. J. Wiesman, Phys. Rev. Lett. 52, 1033 (1984).
[3] L. deArcangelis, S. Redner and A. Coniglio, Phys. Rev. B 31, 4725 (1985).
[4] H. E. Stanley, J. Phys. A 10, L211 (1977).
[5] A. S. Skal and B. I. Shklovskii, Sov. Phys. Semicond. 8, 1029 (1975); P. G. deGennes, La Recherche 7, 919 (1976).
[6] Y. Gefen, A. Aharony, B. B. Mandelbrot and S. Kirkpatrick, Phys. Rev. Lett. 47, 1771 (1981).
[7] A. Coniglio, Phys Rev. Lett. 46, 250 (1981); A. Coniglio, J. Phys. A 15, 3824 (1982).
[8] R. Pike and H. E. Stanley, J. Phys. A 14, L169 (1981); H. J. Herrmann and H. E. Stanley, Phys. Rev. Lett. 53, 1121 (1984).
[9] R. Rammal, C. Tannous. P. Breton and A. M. S. Tremblay, Phys. Rev. Lett. 54, 1718 (1985).
[10] A. Coniglio, in Proceedings of Les Houches Conference on Physics of Finely Divided Matter (1985).
[11] A. Coniglio and S. Redner, unpublished.
[12] P. Meakin, H. E. Stanley, A. Coniglio and T. Witten, Phys. Rev. A 32, 2364 (1985).
[13] L. A. Turkevich and H. Scher, Phys. Rev. Lett. 55, 1026 (1985).
[14] F. Leyvraz, J. Phys. A 18, xxx (1985).
[15] H. G. E. Hentshel and I. Procaccia, Physica D 8, 835 (1983).
[16] T. C. Halsey, P. Meakin and I. Procaccia, to be published.
[17] P. Grassberger, unpublished.

Blackboard Discussion

COMPUTER SIMULATION OF GROWTH AND AGGREGATION PROCESSES

Paul Meakin

Central Research and Development Dept.
E. I. du Pont de Nemours and Company, Experimental Station
Wilmington, Delaware 19898

In recent years considerable interest has developed in the formation of random structures under non-equilibrium conditions. Several factors have contributed to the growth of this area. These include the possibility of substantial practical benefits and impact on other areas of science, the hope that the sort of theoretical methods which have been so successful in advancing our understanding of critical phenomena can be applied to a broader range of problems and the recognition that this is an important area of science which has received little attention in the past. Computer simulations have played an important role in the development of this area. In fact much of what we now know about non-equilibrium growth processes has come from computer simulations. Computer simulations can be used to test ideas concerning the behavior of experimental systems and to pose well defined problems for theoretical analysis. In this way they provide a valuable bridge between theory and experiment.

In these lectures I will focus attention on two models, the diffusion limited aggregation model[1] and the cluster-cluster aggregation model,[2,3] which have received considerable attention during the past few years. I will also discuss more briefly some earlier models which are still of major interest.

EDEN (SURFACE GROWTH) MODELS

The Eden model[4] was developed to simulate the growth of cell colonies (not cancer as is often supposed). In the original version of this model the growth process is started with a single occupied lattice site and unoccupied surface sites (unoccupied sites with one or more occupied nearest neighbor) are occupied randomly with a probability which is proportional to the number of occupied nearest neighbors. In a simpler (now more common) version of this model, unoccupied surface sites are selected randomly and occupied with equal probability. We will refer to this modified version as "the Eden Model." This model is so simple that it is possible to grow large clusters even with very crude algorithms. Figure 1 shows the surface sites for a 200,000 site cluster grown on a square lattice. It is clear from this figure that the cluster is compact (i.e, its fractal[5] dimensionality, D, is equal to its Euclidean dimensionality, d).

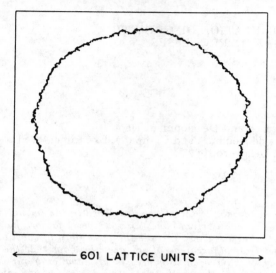

←——— 601 LATTICE UNITS ———→

Fig. 1. The 2372 unoccupied interface sites associated with a 200,000 site cluster grown using the Eden model.

Since the the original work of Eden twenty five years ago, a number of modifications have been developed for various reasons. For example, if enclosed voids (unoccupied sites surrounded by occupied sites) are not allowed to fill in, a compact porous structure with a density of about 0.79 (for the case of a two-dimensional square lattice) is formed.[6] This model is equivalent to the Witten-Sander model for diffusion limited aggregation in the limit of zero sticking probability (see below).

An interesting modification of the Eden model was developed by Williams and Bjerknes[7] to simulate the growth of skin cancer. In this model "cancerous" cells are represented by occupied lattice sites and "normal" cells are represented by unoccupied lattice sites. When a cell divides, one of its nearest neighbors (selected at random) is replaced by a "daughter" cell. If the dividing cell is cancerous and the selected nearest neighbor is normal, the cancer grows, but if the dividing cell is normal and the randomly selected neighbor is cancerous, the cancer retreats. A typical simulation starts with a single cancerous cell and cancerous cells divide with a slightly larger probability than normal cells. Cells (lattice sites) are selected at random and division occurs. The probability with which cancerous cells are selected for division is K times as large as the probability for normal cells. The parameter K is called the "carcinogenic advantage." From relatively small scale simulations (up to 2000 occupied lattice sites) using honey comb lattices Williams and Bjerknes found that the surface size (s) grew with the number of occupied lattice sites (N) according to the power law

$$s \sim N^\sigma, \qquad (1)$$

where the exponent σ had a value of 0.55 for a variety of K values. This result was interpreted in terms of a fractal surface geometry. Figure 2 shows the surface sites for a 100,000 site "cancer" grown using a carcinogenic advantage of 1.1. It is clear from this figure that the interior of the structure is compact ($D = d = 2$). Using clusters of this size we find no evidence that the exponent s in Equation 1 is larger than 0.5 and the two point density-density correlation ($C(r)$) for the surface sites indicates that the fractal dimensionality of the surface is close to 1.0. (i.e., we find $C(r) \sim r^{-\alpha}$ where the exponent α has a value of $\simeq 1.0$ for lengths (r) larger than a few lattice units but smaller than the overall size of the cluster). The effective fractal dimensionality D_α, is then given by $D_\alpha = d - \alpha \simeq 1.0$.[1] Similar results have been obtained for the Eden model itself.[8,9] The conclusion that there is no evidence for a surface fractal dimensionality greater than 1.0 for two dimensional Eden and Williams Bjerknes clusters is supported by the theoretical work of Richardson[10] and more recently by others.

Another interesting model was recently introduced by Sawada et al[11] to simulate dielectric breakdown and other random pattern formation processes.

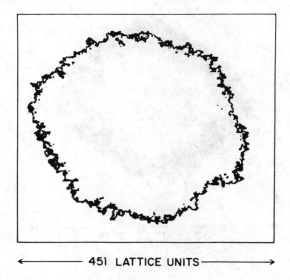

←——— 451 LATTICE UNITS ———→

Fig. 2. The 4135 unoccupied interface sites associated with a 100,000 site cluster grown using the Williams Bjerknes model with a carcinogenic advantage of 1.1.

In this variant of the Eden model a high growth probability is assigned to "tip" sites (unoccupied sites with only one nearest neighbor occupied). The relative growth probability for the tip site, R, is known as the tip priority factor. From two dimensional simulations of clusters containing up to 5,000 occupied sites, Sawada et al concluded that the structure was fractal-like on short length scales with a fractal dimensionality D_1 crossing over to a uniform structure $(D_2 = 2.0)$ on longer length scales. The short length scale dimensionality, D_1, decreased continuously with increasing R up to $R \simeq 35$. Above $R \simeq 40$, D_1 was found to be almost constant with a value of about 1.4. Figure 3 shows the results of a much larger scale simulation (200,000 occupied lattice sites) carried out with a tip priority factor of 80. It is clear from this, and other simulations, that the interior structure of these clusters is compact on all length scales. As the structure grows the surface remains rough but a smaller and smaller fraction of the occupied sites are in the surface region. Although we have not investigated the structure of the surface region it seems probable that it does not have a fractal dimensionality different from 1.0. This does not, of course, exclude an effective fractal dimensionality greater than 1.0 over a limited range of length scales. Recently a more realistic model for dielectric breakdown based on the Witten-Sander[1] model for diffusion limited aggregation (DLA) has been developed by Niemeyer et al[12].

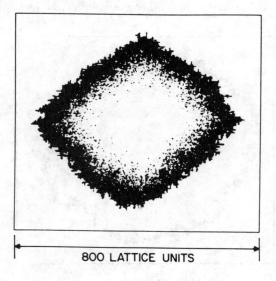

4.25in Fig. 3. The 44111 unoccupied surface sites which border a 200,000 site cluster grown on a square lattice using the model of Sawada et al[11] with a tip priority factor of 80.

Several other Eden type models have been introduced in recent years including Eden Trees,[13] Invading Eden,[14] and the Rickvold(15) model. In all of these models in which the growth probability depends on the local structure in the vicinity of a growth site, compact structures ($D = d$) are obtained. In the screened growth model[16] the growth probability of each surface site depends on the structure of the entire cluster. In this case extensive computer simulations[16] and theoretical considerations show that fractal-like structures are formed.[17,18]

It is clear from the results presented above that large scale simulations and a careful analysis of the results of these simulations are needed to obtain reliable results. Even the most simple versions of the Eden model are subject to large corrections to the asymptotic scaling relationships which can, and often do, produce misleading results.

BALLISTIC AGGREGATION MODELS

In the ballistic aggregation model the growth process is started with a single stationary particle (a hypersphere of unit diameter) and other particles are allowed to follow random ballistic (linear) trajectories in the vicinity of the stationary particle. If a mobile particle contacts the stationary particle it sticks at that point and a stationary cluster is formed. Additional particles are added in the same way, one at a time, until a large aggregate has grown. Figure 4 shows a cluster of 180,000 particles grown in this way during a two dimensional simulation.

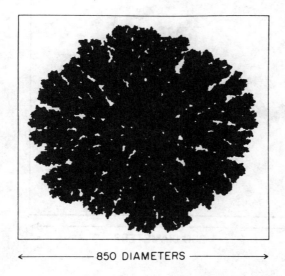

← 850 DIAMETERS →

Fig. 4. A 180,000 particle cluster formed in a two dimensional simulation of off-lattice ballistic aggregation.

The model was first developed more than 20 years ago by M. J. Vold[19] to simulate colloidal aggregation. Today her results from three-dimensional simulations would be interpreted in terms of a fractal dimensionality of about 2.3 (the number of particles, N, within a distance P measured from the original stationary particles was found to be given by $N \sim \ell^{2.3}$). Some time later Sutherland[20] pointed out certain errors in Vold's simulations and on the basis of clusters containing up to about 2,000 particles obtained results which indicate that $D = d$ for both two and three dimensional aggregates. More recently larger scale simulations were carried out by Meakin[21] and by Domany et al.[22] Based on two dimensional clusters containing 90,000 particles Domany et al found a fractal dimensionality of 1.93.

We have now grown clusters containing up to 250,000 particles and find that the fractal dimensionality, D, is larger than 1.95 and probably 2.0 in two dimensional simula- tions and larger than 2.8 and probably 3.0 in three dimensional simulations. Recent theoretical arguments by Ball and Witten[23] support these results and indicate that the fractal and Euclidean dimensionalities are equal. A variety of closely related models have been investigated recently such as ballis- tic aggregation with randomly selected trajectories all in the same direction[24] and a number of investigations of ballistic deposition onto surfaces have been carried out.[25-28]

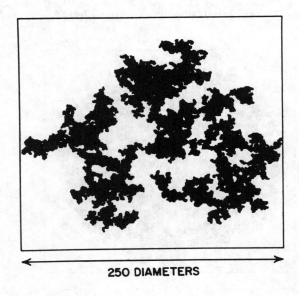

← 250 DIAMETERS →

Fig. 5. A projection of a 32,000 particle cluster grown in a three dimensional simulation using a ballistic aggregation model with random cluster selection.

A few years later Sutherland[29] developed a more realistic model for colloidal aggregation. This model starts with an ensemble of single particles. A pair of particles is selected to form a binary cluster which is returned to the ensemble. At later stages in the simulation objects (particles or clusters) are picked from the ensemble, combined via random linear trajectories and returned. This process is repeated until all of the particles are contained in a large single cluster. In general, the selection of pairs of objects from the ensemble is carried out using probabilities which are dependent on the masses of the two objects. Sutherland found that this model leads to the formation of very low density structures which resemble those observed in a variety of real aggregation processes. He measured both the radius of gyration, R_g, and the maximum radius, R_{max}, of the clusters as a function of the number of particles, N, which they contained. These results indicate that $R_g \sim N^\beta$ and $R_{max} \sim N^{\beta'}$ where the exponents β and β' both have a value of 0.54 for clusters containing up to 256 particles. From these results we can obtain effective fractal dimensionalities (D_β and $D_{\beta'}$ using the relationships $D_\beta = 1/\beta$ and $D_{\beta'} = 1/\beta'$.[30] For these three dimensional simulations Sutherland's results indicate that $D_\beta = D_{\beta'} = 1.85$. This result is in very good agreement with recent, more extensive, larger scale simulations. Figure 5 shows the results of a recent three dimensional simulation carried out using a version of this model in which the selection probabil- ity is assumed to be independent of the masses of the particles or clusters.

DIFFUSION LIMITED AGGREGATION MODELS

The importance of diffusion (Brownian motion) in colloidal aggregation has been recognized for many decades. However, only in recent years has it been practical to carry out computer simulations of aggregation processes using Brownian (random walk) particle trajectories. The first work of this type, which I am aware of, was carried out by Finegold.[31] This model was a cluster-cluster aggregation model which included the effects of both translational and rotational diffusion. However, the simulations were carried out on a rather small scale at high particle densities and no quantitative results were reported. As a result of this, and the focus on biological applications this pioneering work did not have much influence on the general physics community.

The event which contributed most to this area was the discovery by Witten and Sander that a simple diffusion limited growth model in which particles are added, one at a time, to a growing cluster or aggregate of particles leads to a structure which has a fractal-like structure. Because of the interesting scaling and universality properties associated with the structures generated by this model[1,32,33] and its relevance to a wide variety or physical processes including dendritic growth,[1,34] fluid-fluid displacement,[35-38] colloidal aggregation[1] and dielectric breakdown,[11,33] this model has generated considerable enthusiasm and has been extensively investigated during the past few years. Figure 6 shows a large (100,000 site) cluster grown on a two dimensional square lattice using a Witten-Sander model. By analyzing a variety of mass/length scaling relationships associated with these clusters a fractal dimensionality of about 1.70 is obtained in two dimensions.[1,39]

Two almost identical models for diffusion limited cluster-cluster aggregation were developed independently but simultaneously by Meakin[2] and by Kolb, Botet and Jullien.[3] In this model particles are represented by occupied lattice sites on a hypercubic lattice with periodic boundary condi- tions. Particles (or at a later stage in the simulation clusters) are selected at random and moved by one lattice unit in a randomly selected direction. If two objects (particles or clusters) contact via nearest neighbor occupancy after such a move has been made they are joined permanently. The simulation ends when only one large cluster containing all of the parti- cles remains. If the initial particle concentration is low this model leads to structures which have a fractal dimension- ality of 1.4-1.45 for two dimensional simulations and 1.75-1.80 for three dimensional simulations. The results of a typical simulation carried out using 5,000 particles on a 400x400 two dimensional square lattice are shown in Figure 7. Some aspects of the Witten-Sander and cluster-cluster aggregation models are discussed in more detail below.

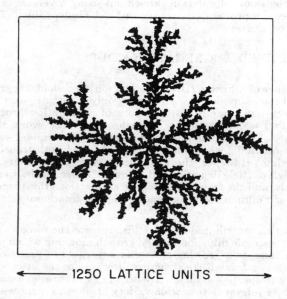

← 1250 LATTICE UNITS →

Fig. 6. A 100,000 site Witten-Sander cluster grown on a square lattice using a semi-lattice model for DLA (off-lattice walks with on-lattice growth).

THE WITTEN-SANDER MODEL FOR DIFFUSION LIMITED AGGREGATION

Figure 8 shows an early version of the Witten-Sander model for diffusion limited aggregation. Particles are launched from a random point on a circle which encloses the cluster. They undergo a random walk on the lattice (a two dimensional square lattice in this case) until they either reach an unoccupied lattice site with an occupied nearest neighbor (trajectory t_1) or they wander a large distance from the cluster (trajectory t_2). Particles which reach a large distance from the cluster (typically $3R_{max}$, where R_{max} is the maximum radius of the cluster) are "killed" and a new particle is started off from the launching circle to reduce computer time requirements. In principal, particle trajectories should only be terminated at a distance which is so large that the particle would return with equal probability to any point on the circle which just encloses the cluster. In practice a distance of $3R_{max}$ is large enough for most purposes and distances shorter than $3R_{max}$ can be used in higher dimensional simulations. To further reduce the computer time needed in these simulations the step length is doubled at distances greater than $R_{max} + 10$, quadrupled at distances greater than $R_{max} + 20$ and increased to eight lattice units at distances greater than $R_{max} + 40$ etc. Using this method and analogous methods for lattices of higher dimensionality

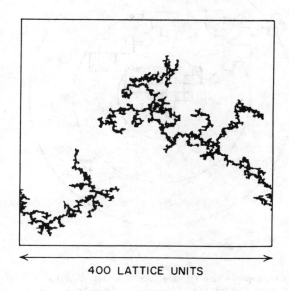

400 LATTICE UNITS

Fig. 7. A 5,000 site cluster grown on a 400x400 two dimensional square lattice using the cluster-cluster aggregation model with mass independent sticking probabilities.

(3d-6d) the results obtained in Table I were obtained from clusters containing up to about 10^4 occupied lattice sites. Apart from the dependence of the fractal dimensionality, D, on the Euclidean dimensionality of the lattice, Table I and similar results obtained by others show that the fractal dimensionality is insensitive to model details such as the lattice type[1,32] and the sticking probability.[32] The results shown in Table I were obtained from the dependence of the radius of gyration, R_g, on cluster size assuming that $R_g \sim N^\beta$ ($b = 1/D$).[30] The results in Table I suggest that $D \simeq 5d/6$. However, this relationship cannot be true in higher dimensional spaces where simple theoretical considerations show that $D > d - 2$ and a more detailed analysis indicates that $D \simeq d - 1$ for $d \to \infty$.[40] This result is, however, in good agreement with the relationship $D = (d^2 + 1)/(d + 1)$ obtained from mean field theories[41,42] for $d = 2, 3$ and 4. In any event the results shown in Table I should not be taken too seriously for $D > 5$. Recently, simple and efficient algorithms have been developed which have allowed large numbers ($> 10^3$) of relatively large clusters ($\simeq 10^5$ particles or clusters) to be generated.[42] This method is illustrated in Figure 9.

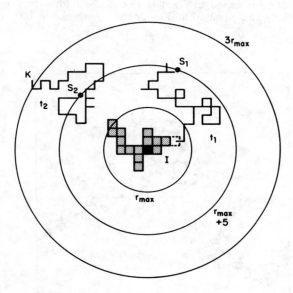

Fig. 8. The figure depicts the simulation of a two dimensional Witten-Sander aggregate using a square lattice. The two particle trajectories (t_1 and t_2) start at random points (s_1 and s_2) on the "launching circle" which has a radius of $R_{max} + 5$ lattice units where R_{max} is the maximum radius for the cluster. Trajectory t_1 reaches an unoccupied interface site and growth occurs into this site.

Table 1

The fractal dimensionality (D) of clusters grown using various diffusion limited particle-cluster aggregation models. The dimensionality of the embedding space or lattice (d) and the sticking probability (S) are indicated. The symbols NN and NNN stand for sticking at nearest neighbor or next nearest neighbor positions and NL indicates off-lattice simulations. These results were obtained from the dependence of the radius of gyration on cluster mass for clusters with 25maximum number of particles.

Model	D	D/d
2d,S(NN)=1.0	1.70± 0.06	0.85± 0.03
2d,S(NNN)=1.0	1.72± 0.05	0.86± 0.03
2d,S(NL)=1.0	1.71± 0.07	0.86± 0.03
2d,S(NN)=0.25	1.72± 0.06	0.86± 0.03
2d,S(NNN)=0.1	1.69± 0.08	0.85± 0.04
3d,S(NN)=1.0	2.53± 0.06	0.84± 0.02
3d,S(NN)=0.25	2.49± 0.12	0.83± 0.04
3d,S(NL)=1.0	2.50± 0.08	0.83± 0.03
4d,S(NN)=1.0	3.31± 0.10	0.83± 0.03
5d,S(NN)=1.0	4.20± 0.16	0.84± 0.03
6d,S(NN)=1.0	~ 5.35	~ 0.89

The random walk has been taken off the lattice and the lattice is used to "inform" the random walker how large a step it may take (in any direction) without "contacting" the cluster. Figure 9 shows the early stages in a small scale simulation in which steps of lengths up to 5 lattice units long may be taken. In more typical practical scale simulations, steps of up to about 40 lattice units would be permitted. In addition if the random walker is outside of the region occupied by the cluster (i.e., at a distance R greater than R_{max}) a step of length $\geq R - R_{max} - 2$ is permitted if this is greater than the step length indicated by the underlying lattice. If the particle is very close to the cluster the step length is one lattice unit and growth occurs if the random walker (which is now considered to be a point rather than a particle of finite size) enters an unoccupied lattice site with one or more occupied nearest neighbors. Using this semi-lattice model clusters containing 10^5 lattice sites can be grown in 10 minutes on an IBM 3081 computer and off lattice aggregates can be grown to a size of 50,000 particles in about 7 minutes using similar methods. In both of these algorithms the radius of the "killing circle" has been extended to 100 R_{max} and it could easily be extended much further if there was any reason for doing so. These methods seem to be comparable in terms of speed to the more elaborate procedures developed by Brady and Ball.[44] However, our algorithms do not make efficient use of the computer storage capabilities and it would be difficult to grow clusters much larger than 10^5 particles or sites. Figure 10 shows a 50,000 particle off-lattice Witten-Sander cluster grown using the approach described above.

The ability to grow large numbers of large aggregates has enabled us to reduce statistical uncertainties by an order of magnitude and is allowing us to investigate some new aspects of the structure of Witten-Sander aggregates. Some aspects of this work are discussed briefly in the next few paragraphs.

In the original work of Witten and Sander the structure of their two dimensional clusters was characterized by the dependence of the radius of gyration on cluster mass and by the two point density-density correlation function. The exponent β which describes how the radius of gyration depends on cluster mass $(R_g \sim N^\beta)$ has no detectable dependence on cluster size for clusters containing more than about 1,000 particles. From 1,000 off-lattice clusters the result $\beta = 0.5832 \pm 0.0008$ or $D_\beta = 1.715 \pm 0.002$ was obtained for clusters in the size range $5,000 < N < 50,000$.[45] If a sticking probability of 0.1 was used instead of 1.0 the result $D_\beta = 1.724 \pm 0.008$ was obtained from 122 clusters in the size range

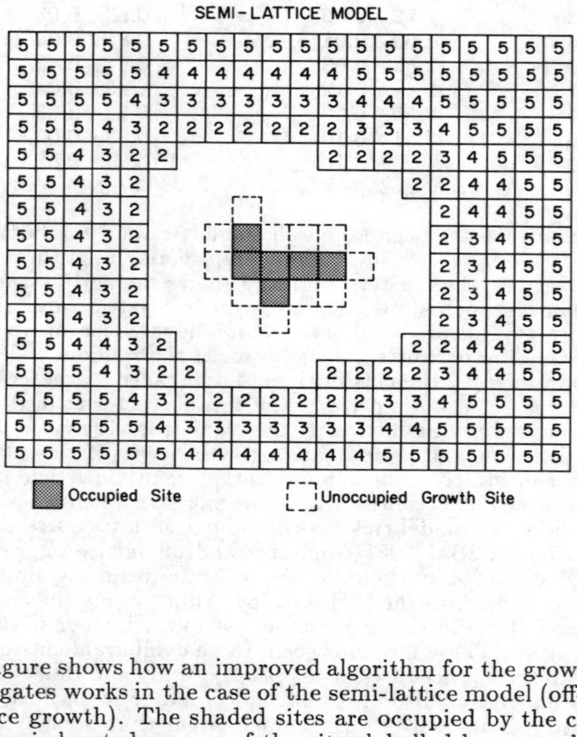

Fig. 9 This figure shows how an improved algorithm for the growth of Witten-Sander aggregates works in the case of the semi-lattice model (off-lattice walks with on-lattice growth). The shaded sites are occupied by the cluster. If the random walker is located on any of the sites labelled by a number (L) it can move by a distance of L lattice units in any direction without "contacting" the cluster. After each growth event the labels in the lattice sites which are in the vicinity of the newly occupied sites are updated. In this small scale illustration steps of length up to 5 lattice units are allowed. In larger scale simulations a more reasonable maximum step length for particles in the region occupied by the cluster would be 40 lattice units.

$50,000 < N < 100,000$. For the semi-lattice model the result $D_\beta = 1.695 \pm 0.002$ was obtained from 452 clusters in the size range $8,000 < N < 80,000$. Although the small difference ($\simeq 1\%$) between the values obtained for the fractal dimensionality D_β is substantially larger than the statistical uncertainties, this difference most probably does not indicate a real difference in the fractal dimensionalities associated with the two models but is an indication that even such large clusters are not close enough to the asymptotic ($N \to \infty$) limit to reduce systematic uncertainties below 1%. Our results do indicate that the fractal dimensionality of two dimensional Witten-Sander aggregates is probably a little larger than the value of 5/3 given by the mean field theories of Muthukumar(41) and of Tokuyama and Kawasaki(42). Curve A in Figure 11 shows the two point density-density correlation functions obtained from 34 50,000 site off-lattice Witten-Sander clusters. For a fractal object we expect that the density- density correlation function, $C(r)$ should have the form[1]

$$C(r) \sim r^{-\alpha} \qquad (\alpha = d - D) \qquad (2)$$

over an intermediate range of length scales long compared to the particle or lattice size and short compared to the overall size of the cluster. The straight line adjacent to the plot of $\ln C(r)$ vs. $\ln(r)$ has a slope of -0.29 which would be expected for a structure with a fractal dimensionality of 1.71. It can be seen from Figure 11 that the expected linear region with a slope of $D-d$ is either very short or absent. The upper curve in Figure 11 shows the two point density-density correlation function obtained by restricting one of the two points in the correlation function to that region of the cluster formed during the first

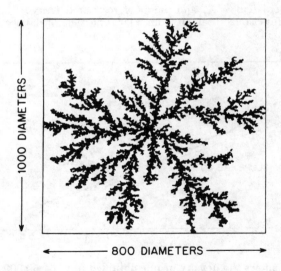

Fig. 10. A 50,000 particle off-lattice Witten-Sander aggregate.

20% of the growth process. The fact that these correlation functions are not well represented by the simple scaling relationship given in Equation (2) may be the result of the fact that the cluster has grown from a single seed which becomes a region of unusually high average density in the cluster.

Figure 12 shows the density profile obtained by averaging the profiles from 64 50,000 site off-lattice Witten-Sander aggregates. Again significant corrections to scaling can be seen at relatively short distances. The density profile shown in Figure 12 is consistent with a structure having a fractal dimensionality of about 1.71.

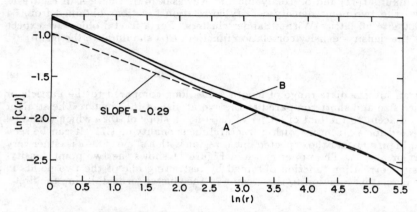

Fig. 11. Unrestricted (curve A) and partially restricted (curve B) density-density correlation functions obtained from 50,000 particle off-lattice Witten-Sander clusters.

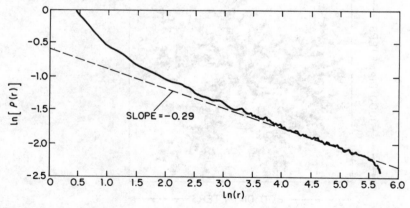

Fig. 12. This figure shows the density profile obtained from 64 50,000 particle off-lattice aggregates.

Some insight into the origin of these rather large corrections to the simple scaling picture for diffusion limited aggregation can be obtained by examining the angular (tangential) correlations.[46] 168 clusters each containing 80,000-100,000 sites were generated using the semi-lattice model and each cluster was divided into circular "shells" with a thickness of $0.01R$ where R is the radius of the annulus or shell. The angular density-density correlation function $(C(\theta))$ was obtained for each shell and then averaged over a small range $(R \pm 0.05R)$ to obtain the angular correlation function at a distance R from the origin or seed site. Figure 13 shows the results obtained from this procedure. The maximum in $C(\theta)$ at $R = \pi/2$ can be related to the formation of a diamond like shape for large lattice based clusters. Over a range of smaller angles we find that $C(\theta) \sim \theta^{\alpha'}$ where the exponent α' has a value of about 0.42. Similar results were obtained using the off-lattice model but there is now no maximum in $C(\theta)$ for $\theta \simeq \pi/2$. Since our results at the larger values of R may be influenced by incomplete growth of the clusters it is difficult to say very much about the asymptotic behavior of $C(\theta)$ in the limit of large R. However, there is some indication that the effective value of the exponent a decreases with increasing R. In any event the fact that the effective value of α' is larger than that of the exponent α which describes $C(r)$ even for quite large clusters illustrates that Witten-Sander clusters have a more complex structure than they were believed to have a short time ago. However, the simple scaling picture for DLA is probably still valid in the asymptotic $(N \to \infty)$ regime. Results similar to those described above have been obtained by M. Kolb.[47]

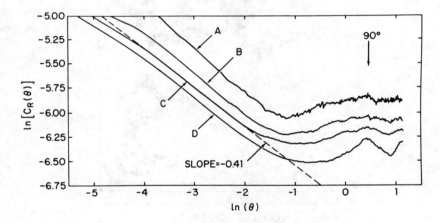

Fig. 13. The tangential (angular) correlation function $c_R(\theta)$ as a function of the angle θ for different values from the origin R. These results were obtained for 168 diffusion limited aggregates grown on a square lattice and containing 80,000-100,000 particles. In order to get curves $A - D$ the data were averaged over the intervals $\delta R = R \pm 0.05R$ with (A) $R = 100$; (B) $R = 200$; (C) $R = 300$ and (D) $R = 400$. The second maximum at $\theta \sim \pi/2$ indicates that the overall shape of the clusters is biased by the underlying square lattice.

The simulation of fractal like structures using DLA models has stimulated experimental work on diffusion limited growth processes. Brady and Ball[48] have found a fractal dimensionality of 2.43 ± 0.03 for the electrode deposition of copper under diffusion limited conditions. Similarly, Matsushita, et al[49] have obtained two dimensional random zinc dendrites with a fractal dimensionality of 1.66 ± 0.03. Also, Kapitulnick[50] has found a fractal dimensionality of about 2.5 in electrodeposited polypyrrole and structures with a fractal dimensionality of about 1.7 have been found to form under what are believed to be diffusion limited conditions in sputter deposited niobium/germanium thin films.[51] All of these experimental observations seem to be in good agreement with the computer simulations described above.

DIFFUSION LIMITED CLUSTER-CLUSTER AGGREGATION

Initially, simulations of diffusion limited cluster-cluster aggregation were carried out using a finite lattice or box with periodic boundary conditions. Figure 14 shows the results of a three dimensional simulation carried out using 2,500 particles (occupied lattice sites) on a 100^3 lattice. The fractal dimensionality of the clusters generated in these simulations has been estimated by

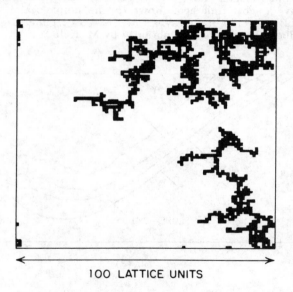

Fig. 14. The results of a simulation of diffusion limited cluster-cluster aggregation using 2,500 particles on a cubic lattice with periodic boundary conditions. A projection of the three dimensional structure onto a plane is shown. In this model only the smallest cluster(s) were allowed to move. Very similar results are obtained if all the clusters are allowed to move with the same probability.

measuring the dependence of the radius of gyration on the cluster mass or from the two point density-density correlation function. Figure 15 shows the density-density correlation functions obtained from similar simulations carried out at different initial particle densities. These results indicate that the aggregates produced in this way have a fractal like geometry on short length scales but are uniform on longer length scales ($C(r) = \rho$ where ρ is the mean density). The crossover from a fractal like geometry to a uniform structure can be extended to longer length scales by carrying out the simulations at lower densities. However, this requires more computer time and statistical uncertainties are larger because there are fewer particles in the structure. An alternative approach is to use the method employed by Sutherland[29] in his simulations of ballistic cluster-cluster aggregation (see above). Particles and/or clusters are brought together two at a time along random walk trajectories. In this way we can avoid altogether the problems associated with finite particle concentrations. These simulations are very similar to the simulations of particle-cluster aggregation (DLA) discussed above. Now we have a mobile cluster instead of a mobile particle. The methods recently developed to improve the efficiency of the simulation of particle-cluster aggregation can also be applied with similar advantages to the simulation of cluster-cluster aggregation via pairwise cluster addition. Another advantage of this approach is that it can readily be extended to higher dimensions. Table 2 shows the results of off-lattice simulations carried out in this manner. Results from two versions of the model are shown. In the hierarchical model[52] (equivalent to the maximum extension model of Sutherland[29]) we start with 2^N particles and form 2^{N-1} binary clusters.

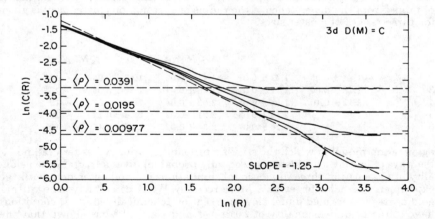

Fig. 15. Density-density correlation functions obtained from four simulations carried out on a 80x80x80 lattice with periodic boundary conditions. The bottom curve was obtained using 2,500 particles [$(\rho) = 0.00488$], the next curve was obtained using 5,000 particles [$(\rho) = 0.00977$], the third curve was obtained using 10,000 particles [$(\rho) = 0.0195$], and the top curve was obtained using 20,000 particles [$(\rho) = 0.0391$]. In these simu- lations the cluster diffusion coefficient was size-independent.

At the next stage 2^{N-2} clusters each containing 4 particles are formed and at the Mth stage 2^{N-M} clusters each containing 2^M particles are generated. The results shown in Table 2 for the hierarchical model are very similar to the results obtained by Jullien, Kolb and Botet[53] using a lattice version of the hierarchical model.

One of the main results from the diffusion limited cluster-cluster aggregation model is the fractal dimensionality of 1.75-1.8 for three dimensional systems. This is in good agreement with results obtained from metal particle aggregates (by Forrest and Witten[54] and by Weitz and Oliveria[55]) formed under conditions under which the aggregation process is probably diffusion limited and reorganization of the aggregate is unlikely. A somewhat higher fractal dimensionality was found by Schaefer et al[56] (2.12 ± 0.05) using colloidal silica (Ludox). However, in this case the aggregation process is probably much slower ("chemically" limited) and reorganization processes are more likely. Jullien and Kolb[57] have shown that a simple modification of the cluster-cluster

Table 2

Values obtained for the effective fractal dimensionalities D_α and D_β from diffusion limited cluster-cluster aggregation simulations. The values for D_α were obtained from the density-density correlation function over a range of length scales larger than a few particle diameters but considerably smaller than the overall size of the aggregate (the range of length scales over which $\ln|C(r)|$ is linearly related to $\ln(r)$). The effective dimensionality D_β (which is obtained from the dependence of the radius of gyration on cluster size) is given for three ranges of cluster sizes.

d	$D_j\alpha$	D_β $5 \leq N \leq 50$	$5 \leq N \leq N_{max}$	$100 \leq N \leq N_{max}$	N_{max}
2	1.44 ± 0.03	1.41 ± 0.01	1.43 ± 0.02	1.46 ± 0.03	$\sim 10^4$
3	1.78 ± 0.06	1.72 ± 0.01	1.75 ± 0.01	1.82 ± 0.08	$\sim 10^4$
4	2.12 ± 0.10	1.94 ± 0.03	2.03 ± 0.04	2.16 ± 0.08	$\sim 10^4$
5	—	2.14 ± 0.03	2.21 ± 0.02	2.36 ± 0.12	$\sim 10^3$
6	—	2.26 ± 0.07	2.38 ± 0.02	2.77 ± 0.17	$\sim 10^3$

aggregation model in which the sticking probability approaches zero (all possible ways of joining the clusters are equally probable) gives a fractal dimensionality of 1.98 ± 0.04 in three dimensional simulations in good agreement with the experiments of Schaefer et al.[56] More recently Weitz et al[58] have shown that gold aggregates formed under slow aggregation (chemically limited) conditions have a fractal dimensionality of 2.05 ± 0.05 and Cannell[59] has shown that the rapid (diffusion limited) aggregation of colloidal silica leads to aggregates with a fractal dimensionality of about 1.7. However, in this case, the mechanism of the aggregation process is uncertain. At this stage the agreement between the simulation results and experiments is quite gratifying. A number of theoretical approaches[60-66] to diffusion limited aggregation have been developed. They all give results which are in reasonably good agreement with the simulation results and they all predict an upper critical dimensionality in the range 6-10. However, they differ in both their approach and their detailed predictions.

In the random selection version of the zero concentration limit cluster-cluster aggregation model a "list" of clusters (including single particle clusters) is maintained throughout the simulation. Pairs of clusters are picked at random from the list and combined to form a larger cluster which is returned to the list which is now shortened. At the final stage of the simulation the list contains just two clusters which are combined to form a single large cluster. This version of the model corresponds to a mass independent addition probability. Since the addition probability in real (three dimensional) systems undergoing diffusion limited colloidal aggregation is also approximately mass independent[67-69] this model is more realistic than the hierarchical model. However, Table 2 shows that both versions of the model give very similar results as is also the case for ballistic cluster-cluster aggregation.

A question related to the possible distinction between the hierarchical and random selection versions of the pairwise addition models is how does the aggregate structure depend on the relationship between the cluster mass and cluster diffusion coefficient (probability of moving by a step length). To address this question we must return to simulations carried out using finite lattices or boxes with periodic boundary conditions. For convenience we have assumed, in most cases, that the dependence of the diffusion coefficient on cluster mass (M) or number of particles is given by $D(M) \sim M^\gamma$. Figure 7 shows the final stage of two dimensional simulations carried out assuming that $\gamma = 0$. Figure

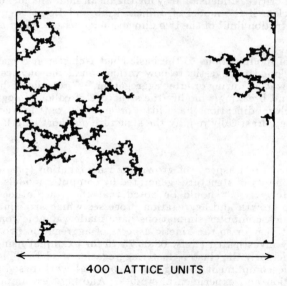

400 LATTICE UNITS

Fig. 16. The final stage in a two dimensional simulation of cluster-cluster aggregation carried out using 5,000 particles (occupied lattice sites) on a square lattice. In this simulation only those clusters with the smallest mass were allowed to move via random walk trajectories. This figure should be compared to Figure 7 which was generated using mass independent diffusion coefficients.

16 shows the results of a similar simulation except that now $\gamma \to \infty$ (only the smallest clusters can move). From many simulations like these we find that the fractal dimensionality is independent of γ (within the accuracy of our simulations) for all $\gamma < 0$. Very similar results have also been obtained from three dimensional simulations.

However, it is clear that in the limits of large (positive) γ and low concentration the cluster-cluster aggregation model becomes equivalent to the Witten-Sander model. Figures 17a and 17b show intermediate stages in simulations carried out with $\gamma = 1.5$ and $\gamma = 2.0$ respectively and Figure 18 shows the final stage of a simulation carried with q set to a value of 1.5. It is difficult to determine the nature of the crossover from cluster-cluster aggregation to particle-cluster aggregation even using relatively large scale simulations such as these. Examination of the density-density correlation function obtained at the end of the simulations for various values of γ suggests a continuous change in the fractal dimensionality from a value of about 1.45 for $\gamma = 1.0$ to about 1.7 for $\gamma = 2.0$. Jullien, Kolb and Botet[70] have investigated the crossover from cluster-cluster aggregation to diffusion limited aggregation using a model in which pairs of clusters are selected with probabilities given by $(M_1 M_2)^\omega$ where M_1 and M_2 are the cluster masses. Based on results obtained from this model and theoretical considerations they find a sudden change from cluster-cluster aggregation to particle-cluster aggregation at $\gamma = 1.0$. However, since 2 seems to be the upper critical dimensionality for the mean field Smoluchowski equations it is not clear if the model of Jullien, Kolb and Botet is equivalent to the zero concentration limit of the two dimensional cluster-cluster aggregation model.

A number of modifications to the basic cluster-cluster aggregation model have been developed to investigate how various plausible physical processes such as reorganization during or after aggregation,[71,72] rotational diffusion(73) and random bond breaking[74] modify the structure of colloidal aggregates. In most cases, such modifications have little or no effect on the fractal dimensionality but may drastically modify the structure on short length scales.

DISCUSSION

Here the structural aspects of growth and aggregation (particularly the fractal dimensionality of structures generated by computer models) have been emphasized. However, it should be noted that there are other aspects of non-equilibrium growth and aggregation processes which are equally important and to which computer simulations have made valuable contributions. During the past year or so the kinetic aspects of aggregation processes have been intensively investigated (Refs. 68,69,75-79 for example) using a variety of models (particularly the cluster-cluster aggregation model). This work provides a valuable compliment to ongoing theoretical work based on Smoluchowski's equation and experimental studies. Another very important area is concerned with developing a better understanding of the physical behavior of fractal aggregates. Simulations have been carried out to explore transport properties,[80-82] chemical reactions,[83] mechanical properties,[84] hydrodynamic properties,[68,85] and scattering properties. A considerable interest has also developed in questions related to how non-equilibrium structures grow and inter-

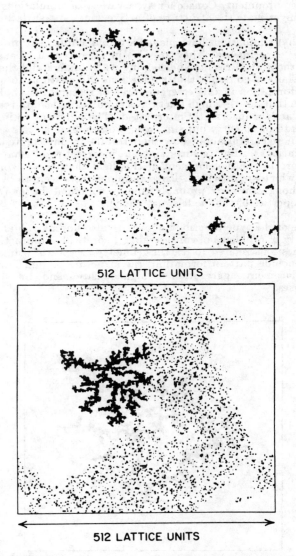

Fig. 17. Intermediate stages in two dimensional simulations of cluster-cluster aggregation carried out using 7,500 occupied lattice sites on 512x512 lattices. Figure 17a was obtained using a value of 1.5 for the exponent γ which describes how the cluster diffusion coefficient depends on cluster mass. Figure 17b was obtained using a value of 2.0 for γ.

act with their environment. Consequently, a variety of simulations have been carried out to investigate the structure of active zones and surfaces.[9,45,86-90]

It is clear that simulations have played a valuable role in stimulating both theory and new experiments. They are also invaluable as a means of checking important theoretical results and helping us to understand the results of experiments. There is no doubt that simulations will continue to provide important new information in this emerging area of research. It should also be noted, however, that a rather high fraction of simulations lead to results which are ambiguous and/or incorrectly interpreted. The only way to improve this situation seems to be to carry out simulations on as large a scale as is practical and to be more careful about how their results are interpreted. It is also true that simulations leading to surprising but incorrect results have probably contributed more to the development of this field than those which lead to correct results which are expected. Programming errors seem to be much less of a problem though they are probably equally common. In many cases the universality properties of the models may save us.

Simulations of the type described above are also finding valuable applications to a variety of problems of considerable theoretical and practical importance where questions related to fractal geometry are not of primary importance. These include pattern formation, the structure of thin film deposits, fluid-fluid displacement, aggregation at high densities and diffusion limited reaction processes.

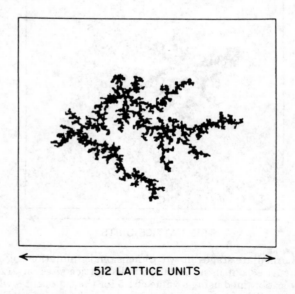

Fig. 18. The final stage in a simulation carried out using the same model parameters which were used to obtain Figure 17a ($\gamma = 1.5$).

[1] Witten, T. A. and Sander, L. M., Phys. Rev. Lett., 47, 1400 (1981).
[2] Meakin, P., Phys. Rev. Lett., 51, 1119 (1983).
[3] Kolb, M., Botet, R., and Jullien, R., Phys. Rev. Lett., 51, 1123 (1983).
[4] Eden, M., Proc. 4th Berkeley Symposium Math., Stat. and Prob., 4, 223 (1961).
[5] Mandelbrot, B. B., "The Fractal Geometry of Nature", W. H. Freeman, San Francisco (1982).
[6] This correspondence between the Eden model with "protected" voids and the Witten-Sander model in the limit of very small sticking probabilities is valid only if the mobile particle in the Witten-Sander model is given an opportunity to stick after each move including those attempted moves which return the particle to its original position because they would result in multiple lattice site occupancy. If the sticking probability is defined as the probability that a particle will stick at a particular unoccupied interface site before moving on to another site then the equivalent Eden model in the limit of small sticking probability is one in which the ratio of growth probabilities for sites with 1, 2 and 3 occupied nearest neighbors is 3:2:1 (with no growth in enclosed voids). This model gives an average density of 0.70-0.71 in good agreement with results obtained from the corresponding Witten-Sander model with a low sticking probability (0.001 at all surface sites).
[7] Williams, T. and Bjerknes, R., Nature, 236, 19 (1972).
[8] Peters, H. P., Stauffer, D., Holters, H. P. and Loewenich, K., Z. Physik B34, 399 (1979).
[9] Meakin, P. and Witten, T. A., Phys. Rev. A28, 2985 (1983).
[10] D. Richardson, Proc Cambridge Philos Soc 74 515 (1973).
[11] Sawada, Y., Ohta, S., Yamazaki, M., and Honjo, H., Phys. Rev. A26, 3557 (1982).
[12] Niemeyer, L., Pietronero, L., and Wiesmann, A. J., Phys. Rev. Lett., 52, 1033 (1984).
[13] Dhar, D. and Ramaswamy, R., Phys. Rev. Lett. 54, 1346 (1985).
[14] Martin, H., Vannimenus, J. and Nadal, J. P., preprint
[15] Rickvold, P. A., Phys. Rev. A26, 647 (1982).
[16] Meakin, P., Phys. Rev. B28, 6718 (1983).
[17] Sander, L. M., in "Kinetics of Aggregation and Gelation", Family, F. and Landau, D. P., Eds. Elsevier-North Holland, Amsterdam, 1984.
[18] Meakin, P., Leyvraz, F. and Stanley, H. E., Phys. Rev. A31, 1195 (1985).
[19] Vold, M. J., J. Colloid Sci. 18, 684 (1963), J. Colloid Sci. 14, 168 (1959), J. Phys. Chem. 63, 1608 (1959), J. Phys. Chem. 64, 1616 (1960).
[20] Sutherland, D. M., J. Colloid and Interface Sci., 22, 300 (1966).
[21] Meakin, P., J. Colloid and Interface Sci. 96, 415 (1983).
[22] Bensimon, D. , Domany, E. and Aharony, A., Phys. Rev. Lett. 51, 1394 (1983).
[23] Ball R. and Witten, T. A., Phys. Rev. A29, 2966, (1984).

[24] Bensimon, D., Shraiman, B. and Kadanoff, L. P., "Kinetics of Aggregation and Gelation", Family, F. and Landau, D. P., Eds. Elsevier, North Holland, Amsterdam (1984).
[25] Ramanlal, P. and Sander, L. M., Phys. Rev. Lett. 54, 1828 (1985).
[26] Leamy, H. J., Gilmer, G. H. and Dirks, A. G., in Vol. 6, "Current Topics in Materials Science", E. Kaldis Ed. (North Holland, Amsterdam, 1980).
[27] Dirks, A. G. and Leamy, H. J., Thin Solid Films 47, 219 (1977).
[28] Meakin, P., unpublished.
[29] Sutherland, D. N., J. Colloid Interface Sci. 25, 373 (1967), Nature 226, 1241 (1970). Sutherland, D. M. and Goodarz-Nia, I., Chem. Eng. Sci. 26, 2071 (1971).
[30] Stanley, H. E., J. Phys. A10, L211 (1977).
[31] Finegold, L. X., Biochim. Bophys. Acta 448, 393 (1976), Donnell, J. H. and Finegold, L. X., Biophys. J. 35, 783 (1981).
[32] Meakin, P., Phys. Rev. A27, 604 (1983); Phys. Rev. A27, 1495 (1983).
[33] Witten, T. A. and Sander, L. M., Phys. Rev. B27, 5686 (1983).
[34] Vicsek, T., Phys. Rev. Lett. 53, 2281 (1984).
[35] Patterson, L., Phys. Rev. Lett. 52, 1621 (1984).
[36] Kadanoff, L. P., preprint.
[37] Nittmann, J., Daccord, G. and Stanley, H. E., Nature 314, 141 (1985).
[38] Tang, C., Phys. Rev. A31, 1977 (1985).
[39] Meakin, P. and Wasserman, Z., Chemical Physics 91, 391 (1984).
[40] Ball, R. C., Nauenberg, M. and Witten, T. A. Phys. Rev. A29 2017 (1984)
[41] Muthukuman, M., Phys. Rev. Lett. 50, 839 (1983).
[42] Tokuyama, M. and Kawasaki, K., Phys. Lett. 100A, 337 (1984).
[43] Meakin, P., J. Phys. A. xx xxxx (1985)
[44] Brady, R. M., and Ball, R. C., CECAM Workshop, Orsay, 1984, (unpublished).
[45] Meakin, P. and Sander, L. M., Phys. Rev. Lett. 54, 2053 (1985).
[46] Meakin, P. and Vicsek, T., Phys. Rev. Axx, xxxx (1985).
[47] Kolb, M., preprint
[48] Brady, R. M. and Ball, R. C., Nature 309, 225 (1984).
[49] Matsushita, M., Sano, M., Hayakawa, Y., Honjo, H. and Sawada, Y., Phys. Rev. Lett. 53, 286 (1984).
[50] Kapitulnik, A., private communication.
[51] Elam, W. T., Wolf, S. A., Sprague, J., Gubser, D. V., Van Vechten, D., Barz, G. L. Jr., and Meakin, P., Phys. Rev. Lett. 54, 701 (1985).
[52] Botet, R., Jullien, R. and Kolb, M., J. Phys. A17, L75 (1984).
[53] Jullien, R., Kolb, M. and Botet, R., J. Physique Lett. 45, L211 (1984).
[54] Forrest, S. R. and Witten, T. A., J. Phys. A12, L109 (1979).
[55] Weitz, D. and Oliveria, M., Phys. Rev. Lett. 52, 1433 (1984).
[56] Schaefer, D. W., Martin, J. E., Wiltzius, P. and Cannell, D. S., Phys. Rev. Lett. 52, 2371 (1984).

[57] Jullien, R. and Kolb, M., J. Phys. A17, L639 (1984).
[58] Weitz, D., Huang, J. S., Lin, M. Y. and Sung, J., Phys. Rev. Lett. 54, 1416 (1985).
[59] Cannell, D. S., private communication
[60] Kolb, M., Phys. Rev. Lett. 53, 286 (1984).
[61] Ball R. C. and Thompson, B. R., J. Phys. A17, L951 (1984).
[62] Hentschel, H. G. E. and Deutch, J. M., Phys. Rev. A29, 1609 (1984).
[63] Ball, R. C. and Witten, T. A., Proc. 3rd Conf. on Fractals, Gaithersburg, MD, 1983.
[64] Matsushita, M. preprint.
[65] Botet, R., J. Phys. A18, 847 (1985).
[66] Obukhov, S. P. preprint.
[67] Feder, J., Jdssang, T. and Rosenqvist, E., Phys. Rev. Lett. 53, 1403 (1985).
[68] Meakin, P., Chen, Z. Y. and Deutch, J. M., J. Chem. Phys. 82, 3786 (1985).
[69] Ziff, R. M., McGrady, E. D. and Meakin, P., J. Chem. Phys. (1985).
[70] Jullien, R., Kolb, M. and Botet, R. in "Kinetics of Aggregation and Gelation, F. Family and D. P. Landau Editors, Elsevier, North Holland, Amsterdam (1984). R. Botet, R. Jullien and M. Kolb, preprint.
[71] Meakin, P. and Jullien, R., J. de Physique xx, xxxx (1985).
[72] Meakin, P. unpublished.
[73] Meakin, P., J. Chem. Phys. 81, 4637 (1984).
[74] Meakin, P. unpublished.
[75] Vicsek, T. and Family, F., Phys. Rev. Lett. 52, 1669 (1984).
[76] Kolb, M., Phys. Rev. Lett. 53, 1653 (1984).
[77] Meakin, P., Vicsek, T. and Family, F., Phys. Rev. B31, 564 (1985).
[78] Botet, R. and Jullien, R., J. Phys. A17, 2517 (1984).
[79] Kang, K. and Redner, S., Phys. Rev. A30 2833 (1984).
[80] Meakin, P. and Stanley, H. E., Phys. Rev. Lett. 51, 1457 (1983).
[81] Witten, T. A. and Kantor, Y., Phys. Rev. B30, 4093 (1984).
[82] Sahimi, M., MacKarnin, M., Nordahl, T. and Tirrell, M. preprint.
[83] Sander, L. M. private communication.
[84] Cantor, Y. and Witten, T. A. preprint.
[85] Chen, Z. Y., Deutch, J. M. and Meakin, P., J. Chem. Phys. 80, 2982 (1984).
[86] Plishke, M. and Racz, Z. Phys. Rev. Lett 53, 415 (1984).
[87] F. Family and T. Vicsek J. Phys A18 L75 (1985).
[88] R. Jullien and R. Botet Phys. Rev. Lett 54 2055 (1985).
[89] P. Meakin, Phys. Rev A xx,xxx (1985).
[90] P. Meakin, H. E. Stanley, A Coniglio and T. A. Witten Phys.Rev. A 32, 2364 (l985).

RATE EQUATION APPROACH TO AGGREGATION PHENOMENA

François Leyvraz

Center for Polymer Studies and Department of Physics
Boston University, Boston, Massachusetts 02215

A general qualitative theory for the mean-field rate equation describing aggregation processes is described. Various exponents describing the cluster-size distribution and its time evolution are derived. It is found that this theory frequently gives a good account of systems with more realistic kinetics, such as cluster-cluster aggregation, but that considerable caution must be exercised to ensure being in the actual asymptotic regime, since very long crossovers are sometimes observed.

INTRODUCTION

Aggregation phenomena have attracted considerable interest in many apparently unrelated field of physics. Just to name a few, one has used the kinetic models of aggregation to be described below in aerosol physics to describe the particle size distribution in aerosols,[1] astrophysics for the distribution of star cluster sizes,[2] cloud physics for the size distribution of droplets;[3] similar equations have also been used to describe polymerization processes.[4,5,6]

All these systems have the following in common: they involve objects that are aggregates of basic units (monomers) and capable of reacting as follows:

$$A_j + A_k \xrightarrow{K(j,k)} A_{j+k},$$

where A_j denotes an aggregate consisting of j monomers and $K(j,k)$ is the rate at which the reaction proceeds. This last depends on the respective sites j and k of the reacting aggregates in a way that is essentially model dependent. It is not the purpose of this article to review the techniques involved in deriving "microscopic" expressions for the rate constants: such expressions must be derived in each experimental situation separately. Rather, the general form of the rate equations shall be given and their qualitative behavior discussed.

Clearly, since there is such a broad range of possibilities for the reaction constants $K(i,j)$, it would be desirable to have a theory that assumes as little as possible about their analytical form. It will be seen that this is indeed possible. The role of the following theory is therefore to make a bridge between the detailed theoretical description of the experimental system under consideration and the qualitative description of general features that can then again be tested against the experimental data.

THE GENERAL FORM OF THE EQUATIONS

Define $c_j(t)$ as the concentration of aggregates of size j at time t. This yields a very partial description of a coagulating system: the geometry and internal structure of the aggregates as well as their spatial distribution are all ignored or, more precisely, averaged over. To obtain a valid description from the functions $c_j(t)$, one must therefore assume that these degrees of freedom can be safely disregarded. This assumes a fairly dilute system, or else a system, the aggregates of which do not readily react with one another. In particular, it is most unlikely to remain valid once an infinite cluster has been formed, since such an infinite cluster of necessity introduces a very complex spatial structure, which presumably cannot be ignored.

Under all these assumptions, however, it appears reasonable to assume that the probability of reaction is given by $K(j,k)c_j c_k$. Hence

$$\dot{c}_j = \frac{1}{2} \sum_{k,l} K(k,l) c_k c_l (\delta_{k+l,j} - \delta_{k,j} - \delta_{l,j}), \tag{1}$$

where the initial condition is usually taken to be that only monomers are present, though general results should not depend on such details.

On physical grounds one expects to have a conservation law for the total number of monomers, i.e., $\sum_{j=1}^{\infty} j c_j(t)$. It is easy to see that this is indeed so, if

$$\sum_{j=1}^{\infty} j K(j,k) c_j < \infty \tag{2}$$

for all k. This implies that $c_j(t)$ starts by a rather slow growth culminating at a maximum at time t_j and then decreasing. The times t_j are then expected to go to infinity as j does. For a given fixed size j, we shall call small those times much less than t_j, large those considerably larger than t_j. We shall say that j is roughly the typical size of the cluster size distribution at time t if $t_j \sim t$.

Some variations add a constant monomer production term and/or a removal term. Various cases have been examined, in particular with respect to stationary solutions, but the time evolution has also been studied.[2,7,8]

SCALING: BASIC CONCEPTS

The basic idea underlying scaling is that the form of the solutions of Eq. (1) is essentially independent of any features of the initial conditions. One then further assumes that there is only one characteristic size $s(t)$. This rather naturally leads to the Ansatz[1,9,10]

$$c_j \sim j^{-\theta}\phi(j/s(t)), \tag{3}$$

where $\phi(x)$ is some arbitrary function with rapid decay at large x. From mass conservation follows quite generally that $\theta = 2$. It is observed as a quite general fact that $s(t)$ increases as a power law. One defines

$$s(t) \sim t^z. \tag{4}$$

The cluster size distribution now can be described in two various limiting cases:

(1) $1 \ll j \ll s(t)$: one then has

$$c_j(t) \sim j^{-2}\phi(jt^{-z}), \tag{5}$$

and the behavior depends crucially on the nature of the singularity of $\phi(x)$ at $x = 0$. If

$$\phi(x) \sim x^{w/z} \quad (x \ll 1),$$

then one obtains

$$c_j(t) \sim j^{-\tau}t^{-w}, \tag{6}$$

with

$$(2-\tau)z = w. \tag{7}$$

If, on the other hand,

$$\phi(x) \sim exp(-x^{-\alpha}) \tag{8}$$

where α is some positive number, then

$$c_1(t) \sim exp(-t^{+\alpha z}) \tag{9}$$

and there is no power law to describe the cluster size distribution for $j \ll s(t)$.

(2) $j \sim s(t)$: one then has

$$c_j(t) \sim j^{-2} \tag{10}$$

independently of any details of the aggregation process. For $j \gg s(t)$, the cluster size distribution decreases exponentially. As an application of this formalism, consider how $\sum_{j=1}^{\infty} j^\rho c_j(t)$ diverges at $t \to \infty$. One obtains, using (3) and changing sum into integrals,

$$\sum j^\rho c_j(t) \cong \sum j^{\rho-2}\phi(j/s(t))$$
$$= \frac{1}{s(t)}\sum (j/s(t))^{\rho-2}\phi(j/s(t))s(t)^{\rho-1}$$
$$= s(t)^{\rho-1}\int x^{\rho-2}\phi(x)dx \propto t^{z(\rho-1)}, \tag{11}$$

if the integral is convergent at the lower limit. If it is not, then the small clusters are the most important and one obtains

$$\sum j^\rho c_j(t) \propto \sum_{j=1}^{\infty} j^{\rho-\tau}t^{-w}$$
$$\propto t^{-w}, \tag{12}$$

since it is easy to see that the sum must then converge. This result can readily be interpreted as follows: the sum $\sum_1^\infty j^\rho c_j$ is either dominated by small clusters or by the typical (large) clusters. In the first case, it will behave just as any $c_j(t)$ with j very small, i.e., it will decay as t^{-w}. In the latter case, however, the weighting factor j^ρ becomes of the order $s(t)^\rho$ and cannot be neglected. In this case, however, the sum can be evaluated as if the whole mass were concentrated in a size j of order $s(t)$, leading to a behavior as $s(t)^{\rho-1}$. Let us now illustrate this by some exactly solved examples.

EXACT SOLUTIONS

Of the many exactly solvable models for $K(i,j)$, we shall discuss two fairly simple ones:

$$K(i,j) = 1$$
$$K(i,j) = i + j, \qquad (13)$$

where, in all cases, the initial condition is taken to be $c_j(0) = \delta_{j1}$ (i.e., only monomer present). These are studied because they show explicitly the kind of behavior to be expected. They are, however, both very specific and sometimes untypical. For the first model, this yields[11]

$$c_j(t) = \frac{4t^{j-1}}{(t+2)^{j+1}} = \frac{4}{(t+2)^2} \cdot \left(\frac{t}{t+2}\right)^{j-1}. \qquad (14)$$

The time t_j are equal to $j - 1$, so that the typical size increases linearly with time. This is borne out by a more conventional definition of typical size as

$$s(t) = \frac{\sum_{j=1}^\infty j^2 c_j(t)}{\sum_{j=1}^\infty j c_j(t)} = t + 1. \qquad (15)$$

Furthermore one sees that for large times but fixed j

$$c_j(t) \sim t^{-2}. \qquad (16)$$

More accurately yet, if one considers the limit $j \to \infty$, $t \to \infty$, $j/s(t) = const.$, one obtains

$$c_j(t) \cong \frac{1}{t^2} exp(-2j/s(t))$$
$$\cong j^{-2} (j/s(t))^2 exp(-2j/s(t)), \qquad (17)$$

which has the correct scaling form given by Eq. (5). For the case $K(i,j) = i+j$, one obtains[12]

$$c_j(t) = \frac{j^{j-2} e^{-j}}{(j-1)!} (1 - e^{-t})^{j-1} exp(je^{-t}) e^{-t}, \qquad (18)$$

and hence

$$t_j = \frac{1}{2} \ln j, \qquad (19)$$

so that the characteristic size increases exponentially with time. This means that a large aggregate is far sooner built up in this case than at a constant reaction rate, which is not surprising. One also obtains

$$s(t) = \sum_{j=1}^{\infty} j^2 c_j(t) \sim e^{2t}. \tag{20}$$

Further
$$c_j(t) \sim e^{-t} \tag{21}$$

for any finite j; moreover, in the limit $j \to \infty$, $s(t) \to \infty$, $j/s(t) = const.$, one obtains

$$c_j(t) \cong \frac{j^{j-2}e^{-j}}{(j-1)!} e^{-t} exp(j \ln(1-e^{-t}) + je^{-t})$$
$$\simeq \frac{j^{j-2}e^{-j}}{(j-1)!} e^{-t} exp(-aj/s(t)), \tag{22}$$

where a is some positive constant. Note that the constant prefactor in this scaling representation, i.e.,

$$\frac{j^{j-2}e^{-j}}{(j-1)!}, \tag{23}$$

has a power-law decay, namely it goes as $j^{-3/2}$ as $j \to \infty$. This implies

$$c_j(t) \cong j^{-3/2} e^{-t} exp(-aj/s(t))$$
$$\cong j^{-2}(j/s(t))^{1/2} exp(-aj/s(t)). \tag{24}$$

SCALING EXPONENTS FOR NON-GELLING SYSTEMS

After having considered some quite specific cases exactly, it is now necessary to look at far more general cases. The reaction rates $K(i,j)$ will be assumed to be completely general, characterized by only two exponents λ and ν, defined as follows:[15]

$$K(aj, ak) = a^\lambda K(j, k)$$
$$K(1, j) \sim j^\nu. \tag{25}$$

On physical and mathematical grounds it is necessary that ν should be less than one and λ less than two.[14] Further, it will be found that if $\lambda > 1$, then gelation will occur.[5,6,13] Thus for the time being, we only consider $\lambda \leq 1$, $\nu \leq 1$.

Assume now the existence of a typical size $s(t)$. The rate of increase of the typical size can be assumed to be the rate at which two aggregates of typical size react with one another. This is so because the definition of typical size is such that it is not influenced by the presence of a background of small clusters. Thus it follows that the increase of $s(t)$ with time goes as

$$\frac{ds}{dt} \sim K(s, s) \sim s^\lambda, \tag{26}$$

and hence

$$s(t) \sim t^{+1/(1-\lambda)}$$
$$z = \frac{1}{1-\lambda}. \tag{27}$$

To determine τ and w, it is necessary to know how the monomer concentrate $c_1(t)$ behaves. Physically this depends crucially on whether it reacts mainly with itself or mainly with clusters of typical size. This information, however, is contained in the exponent ν. Defining $\mu = \lambda - \nu$, one has three cases:

Case 1: $\mu > 0$. This means that large aggregates react preferentially with equally large ones. Using Eq. (11) we see that

$$\sum_{j=1}^{\infty} K(1,j)c_j \sim \sum_{j=1}^{\infty} j^\nu c_j$$
$$\sim s(t)^{\nu-1}, \tag{28}$$

if $\nu - \tau > -1$. Let us consider the implications. Since $s(t) \sim t^{1/(1-\lambda)}$, one obtains

$$\dot{c}_1 \cong -c_1 t^{-(1-\nu)/(1-\lambda)}, \tag{29}$$

and hence

$$c_1 \sim exp\left(-At^{(\nu-\lambda)/(1-\lambda)}\right)$$
$$\to 1 \qquad (t \to \infty), \tag{30}$$

since $\nu < \lambda$. This is, however, a patent absurdity, as a monomer can always react at least with itself, so that $\lim_{t\to\infty} c_1(t) = 0$. We are therefore led back to assuming that

$$\nu - \tau < -1. \tag{31}$$

Hence, since τ is well-defined, w must exist and be finite. This means that

$$\frac{\dot{c}_1}{c_1} = \frac{d}{dt}(\ln c_1) \simeq -w\frac{d}{dt}\ln t \simeq \frac{1}{t}. \tag{32}$$

But

$$\frac{\dot{c}_1}{c_1} = \sum K(1,j)c_j \sim \sum j^\nu c_j \sim t^{-w}. \tag{33}$$

Thus

$$w = 1$$
$$\tau = 1 + \lambda. \tag{34}$$

Case 2: $\mu = 0$. In this case, the value of the exponents depends intricately on the nature of the constants $K(i,j)$. A result has been proposed by Leyvraz for $K(i,j) = i^\omega + j^\omega$, but has been shown to be erroneous by van Dongen and Ernst.[15] See their work for further details.

Case 3: $\mu < 0$. Using the same reasoning as in Case 1, one is led to say that if

$$\sum j^\nu c_j \sim s(t)^{\nu-1}, \tag{35}$$

then

$$c_1(t) \sim exp\bigl(-t^{-(\lambda-\nu)/(1-\lambda)}\bigr)$$
$$\to 0 \quad (t \to \infty), \tag{36}$$

so that this does **not** lead to a contradiction any more. The alternative hypothesis, however, namely that $\nu - \tau < -1$ and hence

$$w = 1$$
$$\tau = 1 + \lambda \tag{37}$$

leads to

$$\nu - \tau = \nu - 1 - \lambda > -1, \tag{38}$$

and is therefore contradictory. The long-time behavior of such kernels is characterized by a typical size growing as t^z, where all clusters are of approximately that size and all small clusters decay as exponentials of some power of time. The difference between these and kernels with $\mu > 0$ is understandable. In the case $\mu > 0$, large clusters react primarily with other large clusters, thus leaving behind a very polydisperse system. In the case $\mu < 0$, however, large clusters grow first by eliminating whatever remains of small clsuters, so that the system remains fairly monodisperse.

NUMERICAL SIMULATIONS

So many simulations of aggregating systems have been performed that it is impossible within this article to go into any detail. I shall merely consider two well-studied models.

In the first model[16,17] a dilute solution of monomers on a lattice is allowed to diffuse. Each time two clusters become adjacent to one another, they combine irreversibly and proceed to diffuse as a rigid whole. The diffusion constant is allowed to depend on cluster size as follows:

$$D(i) \sim i^{-\gamma}. \tag{39}$$

It is then observed that the objects thus created are fractals, so that the radius $R(i)$ depends on cluster size as:

$$R(i) \sim i^{1/d_f}, \tag{40}$$

where d_f is the clusters' fractal dimension (numerically ~ 1.45 in two dimensions). The theory for aggregation of Brownian particles gives[11]

$$K(i,j) = \bigl(R(i) + R(j)\bigr)^{d-2}\bigl(D(i) + D(j)\bigr)$$
$$\cong (i^{1/d_f} + j^{1/d_f})^{d-2}(i^{-\gamma} + j^{-\gamma}). \tag{41}$$

This implies, in the notation of Sec. 4,

$$\lambda = \frac{d-2}{d_f} - \gamma$$
$$\nu = \frac{d-2}{d_f}, \tag{42}$$

so that as long as $\gamma > 0$ and $d > 2$, one obtains a sharply peaked, bell-shaped cluster-size distribution, as in Case 3 above. If, however, γ is taken to be negative, one obtains the situation in Case 1, with

$$\tau = 1 - \gamma + \frac{d-2}{d_f}$$
$$w = 1, \tag{43}$$

whereas the case $\gamma = 0$ is difficult. However it is known in this case (15) that $\tau > 1$ and hence $w < z$. In any case, however, $\tau = \frac{1}{1-\lambda}$, with λ given as above, in accordance with the result of Kolb.[19]

Qualitatively, this picture appears to be correct.[18] There is a critical value γ_c of γ below which the distribution is bell-shaped, above which it shows power-law behavior. Many significant details, however, do not fit quantitatively: the observed value of γ_c is not zero; the value of w above γ_c is not one. However, it is not easy to say what exactly is wrong with the theory, since the form of the kernel itself is open to doubt as well as the value of d_f. Moreover it is not clear whether the mean-field like approximations involved in the Smoluchowski equations are justified. Finally, there is the possibility that the asymptotic behavior described above takes an anomalously long time to set in completely.

To settle this issue, another model has been studied, where two clusters only stick together when they occupy the same site and the clusters therefore remain pointlike, although their mass increases. As above, the diffusion constants $D(i)$ go as $i^{-\gamma}$. This model has been extensively studied by Kang, Redner and Meakin, leading to the following conclusions:

¶The critical dimension d_c below which spatial fluctuations make the rate equation approach invalid is two.

¶Critical behavior is usually quantitatively consistent with theoretical predictions, but exceedingly long crossovers have been observed, particularly if $K(i,j)$ is close to being the constant kernel. Numerical integration of the Smoluchowski equations confirm the existence of these crossovers, though from a theoretical point of view they remain little understood.

SUMMARY

In summary, we have found that aggregation phenomena are qualitatively well-described by rate equations for the cluster size distribution. The behavior of these equations depends on the precise form of the reaction constants, but for large times are found to fall into two categories: (1) bell-shaped distributions sharply peaked at some typical size and (2) highly polydisperse power-law cluster-size distributions, with a slow exponential decay setting in beyond some typical size. Of these two cases, the first was found to hold if reactions between large and small aggregates had larger rate constants than reactions between aggregates of comparable size, the second otherwise. Quantitative expressions for these exponents were derived from a scaling theory which has been conformed by real[2] and numerical[18] experiments as well as by many exactly-solved models.

[1] S. K. Friedlander, Smoke, Dust and Haze (Wiley-Interscience, New York, 1977).
[2] G. B. Field, W. C. Saslaw, Astrophys. J. **142**, 568 (1965); G. B. Field and J. Hutchins, Astrophys. J. **153**, 737 (19xx).
[3] R. L. Drake, in Topics in Current Aerosol Research, vol. 3, eds. G. M. Midy and J. R. Brock.
[4] W. H. Stockmayer, J. Chem. Phys. **11**, 45 (1943).
[5] F. Leyvraz and H. R. Tschudi, J. Phys. A **15**, 1951 (1982).
[6] E. M. Hendriks, M. H. Ernst and R. M. Ziff, J. Stat. Phys. **31**, 519 (1983); Phys. Rev. Lett. **49**, 593 (1982).
[7] E. M. Hendriks, J. Phys. A **17**, 2299 (1984).
[8] W. White, J. Coll. Interface Sci. **87**, 204 (1982).
[9] A. A. Lushnikov and V. N. Piskunov, Dolklady Phys. Chem. **231**, 1166 (1976).
[10] T. Vicsek and F. Family, Phys. Rev. Lett. **52**, 1669 (1984).
[11] M. v. Smoluchowski, Phys. Z. **17**, 585 (1916).
[12] A. M. Golvin, Izv. Geophys. Ser. **1963**, 783 (1963) [English translation: Bull Acad. Sci. USSR, Geophys. Ser. No. 5, 482 (1963).
[13] R. M. Ziff, J. Stat. Phys. **23**, 241 (1980).
[14] F. Leyvraz and H. R. Tschudi, J. Phys. A **14**, 3389 (1981).
[15] P. van Dongen and M. H. Ernst (preprint).
[16] P. Meakin, Phys. Rev. Lett. **51**, 1119 (1983).
[17] M. Kolb, R. Botet and R. Jullien, Phys. Rev. Lett. **51**, 1123 (1983).
[18] P. Meakin, T. Vicsek and F. Family, Phys. Rev. B **31**, 564 (1985).
[19] M. Kolb, Phys. Rev. Lett. **53**, 1653 (1984).

EXPERIMENTAL METHODS FOR STUDYING FRACTAL AGGREGATES

José Teixeira

Laboratoire Léon Brillouin, * CEN-Saclay
91191 Gif-sur-Yvette Cedex, France

Porous materials, aggregates and ramified structures frequently display self-similarity, i.e., their structure is associated with power-law density-density correlation functions. In a real physical system, such behavior has two natural limits: the particle or pore size and the cluster size. Within these limits, the system is well described by a fractal or Hausdorff dimension. For the experimentalist, it is important to find methods which can isolate this behavior, i.e., to distinguish it from the particle and cluster contributions, in order to determine the fractal dimension unambiguously. Depending on the characteristic size of the fractal object, different experimental methods can be used. We discuss the use of microscopic photographs and give the theory of scattering applied to fractal objects and fractal surfaces. We present, in particular, the method of separating the particle and the cluster size contributions. Finally we give a short introduction to the dynamical aspects of fractals and to the interpretation of low frequency vibrational spectra of fractal systems in terms of fractons.

For the experimentalist, a fractal structure[1] appears as a self-similar structure within some observable length scale. In general the purpose of an experiment is to determine the parameters that characterize the fractal structure, namely the fractal (Hausdorff) dimension, and the limits within which the definition applies.

We present in the following sections the more current techniques used for this purpose. Part 2 describes the observations performed in real space and Part 3 is devoted to scattering techniques. Finally, Part 4 gives an introduction to the observation of excitations in fractal objects.

---

* Laboratoire commun CEA-CNRS

OBSERVATION OF FRACTAL OBJECTS IN REAL SPACE

Aggregates often have large sizes and can be observed by direct techniques such as optical microscopy. The fractal character of these aggregates can be determined by detailed analysis of photographs. Preliminary studies have been established in this way, power law relations between the number of particles constituting the aggregate and its projected area on a plane.[2]

In a general way, the fractal dimension D is defined by

$$N(r) = \left(\frac{r}{r_0}\right)^D, \tag{1}$$

where $N(r)$ is the quantity obtained by measuring a fractal medium with a gauge r_0. For instance, $N(r)$ can be the number of particles of radius r_0 which lie within a sphere of radius r centered on an arbitrary particle.

Niemeyer et al[3] have shown that the discharge pattern on dielectric breakdown is fractal. This patterns has, in two dimensions, a shape analogous to the structures generated by the diffusion limited aggregation model of Witten and Sander.[4] In the experiment of Niemeyer et al, the determination of the fractal dimension is done by counting the number of branches of the pattern as a function of their distance, r, from the central point where the discharge started. The number of branches $n(r)$ at a given distance r from the center is then obtained from the relation

$$n(r) \sim \frac{dN(r)}{dr} \sim r^{D-1}. \tag{2}$$

A more general procedure for determining the fractal dimension of an aggregate observed in real space uses the notion of the space density–density correlation function:

$$G(r) = \frac{1}{V\phi} \int_V dr' \langle \phi(r')\phi(r+r') \rangle, \tag{3}$$

where $\phi(r)$ is the density at point r, $\langle \ldots \rangle$ means an ensemble average, the integral extends over the total volume V of the sample and $\phi = \langle \phi(r) \rangle$ is the average density.

In practice, either the aggregate is formed by elements that can be identified one by one, or the image of the aggregate is digitized, for instance by a square lattice of size r_0. Then, each square contributes to the evaluation of the correlation function $G(r)$ by $\phi(r) = 0$ or 1, depending whether the square is empty or occupied. The following approximate expression is then used to obtain the correlation function of an N-particle aggregate:[5]

$$C(r) = \frac{1}{N} \sum_{r'} \phi(r')\phi(r+r'). \tag{4}$$

The space correlation function $G(r)$ is directly related to the pair-correlation function $g(r)$, which is discussed in detail in Part 3, through the relation:

$$G(r) = \delta(r) + \phi g(r) = G_s(r) + G_d(r). \tag{5}$$

The meaning of $g(r)$ is the following: take an arbitrary occupied point in the cluster; $g(r)$ represents then the probability that another point at distance r is occupied. Equation (5) separates a self part (points for which $r = 0$) and a distinct part $(r \neq 0)$ and implies that $g(0) = 0$ and $\lim_{r \to \infty} g(r) = 1$.

It is easy to prove that the spatial dependence of $C(r)$ gives directly the fractal dimension D. For a fractal object, the distinct part $g(r)$ is homogeneous:[5]

$$\langle \phi(\lambda r_1)\phi(\lambda r_2)\rangle = \lambda^{-A}\langle\phi(r_1)\phi(r_2)\rangle, \tag{6}$$

which implies

$$\langle\phi(r_1)\phi(r_2)\rangle - \phi^2 \sim |r_2 - r_1|^{-A}. \tag{7}$$

From the definition of $g(r)$, the number $N(r)$ of particles within a radius r is given by

$$N(r) = \int_0^r \phi g(r) d^d r, \tag{8}$$

where the exponent d represents the Euclidean space dimension.

The spatial dependence of $N(r)$ is directly related to that of the correlation function $G(r)$

$$\int_0^r \left[\frac{\langle\phi(0)\phi(r)\rangle}{\phi} - \phi\right] d^d r \sim \int_0^r r^{-A} d^d r \sim r^{d-A}. \tag{9}$$

The comparison with Eq. (1) gives

$$D = d - A. \tag{10}$$

This means that a logarithmic representation of the correlation function $C(r)$ as a function of r will be a straight line with a slope A related to the fractal dimension D through Eq. (10).

This technique has been used to study several systems of aggregates. However, it has some important limitations. In particular, it is in general applicable only at two dimensions because of the use of photographs. Weitz and Oliveria[6] have shown, however, that the geometric projection of a 3-dimensional aggregate onto a 2-dimension space does not change the fractal dimension D, if $D \leq 2$.

The evaluation of Eq. (4) from the image of a real cluster must be done with some precautions, in particular, avoiding the edge effect when evaluating the ensemble averages. The normal way to proceed is to take the smallest circle that encloses each cluster and then to evaluate the correlation function inside the largest circle that is both centered at each particle and tangent to the outer circle.[6]

Another limitation comes from edge effects which strongly limit the operational region of the aggregate. One possible way of reducing such effects is to take the Fourier transform of the density profile of the aggregate in order to obtain its power spectrum. The density-density correlation function $g(r)$ is then obtained by inversion of the Fourier transform. This procedure has been used by Matsushita et al[7] in the study of the fractal structure of zinc metal leaves grown by electrodeposition.

Finally, this technique appears to be appropriate essentially in the study of aggregates of semi-microscopic particles for which the current amplification methods are easy to apply. This is the reason very often why the direct space observations have been used in *ad hoc* experiments to test theoretical models or, more simply, to put computer simulations on a more solid base.

SCATTERING TECHNIQUES

Scattering techniques are a more powerful way of studying fractal structures. Depending on the characteristic length scale to be observed and the nature of the aggregate, it is possible to use the scattering of light, X-rays or neutrons. The first part of this section summarizes some relations of the general theory of scattering which are useful in the interpretation of the scattering pattern of fractal objects. The second part presents the scattering function of a fractal object and the third part that of a fractal surface. Finally, the fourth part summarizes some of the experimental features which influence the determination of the fractal parameters.

Generalities. In a scattering experiment, a beam of electromagnetic radiation (light, X-rays) or neutrons of intensity I_0 is directed on a sample and the scattered intensity is measured as a function of an angle θ to the incident direction (Fig. 1). The incident and scattered beams are characterized by their wave numbers $\vec{k}_0$ and $\vec{k}_s$, respectively (Fig. 1). The momentum transfer, $\vec{Q}$, is obtained from the conservation relation

$$\vec{Q} = \vec{k}_0 - \vec{k}_s. \tag{11}$$

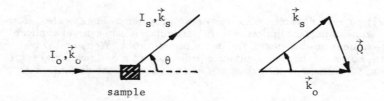

FIG. 1: Schematic representation of the incident and scattered beam in direction θ, characterized by the wave numbers $\vec{k}_0$ and $\vec{k}_s$ respectively and vectoral representation of Eq. (11).

Most of the time, in a scattering experiment, the quasi-elastic scattering is dominant, i.e., inelastic contributions to the scattered intensity can be neglected especially at small scattering angles θ. Then, the scattered intensity is mainly due to processes with $|\vec{k}_1| = |\vec{k}_2|$ and consequently

$$Q = |\vec{Q}| = 2k\sin(\theta/2), \quad (12)$$

where

$$k = |\vec{k}_1| = |\vec{k}_2| = 2\pi/\lambda = 2\pi n/\lambda_0, \quad (13)$$

λ being the wavelength of the incident beam in the sample, λ_0 the same quantity in the vacuum and $n = \lambda_0/\lambda$ the refractive index, which can be considered equal to 1 for X-rays and neutron scattering.

The general theory of scattering[8] relates the scattered intensity and the density distribution in real space through a Fourier transform. We come later to such relations. For the moment, it is sufficient to note that measuring the scattered intensity at momentum transfer Q corresponds to analysing the real space density distribution with a resolution $2\pi/Q$.

Table I gives the typical wavelengths and wave vectors associated with different types of radiation, taking into account some experimental constraints such as the minimum accessible scattering angle in conventional instruments. In the case of neutron scattering, the possibility of using long wavelength neutrons from cold sources in reactors is assumed.

From this table, one sees that an extremely large Q range is covered using several techniques. However, we must emphasize that each type of radiation couples with the sample in a different way, and the addition of spectra obtained by different techniques must be done with caution, taking into account the chemical composition of the sample.

TABLE I

Type of radiation	Typical wavelength ($\mathring{A}$)	Q range ($\mathring{A}^{-1}$)	Spatial resolution ($\mathring{A}$)
Light	4000 – 6000	$5 \times 10^{-5} - 3 \times 10^{-3}$	2000 – 100,000
X-rays	1 – 4	$10^{-2} - 15$	0.5 – 500
Neutrons	1 – 30	$10^{-3} - 15$	0.5 – 5000

Because of the characteristic size of fractal objects one sees moreover that, in the case of X-rays and neutrons, we will be concerned only with small angle scattering, i.e., with experiments where the scattering angle θ does not exceed typically 5°. This corresponds to situations for which the product of Q by the size is smaller than 1.

The important quantity to be considered in light scattering, small angle x-ray (SAXS) or small angle neutron (SANS) experiments, is the contrast. The contrast can be defined as the difference between the scattering properties of the particles, measured by a density $\rho(r)$, and the equivalent quantity ρ_0 of the "solvent," treated as a continuum because of the low spatial resolution of this kind of experiment.

The densities ρ introduced in this definition are quite different for light, X-rays and neutron scattering. In light scattering it is identified with the refractive index.[9] For X-ray, because the interaction is between the electromagnetic field and the electrons, the density, ρ, is proportional to the atomic number Z of the atoms. For neutrons, one defines a coherent scattering length, b, a quantity that is dependent on the nucleus and is different for each isotope. Such a situation is often used, for instance, to isolate the contribution of parts of a molecule by isotopic substitution. Table II gives the scattering lengths for some common atoms.[10] In the case of X-ray, the scattering length at zero-angle is proportional to the atomic number Z. In the case of neutrons, we indicate also the coherent cross section, $\sigma_c = 4\pi b^2$ and the incoherent cross section, σ_i, which constitutes a constant background to be subtracted from the total scattered intensity. Note the particularly intense incoherent scattering of hydrogen atoms, but, on the other hand, the large difference between the coherent scattering lengths of hydrogen and deuterium.

In a small angle scattering experiment, one measures the scattered intensity I_s at a distance R from the sample, on a detector surface subtending a solid angle $d\Omega$ at a scattering angle θ. The total differential cross-section of the sample, $d\sigma$, is given by

$$d\sigma = \frac{I_s}{I_0} R^2 d\Omega, \qquad (14)$$

where

$$I_0 = \text{Intensity of the incident beam.} \qquad (15)$$

The scattering cross-section per unit volume of the sample is

$$I(Q) = \frac{1}{V} \frac{d\sigma}{d\Omega} = \frac{R^2}{V} \frac{I_s}{I_0}, \qquad (16)$$

where V is the volume of the sample. Equation (16) represents the probability of a neutron being scattered in a given direction θ, per unit solid angle around θ, when it traverses a unit length of that sample.

Because the spatial resolution in a typical SAXS or SANS experiment is very small compared to interatomic distances, it is often useful to identify an individual scatterer, such as a particle, a micelle, a monomer, etc. and attribute to it a form factor $P(Q)$. The intensity $I(Q)$ is then separated into two factors $P(Q)$ and $\bar{S}(Q)$

$$I(Q) = \phi P(Q) \bar{S}(Q), \qquad (17)$$

where $\phi = N/V$ is the number density of particles or individual scatterers in the sample.

TABLE II:

Atoms*	Z	b (X-rays) Q=0 (10^{-12}cm)	b(neutrons) (10^{-12}cm)	σ_c (barn)	σ_i (barn)
^{1}H	1	0.281	-0.37423	1.7599	79.91
^{2}H or D	1	0.281	0.6674	5.597	2.04
Li	3	0.843	-0.190	0.454	0.91
C	6	1.686	0.66484	5.554	0.001
N	7	1.967	0.936	11.01	0.49
O	8	2.249	0.5805	4.235	0.000
F	9	2.530	0.5654	4.017	0.0008
Na	11	3.092	0.363	1.66	1.62
Mg	12	3.373	0.5375	3.631	0.077
Al	13	3.654	0.3449	1.495	0.0085
Si	14	3.935	0.4149	2.163	0.015
P	15	4.216	0.513	3.307	0.006
S	16	4.497	0.2847	1.0186	0.007
Cl	17	4.778	0.95792	11.531	5.2
K	19	5.340	0.371	1.73	0.25
Ca	20	5.621	0.490	3.02	0.03
Fe	26	7.308	0.954	11.44	0.39
Cu	29	8.151	0.7718	7.486	0.52
Zn	30	8.432	0.5680	4.054	0.077
Pb	82	23.0	0.94003	11.104	0.0030

(*) - Natural isotope mixture, except for hydrogen and deuterium.

In Eq. (17), $\tilde{S}(Q)$ plays the role of an effective structure factor[11] and is given as a function of the interparticle structure factor, $S(Q)$, by

$$\tilde{S}(Q) = 1 + \frac{|\langle F(Q)\rangle|^2}{\langle|F(Q)|^2\rangle}[S(Q) - 1], \tag{18}$$

where $F(Q)$ is the form factor of the particle, related to $P(Q)$ by[12,13]

$$P(Q) = \langle|F(Q)|^2\rangle. \tag{19}$$

Here $F(Q)$ is directly related to the shape of the particle and to the contrast $\rho(\vec{r}) - \rho_0$. The densities are obtained from the scattering length b_i of each atom constituting the particle

$$\rho(\vec{r}) = \sum_i b_i \delta(\vec{r} - \vec{r_i}), \tag{20}$$

where $\vec{r_i}$ is the position of atom i. Now $F(Q)$ is the Fourier transform of the particle density distribution and is given as a function of the contrast by

$$F(Q) = \int_{\text{volume of particle}} [\rho(\vec{r}) - \rho_0] \exp(i\vec{Q} \cdot \vec{r}) d\vec{r}. \tag{21}$$

For a centrosymmetric particle, $\langle|F(Q)|^2\rangle = |\langle F(Q)\rangle|^2$ so that $\tilde{S}(Q) = S(Q)$.

A particularly simple case is that of a monodisperse system of spherical and homogeneous particles of radius r_0 and density $\rho(r) = \rho$. Equation (21) is then proportional to the first-order spherical Bessel function $j_1(Qr_0)$

$$F(Q) = V(\rho - \rho_0)\frac{3j_1(Qr_0)}{Qr_0}, \tag{22}$$

with

$$j_1(Qr_0) = \frac{\sin(Qr_0) - Qr_0\cos(Qr_0)}{(Qr_0)^2}. \tag{23}$$

Replacing (22) and (23) in (19) gives

$$P(Q) = V^2(\rho - \rho_0)^2 \left[3\frac{\sin(Qr_0) - Qr_0\cos(Qr_0)}{(Qr_0)^3}\right]^2. \tag{24}$$

The function $P(Q)/(V(\rho - \rho_0))^2$ is plotted in Fig. 2.

It is interesting to investigate the behavior of $P(Q)$ at small and large values of Q. For $Qr_0 \ll 1$, an expansion of Eq. (24) to the fifth order in Qr_0 gives

$$P(Q) \simeq V^2(\rho - \rho_0)^2 \left(1 - \frac{Q^2 r_0^2}{5}\right) \quad (Qr_0 \ll 1). \tag{25}$$

This is a particular application to spheres ($R_g^2 = 3r_0^2/5$) of the well-known Guinier approximation

$$P(Q) = V^2(\rho - \rho_0)^2 \left(1 - \frac{Q^2 R_g^2}{3}\right)$$
$$\simeq V^2(\rho - \rho_0)^2 \exp\left(-\frac{Q^2 R_g^2}{3}\right) \quad (QR_g \ll 1), \tag{26}$$

which applies for a system of monodisperse particles with radius of gyration R_g and density ρ in a medium of density ρ_0.

At large Q ($Qr_0 \gg 1$), the average of $P(Q)$ decreases as Q^{-4}. From Eq. (24), putting $\sin^2 x = \cos^2 x = 1/2$, one obtains

$$P(Q) \simeq \frac{9V^2(\rho - \rho_0)^2}{2} \frac{1}{(Qr_0)^4} \qquad (Qr_0 \gg 1). \qquad (27)$$

This is called the Porod law which characterizes the scattered intensity of a system where the two densities ρ and ρ_0 are separated by a sharp boundary. A more detailed treatment gives the general Porod law

$$P(Q) = \frac{2\pi(\rho - \rho_0)^2 S}{Q^4}, \qquad (28)$$

where S is the total surface of the boundary.

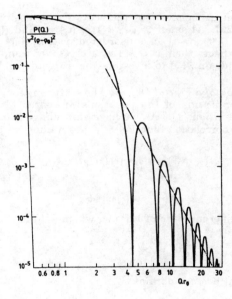

FIG. 2: A log-log diagram showing the form factor $P(Q)$ of a homogeneous sphere of radius r_0 and density ρ. Note that the function is almost constant for $Q < 1/r_0$. The dot-dashed line represents the Porod law and has a Q^{-4} dependence.

For the particular case of spheres, Eq. (27) and Eq. (28) coincide, of course, putting $V = 4\pi r_0^3/3$ and $S = 4\pi r_0^2$. Equation (27) is plotted also in Fig. 2. This figure shows that $P(Q)$ is not significantly different from $V^2(\rho - \rho_0)^2$ for $Q < 1/r_0$. The oscillations beyond $Qr_0 = 4$ are very often not seen because, even a small polydispersity smears the curve and only the Q^{-4} dependence is apparent.

The structure factor introduced by Eq. (18) describes the spatial arrangement of the elementary scatterers or particles. The theory of scattering (8) relates $S(Q)$ with the pair-correlation function introduced in Part 2, Eq. (5)

$$S(Q) = 1 + \phi \int [g(r) - 1] \exp(i\vec{Q} \cdot \vec{r}) d\vec{r}. \tag{29}$$

In an isotopic system in three dimensions, Eq. (29) takes the form

$$S(Q) = 1 + 4\pi\phi \int_0^\infty [g(r) - 1] r^2 \frac{\sin(Qr)}{Qr} dr. \tag{30}$$

Consider the limiting behavior of the scattered intensity, $I(Q)$, given by Eqs. (17), (24) and (29). At small Q ($Qr_0 \ll 1$), $P(Q) \simeq 1$ and $I(Q) \simeq \phi S(Q)$. On the contrary, at large Q ($Qr_0 \gg 1$), in a disordered system, $g(r) \simeq 1$, $S(Q) \simeq 1$ and $I(Q) \simeq \phi P(Q)$. One expects then at $Q \sim 1/r_0$ a crossover from a region where $I(Q)$ essentially depends on $S(Q)$ to a region where $I(Q)$ depends on $P(Q)$.

Scattering Function of a Fractal Object. The scattering function of a fractal object is a direct application of the equations presented in the preceeding paragraph and of definition (1). As we have seen above [Eq. (8)], from the definition of the pair correlation function $g(r)$, the number of particles within a sphere of radius r is

$$N(r) = \phi \int_0^r g(r) 4\pi r^2 dr, \tag{31}$$

or

$$dN(r) = \phi g(r) 4\pi r^2 dr. \tag{32}$$

On the other hand, differentiation of Eq. (1), which defines the fractal object, gives

$$dN(r) = \frac{D}{r_0} \left(\frac{r}{r_0}\right)^{D-1} dr. \tag{33}$$

Combining both equations one obtains

$$\phi_g(r) = \frac{D}{4\pi} \frac{1}{r_0^D} r^{D-3}. \tag{34}$$

Equation (29) shows that $S(Q)$ is obtained from the Fourier transform of $g(r) - 1$. The term 1 is introduced to remove the $\delta(Q)$ function corresponding to the limiting value of $g(r)$ at large distances. In the case of a fractal object, this problem is solved by introducing a cutoff ξ in a manner analogous to that used in critical phenomena[14]

$$\phi[g(r)-1] = \frac{D}{4\pi}\frac{1}{r_0^D}r^{D-3}\exp(-r/\xi). \tag{35}$$

We can now apply Eq. (30) and finally obtain an expression for $S(Q)$

$$S(Q) = 1 + \frac{D}{r_0^D}\int_0^\infty r^{D-1}\exp(-r/\xi)\frac{\sin(Qr)}{Qr}dr. \tag{36}$$

This Fourier transform can be evaluated from the mathematical identity[15]

$$\int_0^\infty x^{\mu-1}\exp(-\beta x)\sin\delta x\,dx = \Gamma(\mu)(\beta^2+\delta^2)^{-\mu/2}\sin[\mu tg^{-1}(\delta/\beta)]$$

$$(Re\mu > -1, Re\beta > |Im\delta|). \tag{37}$$

Hence[16]
$$S(Q) = 1 + \frac{1}{(Qr_0)^D}\frac{D\Gamma(D-1)}{\left(1+\frac{1}{Q^2\xi^2}\right)^{(D-1)/2}}\sin[(D-1)tg^{-1}(Q\xi)], \tag{38}$$

where $\Gamma(x)$ is the gamma-function. The scattered intensity of a fractal object is then given by Eqs. (17), (18), (19), (21) and (38).

It is important now to give some properties of $S(Q)$. At large Q ($Qr_0 \gg 1$) we obtain, as expected, $S(Q) \simeq 1$. As noted above, one sees only the scattering due to individual particles: $I(Q) \simeq \phi P(Q)$. At small Q compared with $1/r_0$, but large compared with $1/\xi$ ($1/\xi \ll Q \ll 1/r_0$), $P(Q) \simeq 1$, $I(Q) \simeq \phi S(Q)$ and $S(Q)$ has the limiting value

$$\lim_{\xi \to \infty} S(Q) = 1 + \frac{1}{(Qr_0)^D}D\Gamma(D-1)\sin\left[\frac{(D-1)\pi}{2}\right]$$
$$= 1 + \frac{const}{(Qr_0)^D} \simeq (Qr_0)^{-D} \qquad (Qr_0 \ll 1). \tag{39}$$

This is a result very often directly used to analyse the scattered intensity by a fractal object. It is clear, from the way it is obtained, that it applies only in an intermediate Q region for which both inequalities, $Q\xi \gg 1$ and $Qr_0 \ll 1$ apply. The application of Eq. (39) outside these limits can be a source of important errors in the determination of the fractal dimension D.

Of course, in the non-real situation of a mathematical fractal object, such as for instance, the Sierpinski gasket or in percolation,[1,17] Eq. (19) is again exact. Actually, for such objects, r_0 can be infinitely small (even smaller than the atomic distances!) and the condition $Qr_0 \ll 1$ is always satisfied, as well as the condition $Q\xi \gg 1$.

Note that one can evaluate the behavior of $S(Q)$ at small Q ($Q\xi \ll 1$). The development of Eq. (38) gives

$$\lim_{Q \to 0} S(Q) = \Gamma(D+1)\left(\frac{\xi}{r_0}\right)\left[1 - \frac{D(D+1)}{6}Q^2\xi^2\right]. \quad (40)$$

The significance of the cutoff ξ appears now more clearly. It reflects the finite overall size of the system. For an aggregate it can correspond to its size. In other systems, such as silica-particle aggregates, ξ represents the length scale at which the cluster density approaches the average density.[18]

Finally, it is worth noting that Eq. (38) gives a particularly simple result when $D = 2$

$$[S(Q)]_{D=2} = 1 + 2\left(\frac{\xi}{r_0}\right)^2 \frac{1}{1 + Q^2\xi^2}. \quad (41)$$

At small Q, we obtain the well-known result of the Ornstein-Zernike theory[14]

$$S(Q) \sim \frac{\xi^{-2}}{\xi^{-2} + Q^2}. \quad (42)$$

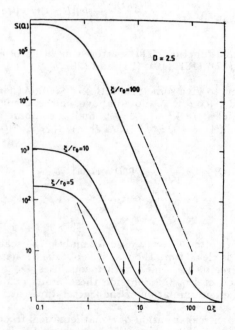

FIG. 3: Representation of the scattering function $S(Q)$ given by Eq. (38), for $D = 2.5$ and different ratios ξ/r_0. The arrows indicate the points where $Q = 1/r_0$. The dot-dashed lines show the $Q^{-2.5}$ dependence.

Note, from the development given by Eq. (40) that the meaning of ξ is conserved at all dimensions. Figure 3 is a plot of the structure factor $S(Q)$ given by Eq. (38), for the particular case $D = 2.5$. Three situations are represented corresponding to different ratios ξ/r_0. The points for which $Qr_0 = 1$ are indicated by arrows. The scattered intensity, in agreement with Eq. (17), is the product of this function and the form factor $P(Q)$ which as we noted above is almost equal to 1 for $Qr_0 < 1$. From this figure, it is clear that the scattered intensity is almost linear on a log-log plot, between the points $Q = 1/\xi$ and $Q = 1/r_0$. However, it must be noted that this linear part has a slope equal to the fractal dimension D only for very large values of the ratio ξ/r_0. For example, the slope of the straight part of $S(Q)$ for $\xi/r_0 = 5$ is equal to 2.15.

A complete treatment of the experimental data can in principle give ξ, D and r_0 if a form factor, $P(Q)$, can be assumed with some confidence. As an example, we take recent results obtained with aqueous solutions of proteins and the surfactant LDS.[16] In this case, the individual particles are micelles formed around the hydrophobic sites of the protein (bovine serum albumin or BSA). The system contains 1% by weight of BSA and 1, 2 or 3% by weight of surfactant in a buffer solution with a pH adjusted to 5.2 in order to avoid the interactions. The experimental results obtained at spectrometer PACE, at Saclay, are shown in Fig. 4. The fit is obtained through a non-linear least-square technique of equations (17), (24) and (38). It gives simultaneously the three parameters r_0, ξ and D. It is significant that only r_0 is independent of the surfactant concentration as expected. ξ increases with the surfactant concentration and its effect is almost negligible for the highest concentration, in the measured Q-range. The values of D are accurately determined and one notices again that it is different from what would be determined from the slope of the linear part of the curves.

Scattering Function of a Fractal Surface. The scattering due to a fractal surface can be derived in a way analogous to that used to derive the Porod equation,[12] i.e., evaluating the probability that a point at a distance r from a given point in the particle will also be in the particle. Based on concepts of absorption at fractal surfaces by Pfeifer et al,[19] Bale and Schmidt[20] have derived a general expression for the fractal D_s-dimensional surface of lignite coals.

The pair correlation function can be expressed by

$$g(r) = 1 - \frac{N_0 r^{3-D_s}}{4c(1-c)V}, \qquad (43)$$

where c is the fraction of dense material and $1 - c$ the fraction of pores in the sample volume V. N_0 represents a constant in the measurement of the total surface through the equation

$$n = N_0 r^{-D_s}. \qquad (44)$$

The meaning of this equation is the following: n is the number of cubes of edge r needed to make a layer covering all points of the boundary surface.[20] Fourier transformation of Eq. (43) gives

$$S(Q) = \pi N_0 \rho^2 \frac{\Gamma(5-D_s)}{Q^{6-D_s}} \sin \frac{\pi(D_s-1)}{2}. \qquad (45)$$

It is not obvious in a real system when one passes from the scattering by a fractal object to the scattering by a fractal surface. In a porous and ramified sample, where both the object and its surface can be thought as having a fractal structure, one can suppose that at low Q the scattering follows a Q^{-D} law [Eq. (39)], where D is the fractal dimension of the object. At larger values of Q, the neutron or X-ray probe is more dependent on the surface structure and the scattered intensity follows a Q^{D_s-6} dependence, where D_s is the fractal dimension of the surface.[21] Clearly, if D is between 2 and 3, as is often the case, the cross over between the two regimes is not seen. We must also note that, in contrast to theoretical models,[22] there is no a priori reason to have $D = D_s$.

Another difficulty is that, if the two cross-overs at high and low Q that we have discussed in the preceeding part are present too, the usual evidence of a fractal behavior, i.e., the power law Q-dependence of the scattered intensity, will be restricted to regions so small that the concept itself will be useless.

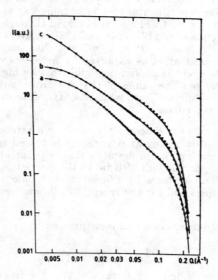

FIG. 4: Scattered intensity of a buffer solution of 1% BSA (protein) and x% LDS (surfactant).[16] Dots: experimental points. Solid lines: best fits with Eqs. (17), (24) and (38). The concentration x and the fitted parameters are:

(a) $x = 1\%$, $r_0 = 18 \text{Å}$, $D = 2.30$, $\xi = 84 \text{Å}$;
(b) $x = 2\%$, $r_0 = 18 \text{Å}$, $D = 1.91$, $\xi = 108 \text{Å}$;
(c) $x = 3\%$, $r_0 = 19 \text{Å}$, $D = 1.76$, $\xi = 338 \text{Å}$.

Experimental Constraints. In this last part we discuss some of the more important corrections of the data which eventually influence the determination of the parameters that characterize a fractal structure. As we have seen in paragraph 3.1 (see Table I), neutron and light scattering domains overlap around 10^{-3} A^{-1}. It can be interesting to explore this possibility to cover a large Q range. However, the measured quantities are in general very different for each technique and the "addition" of spectra obtained in this way has to be done with caution. Schaefer et al[23] measured the fractal dimension of colloidal aggregates of silica particles, using both light and SAXS and covering, in this way, almost 3 decades in Q space and 7 decades in intensity. We have mentioned already the influence of the limited size of the object or the aggregate in the determination of the fractal dimension from the slope of the log I vs. log Q plot. It can be noticed also that the presence of clusters of small size, for instance in experiments performed at different stages of aggregation,[23] can drastically affect the scattered intensity. This problem can be relevant in the study of percolating systems, such as gels, where the distribution of cluster sizes is very broad.

Polydispersity of the particles forming the aggregate can also modify the large Q behavior of the scattered intensity. However, if the interface between the particle and the medium is sharp, the Q^{-4} dependence characteristic of the Porod law [Eq. (28)] is observed. Even for a monodisperse system, there is a broadening due to the instrumental resolution.

For systems where the electrostatic interactions between aggregates or between particles can create a superstructure masking the fractal structure, it can be useful to screen such interactions, for instance, by modifying the pH[16] or adding salt to the solvent.

Particularly important in the case of SANS are multiple scattering and background corrections. To avoid the first, thin enough samples have to be used. The transmission (ratio between the intensity of the transmitted beam and the intensity of the incoming beam) must be of at least of the order of 0.7. Experimental results[18,24] have shown that the determined fractal dimension can be reduced if the amount of multiple scattering is important. Background corrections serve the purpose of subtracting the contributions of both the solvent and the incoherent scattering to the total scattered intensity. This subtraction can be very critical at large Q when the scattering due to the fractal object approaches zero. However, this correction, in principle, does not substantially affect the low-Q domain from which the fractal dimension is essentially determined. When a fractal structure can be observed over a very large Q range, the high-Q region eventually imposes relatively long time measurements. The use of synchrotron radiation can substantially increase the quality of SAXS data.

FRACTONS

In a series of papers Alexander, Orbach and collaborators have developed a general theory for the excitations, or "fractons," in fractal systems.[25-29] They introduced a fracton dimension, $\tilde{d}$, related to the static Hausdorff dimension D through the relation[25]

$$\tilde{d} = \frac{2D}{2+\theta}, \qquad (46)$$

where θ is the diffusion constant scaling exponent. In the lecture by Stanley, the symbol $\tilde{d}$ is replaced by d_s and $(2+\theta)$ is replaced by d_w.

It can be shown that the fracton dimension, $\tilde{d}$, is always smaller than D and that, in general, the following double inequality applies[30]

$$d \geq D \geq \tilde{d}, \qquad (47)$$

where d is the dimension of the Euclidean space.

The density of vibrational states in a fractal, $N_{fr}(\omega)$, is evaluated in a "Debye-like" approximation and, like the phonon density of states in a classical system, $N_{ph}(\omega)$, is proportional to $\omega^{\tilde{d}-1}$.[25] These results apply for times or frequencies such that the associated distances are much larger than the size of the individual bonds which characterize the fractal structure, but are much smaller than the total size of the network. Consequently, for length scales larger than the size ξ of the fractal object, the density of vibrational states will be again the classical $N_{ph}(\omega)$, essentially because the self-similarity which characterizes the fractal structure is then replaced by the translational symmetry.

As a function of the frequency ω, the vibrational density of states changes its behavior at some crossover frequency ω_c. At frequencies below ω_c there is a phonon-like behavior and at frequencies above ω_c, a fracton-like behavior.

Derrida et al[28] have applied the theory to percolation networks. The length scale ξ at which the crossover occurs in these systems is the percolation correlation length. It is then possible to derive expressions for the vibrational density of states, $N_{fr}(\omega)$ and $N_{ph}(\omega)$ as a function of the distance Δp from the percolation threshold. It is also shown that the velocity of sound varies with $\Delta p^{1/2}$.

Fracton theory applies in principle to glassy and disordered materials because in these systems there is a characteristic length scale larger than the lattice parameters. Anomalies in the density of the vibrational states of glasses can be eventually understood within the fracton theory. This has been done for the interpretation of low-temperature thermal conductivity and heat capacity data of epoxy resin.[31] The density of states in such experiments is proportional to ω^2, as expected, for temperatures below 8 K, but is proportional to ω above this temperature. This behavior suggests that the crossover frequency in epoxy resin is $\omega_c = 8k_B/\hbar \sim 0.17$ THz, corresponding to a length scale of 30Å which turns out to be about the length of the epoxy molecule.[26] Results for the low-frequency vibrations in vitreous silica, obtained by inelastic neutron scattering,[32] can also be interpreted within this theory.

The application of the theory to percolating systems, such as polymers and gels, would be interesting because the length scale, in principle, can be well-controlled. However, contributions from the solvent, and the broad distribution of cluster sizes can affect the final result. Results of Helman et al[33] show moreover that the description of the spin-lattice relaxation rate of hemoproteins and ferrodoxin in terms of the fracton model implies $\tilde{d} = D$, instead of $\tilde{d} = 1$, as expected, for the linear chain representing the protein backbone.[25] The explanation could be the presence of bridges (chemical bonds) between different points of the backbone. If these bridges can be considered as massless springs, the static fractal dimension D is not changed but, for the dynamical properties $\tilde{d}$ will be different of 1 and approach the dimension D.[33]

CONCLUSIONS

We have presented above some procedures to measure the characteristic features of fractal structures and dynamics, with a particular emphasis on scattering techniques.

Fractal objects are, in general, relatively easy to observe. In spite of the very simple mathematical concept leading directly to the well-known power-law dependences, real systems have to be treated in such a way that the natural geometric limits of the object and the experimental conditions are taken into account.

Theoretical models of growth and aggregation predict a limited set of fractal dimensions. However, experimental results are spread over a very large domain. For 2-dimensional systems, fractal dimensions of 1.5 and 1.6 were found for clusters of iron, zinc, silica[34] and wax balls,[35] and 1.7 for gold colloids.[7] In 3-dimensional systems, the results extend from 1.65 in proteins[36] to 2.5 to 2.6 in silica[18] or immunoglobulin aggregates.[37]

More experimental studies with appropriate systems are certainly necessary before a complete picture of the structural and dynamical aspects of fractals will be established.

I would like to thank J. P. Cotton for many helpful discussions and B. Carvalho and D. Mildner for their critical reading of the manuscript.

[1] B. Mandelbrot, *Fractals, Form and Dimension* (Freeman, San Francisco, 1977).
[2] A. I. Medalia and F. A. Heckman, J. Colloid Interf. Sci. **36**, 173 (1971).
[3] L. Niemeyer, L. Pietronero and H. J. Wiesmann, Phys. Rev. Lett. **52**, 1033 (1984).
[4] T. A. Witten and L. M. Sander, Phys. Rev. B **27**, 5686 (1983).
[5] T. A. Witten and L. M. Sander, Phys. Rev. Lett. **47**, 1400 (1981); R. Richter, L. M. Sander and Z. Cheng, J. Colloid Interf. Sci. **100**, 203 (1984).
[6] D. A. Weitz and M. Oliveria, Phys. Rev. Lett. **52**, 1433 (1984).
[7] M. Matsushita, M. Sano, Y. Hayakawa and Y. Sawada, Phys. Rev. Lett. **53**, 286 (1984).
[8] See, for example, G. L. Squires, *Introduction to the Theory of Thermal Neutron Scattering* (Cambridge University Press, London, 1978).

[9] B. I. Berne and R. Pecora, *Dynamic Light Scattering* (Wiley, New York, 1976).
[10] V. F. Sears, *Thermal Neutron Scattering Lengths and Cross-Sections for Condensed Matter Research* (Atomic Energy of Canada, Chalk River, 1984).
[11] S.-H. Chen and D. Bendedouch, in *Enzyme Structure, Method in Enzimology*, eds. C. H. W. Hirs and S. N. Timasheff (Academic, New York, 1985).
[12] A. Guinier and G. Fournet, *Small Angle Scattering of X-Rays* (Wiley, New York, 1955).
[13] J. P. Cotton, in *Introduction à la spectrométrie neutronique*, Cours Saclay (1974).
[14] P. A. Egelstaff, *An Introduction to the Liquid State* (Academic, London, 1967), p. 211.
[15] I. S. Gradshteyn and I. M. Ryzhik, *Table of Integrals, Series and Products* (Academic, New York, 1980), p. 490.
[16] S.-H. Chen and J. Teixeira, to be published.
[17] H. E. Stanley, J. Phys. Soc. Jpn. **52**, 151 (1983).
[18] S. K. Sinha, T. Freltoft and J. Kjems, *Kinetics of Aggregation and Gelation*, eds. F. Family and D. P. Landau (Elsevier, Amsterdam, 1984).
[19] P. Pfeifer and D. Avnir, J. Chem. Phys. **79**, 3558 (1983); D. Avnir, D. Farin and P. Pfeifer, J. Chem. Phys. **79**, 3566 (1983).
[20] H. D. Bale and P. Schmidt, Phys. Rev. Lett. **53**, 596 (1984).
[21] D. F. R. Mildner and P. L. Hall, to be published.
[22] P. Meakin, Phys. Rev. B **30**, 4207 (1984).
[23] D. W. Schaefer, J. E. Martin, P. Wiltzius and D. S. Cannell, Phys. Rev. Lett. **52**, 2371 (1984).
[24] J. A. Helsen and J. Teixeira, to be published.
[25] S. Alexander and R. Orbach, J. Phys. Lett. (Paris) **43**, L625 (1982).
[26] S. Alexander, C. Laermans, R. Orbach and H. M. Rosenberg, Phys. Rev. B **28**, 4615 (1983).
[27] P. F. Tua, S. J. Puttermann and R. Orbach, Phys. Lett. **98A**, 357 (1983).
[28] B. Derrida, R. Orbach and K.-W. Yu, Phys. Rev. B **29**, 6645 (1984).
[29] A. Aharony, S. Alexander, O. Entin-Wohlman and R. Orbach, Phys. Rev. B **31**, 2565 (1985).
[30] R. Rammal and G. Toulouse, J. Phys. Lett. (Paris) **43**, L625 (1982).
[31] S. Kelham and H. M. Rosenberg, J. Phys. C **14**, 1737 (1981).
[32] U. Buchenau, N. Nücker and A. J. Dianoux, Phys. Rev. Lett. **53**, 2316 (1984).
[33] J. S. Helman, A. Coniglio and C. Tsallis, Phys. Rev. Lett. **53**, 1195 (1984).
[34] S. R. Forrest and T. A. Witten, Jr., J. Phys. A **12**, L109 (1979).
[35] C. Allain and B. Jouhier, J. Phys. Lett. (Paris) **44**, L421 (1983).
[36] H. J. Stapleton, J. P. Allen, C. P. Flynn, D. G. Stinson and S. R. Kurtz, Phys. Rev. Lett. **45**, 1456 (1980).
[37] J. Feder, T. Jossang and E. Rosenqvist, Phys. Rev. Lett. **53**, 1403 (1984).

ON THE RHEOLOGY OF RANDOM MATTER

Etienne Guyon

L.H.M.P., E.S.P.C.I
10, rue Vauquelin 75231 PARIS Cedex 05

The mechanical properties of random matter are of importance in two different ways: on one side, one considers the form of matter as granted and one raises questions concerning the effect of the structure on these properties. On the other side, it is of considerable importance to study how the constraints imposed by the mechanics control the problems of growth, in particular for large non-brownian objects. Such a teleological approach was used through D'Arcy Thompson's classical masterwork.[1] We will focus mostly on granular media and will consider properties of both grain and pore phases starting from homogeneous, well-connected media (Sec. 1). In Sec. 2, we will consider the extreme case of very heterogeneously-connected media where bond percolation is often found to apply. In Sec. 3, we will mention effects of macroscopic heterogeneities controlled by hydrodynamic processes, such as in two phase flow in porous media or in flowing suspensions.

THE FORMS OF TISSUES OR CELL AGGREGATES

In this section we want to show that, even in well-connected media, there are strong heterogeneities that deserve descriptions in the spirit of the present book.

The dense packing of monodisperse spheres leads to periodic structures[2] in $2D$ and random ones in $3D$. The well-known variation of porosity in the latter case (between ~ 0.70 and 0.66, depending on the packing conditions) indicates the strong influence of the mechanics on the nature of dense random packings.[3] This is even more crucial for polydisperse assemblies: a mixture of two sizes of spheres (radii R_1, R_2 with $R_1 < R_2$) is a simple but quite representative case of a multiply-connected system in both the grain and pore space with strong local heterogeneities (three classes of contacts; five classes of tetrahedron voids that are the basic representative structures of such random porous structures).

[1] Mixtures of conducting (percentage p) and insulating spheres have been used for a characterization of the *grain space* in the above systems. If we restrict ourselves to the monodisperse problem, we expect a percolation conductance threshold for a site probability p_c, which can be expressed in terms of a "quasi-invariant" V_c, the critical volume fraction of conducting matter.

$$V_c \sim 0.15 \quad \text{in} \quad 3D \tag{1}$$

fairly independently of grain shape, pressure applied to the distribution,[4] (but variable when polydisperse assemblies are used[5]). On the other hand, for $p = 1$, only the "good" *bonds* between spheres are of importance. We can relate the initial slope of the conductance curve $\Sigma(p)$

$$K = \frac{1}{\Sigma}\left(\frac{d\Sigma}{dp}\right)_{p=1}, \tag{2}$$

the "vulnerability" of the packing, to the number of such good contacts z by the formula

$$K = (2z - 2)/(z - 2).$$

This formula is exact for a particular class of lattices in different dimensions and is also a good interpolation for effective medium approximation results obtained on various $2D$ and $3D$ lattices.[4]

The lowest coordinance value for very weakly compacted packing is consistent with a *bond* percolation threshold

$$\begin{aligned} z_c &\sim 1.5 \quad \text{in} \quad 3D \\ &\sim 2.0 \quad \text{in} \quad 2D, \end{aligned} \tag{3}$$

i.e., a number of active bonds per site of 1.5 or 2.0, this "quasi invariant" result was also obtained from estimates on various $2D$ and $3D$ lattices.

The Hertz-Medlin theories describing the flattening of individual contacts under pressure and the (viscous-like) resistance to shear are a necessary *Ansatz* for a local description of electrical and mechanical properties of unconsolidated assemblies; tribology, the science of contact, adhesion, wetting effects, should complement the statistical descriptions.[7] Let us just mention that, due to the distribution of contacts, particularly for polydisperse assemblies, the stresses are very heterogeneously distributed: this can be seen spectacularly in photoelastaic pictures of packings of parallel transparent cylindrical rods under pressure[8] which show that only a small percentage of rods forming continuous weakly-branched lines are submitted to strains. We have there a first image of a "mechanical backbone."

[2] Let us pass now to the properties of the *pore space* of such assemblies. It can be characterized by a few parameters:

(a) The *porosity* Φ can be decreased by pressure or by thermal sintering. By varying the concentration in a mixture of spheres (R_1, R_2), one gets a non-monotonic variation of Φ with a minimum around $C_1/C_2 \sim 0.25$ corresponding to an optimal packing of the small spheres within the large ones[9] (this would be the first step in an appolonian iterative filling of spheres when each class of spheres fills the space between the voids of the previous one—for a ratio $R_1/R_2 \sim 0.22$).

(b) We next define the *formation factor*, which measures the effective conductivity σ of an insulating porous medium filling the brine (conductivity σ_f). It is empirically related to ϕ by

$$\frac{\sigma}{\sigma_f} \propto \frac{\phi^m}{\alpha} \qquad (4)$$

(α is a measure of the tortuosity of the medium[10]).

(c) *Darcy's permeability*: K measures the ratio between the average flow rate of a fluid of viscosity η and the applied pressure gradient

$$\frac{\vec{Q}}{\vec{\nabla}_p} \propto \frac{K}{\eta}. \qquad (5)$$

The linear relation comes from a complex average of the microscopic linear Stokes law for the flow, the average being made on a suitable representative elementary volume (REV) which should be of a few grain sizes for monodisperse assemblies but which becomes critically large (like ξL) near a percolation threshold ... precluding any homogenization-type calculations! We may also note that the range of linear Darcy's law could be a very useful tool for the study of heterogeneous flows. Non-linear effects due to convection terms $(\vec{v}\vec{\nabla})\vec{v}$ take place above a critical Reynolds number of the local flow. In heterogeneous porous media, this should occur more rapidly in terms of a mean flow Re number because of the simultaneous effect of the distribution of length scales and of local velocities. K is a function of the geometry (typically K is of the order of the pore size[2]) and scales roughly as $K \propto \phi^m$.

A ratio

$$\frac{m'}{m} = \frac{4}{2} = 2$$

was found in a series of experiments on consolidated monodisperse spheres with a variable degree of sintering, the ratio 2 being related to the difference between the average on the parabolic flow profile (*dissipative* problem) and the *potential* electrical problem.

Different values of m that, together with α, express the potential scattering by the solid medium are obtained: for a dilute fixed bed, one theoretically obtains $m = 3/2$,[12] whereas for an unconsolidated packing experiments give $m \simeq 2$.[13]

Because laws (4) and (5) are of practical importance in site ("case") studies of rocks, the present effort extends them to various pore distributions.

(d) *Dispersion* should be an even more sensitive tool of heterogeneities in porous media. It deals with the spread of tracer particles in a fluid flowing through a porous media. The importance of convection effects with respect to those due to molecular diffusivity D_m is measured by the Peclet number $Pe = Ud/D_m$ where d is a characteristic intergrain space. For high Pe and large enough samples for the central limit theorem to apply, the mixing due to the distribution of flow lines is gaussian (Fig. 1, curve A); longitudinal (along mean flow) dispersion is given by a diffusion-like coefficient

$$D_\| \sim Ud. \tag{6}$$

The linear dependence expresses the reversibility of the flow at low Re numbers.

However the situation is already not so simple for packings of 2 size spheres. The curveve B of Fig. 1 is obtained for a ratio of 24% of small spheres (the minimum of porosity). The skewness of the dispersion is due to the heterogeneous distribution of current in the various channels (or of their necks which control the flow resistance). The fast-exiting dye uses the first passage channels (statistically interconnected large necks).

But there are also very slow (nearly stagnant) paths. In addition, the wall regions are nearly stagnant zones.[14] Such are the stagnation points for flow around spherical obstacles.[15] These features lead to log corrections of

$$D_\| \sim Ud \log(Ud/D_m),$$

where the molecular diffusion limits the long time divergence by permitting the dye to exchange between these zero velocity regions and the main flow.

An extreme system for multiple scale packing is the iterative appolonian packing of spheres (more generally, a self-similar space-filling packing). The problem is reminiscent of a Sierpinski gasket geometry: the pores have the scale of the smallest grains; there are loops of all sizes between two extreme radii R_{min}, R_{max}. Thus the rate of separation for two points distant of R ($R_{min} \ll R \ll R_{max}$) should be a function of R,

$$\frac{dR^2}{dt} = D(R),$$

increasing at least linearly with R.

A correspondance can be made with turbulent diffusion where, in the inertial range of turbulent scales (between large eddy scales characteristic of the main flow and Kolmogorov smallest scales), the pair-diffusion coefficient at scale R is given by $U(R)R \propto R^{4/3}$;[16] thus steady flow in a porous medium appears as a frozen turbulent flow; but a particle moving through it experiences the velocity fluctuating field as it moves with the fluid. In fact, forced Rayleigh techniques we had used to study pair diffusion should also apply to dispersion in transparent porous media.[17]

Such appolonian packings should not be easy to obtain using spheres except possibly in zero gravity experiments. However granular assemblies with large distribution of grains, as found in well-mixed grain mixtures used for concrete, have porosities that go to zero $(R_1/R_2)^n (n \sim 1/5)$ as the ratio of the smallest to largest grains. It is possible that diagenesis processes (dissolution + redeposition within the medium) also produce space-filling fractal structures[18] but the problem of fractal pore size should not be confused with that of fractal surface area![19]

ON FORMS AND MECHANICAL EFFICIENCY

We consider now poorly-connected geometries. We will see that the percolation concepts which can be applied to these systems can, in fact, also been used for some (Ref. 1) systems of II provided the heterogeneities in the quality of the connections are large enough.

Weakly-connected systems as modelled by percolation near threshold have usually a large length scale which gives the size of the REV over which local properties should be averaged if homogeneization techniques were to be obtained. Usually, in $3D$ geometries, a porous medium will not be weak at the same time in its solid and pore space and we may separate the description of the two problems (in particular we will only present schematically the mechanics of weak objects which is discussed elsewhere[20] in this book). For example, let us consider the sol-gel transition of a polymeric system.[21] At threshold, the elastic modulus goes to zero as

$$E \propto (p - p_c)^T, \qquad (7)$$

where p measures the number of branches between linear polymer chains. On the other hand, the permeability ratio K/η (see Eq. (5)) should diverge as

$$\xi^2/\eta \propto (p - p_c)^{-2\nu+S}, \qquad (8)$$

where

$$\eta \propto |p - p_c|^{-S} \qquad (9)$$

describes the critical divergence of the viscosity of the sol phase near p_c and ξ gives the mesh size of the $3D$ polymeric lattice holding the polymer solution. This behavior is studied in our laboratory as well as the corresponding finite frequency behavior due to inter- and intra-cluster interactions.

ELASTICITY OF WEAK OBJECTS

A correspondence between the critical behavior of Eq. (7) and that of electrical conductance above threshold $(\sigma \propto (p - p_c)^t)$ had been suggested by de Gennes[22] and justified for strongly strained lattices by Alexander[23] but this is still an object of scientific controversy in view of the experimental results ($T = 1.8$ to 2.4 for acrylamide gels; ~ 3.0 to 3.5 for gels obtained by polycondensation as $t \sim 1.8$) and of bound estimates (see S. Roux and E. Guyon, this book). The correspondance between the divergence of η (Eq. (9)) and that of dielectric conducting mixtures is also questionable and the experiments in the sol phase show the need to take into account finite frequency rheology[24] (a finite shear rate s means a finite time constant $\sim s^{-1}$ of the experiment). A number of models have been published recently using *unstrained* lattices in $2D$ and $3D$, which also indicate values of T markedly larger than t, and show the crucial role of bending elastic terms (due to changes of bond angle in unstained lattices of springs) in addition to compressive ones. We have recently used a square lattice of beams rigidly fixed at the nodes, which can deform under constraint. This reduces the problem to a bond percolation problem.[25] The problem is similar to one of Feng (preprint) which characterizes the elasticity of an incomplete array of spheres (or a random mixture of hand + soft spheres similar to the initial mixture of conducting plus insulating ones) by taking into account compressive and shear forces between grains but also local rotations. In all problems, the rotational degrees of freedom characterizes the critical elasticity at threshold because the effective moments applied to the weakly connected structure are large (they apply over lever arms $\sim \xi$). On the experimental side, a number of 2D experiments have been undertaken in our group and elsewhere with only qualitative results due to the small size of the lattices used. The work of Deptuck et al[26] using large porosity sintered Ag grains and comparing directly Σ and E giaves values $T = 3.8 \pm 0.5$ and $t = 2.15 \pm 0.25$ in agreement with a theoretical estimate $T = t + 2\nu$.

WEAK POROUS MEDIA

Weak porous media are obtained in several instances:

Clogged filters. The example was used in the original work of Broadbent and Hammersley[27] but, in depth filtration operations, there is usually a gradient in the concentration of particles along the length of the filter. The swelling of clays in sandstones can be modified by the salinity of water and thus provides a way to control the porosity; this is being used in present simulation experiments by C. Baudet in our group.

Heavily sintered materials. Below a porosity of the order of a few percent, experiments and models on regularly-sized particles give a zero permeability (the porosity is in closed pockets). Near threshold, it is likely that a percolation model applies[28]; similar effects are obtained in very low porosity (less than 5%) sandstones.

Fractured rocks. In granitic sites, the permeability due to large arrays of open fractures can vary from zero to a finite value depending on their density. The problem has been modelled by random arrays of overlapping disks in $3D$ (needles in $2D$) and the overlapping threshold estimated from the dimensionless invariants

$$\pi^2 N_3 R^3 \sim 1.8 \qquad (10)^{29}$$

$$(2/\pi) N_2 L^2 \sim 3.6 \qquad (11)^{30}$$

where $N_3(N_2)$ is the number of centers of spheres (needles) per unit volume (area), R the disk radius, L the needle length; the numbers were obtained from numerical calculations. These invariants express in fact an average number of contacts between disks. The excluded volume of a disk ($V_{exc} = \pi^2 R^3$) is an average over all relative orientations and positions of the volume around one fixed disk center for another disk center to intercept the first one. Thus the criterion for disks is expressed as $N_c \pi^2 R^3 \sim z_c$, where z_c (defined in Eq. (3)) is of the order 1.5 (the quasi-invariant value must be modified to take into account effects of overlapping objects).

These threshold effects are of importance for practical problems of permeability: determination of sites for storage of wastes (in particular nuclear ones), hydrogeology and oil exploration in granitic sites. However a larger class of problem deals with richly-connected fracture lattices, which, however, show a large range of aperture (δ) and extent (R) distributions (often log normal). In such cases it can be easily guessed that the flow within the largest size fractures will completely dominate that in the array of small ones (the viscous losses between two parallel planes vary as δ^3).

We are going to see from this particular example *how to apply percolation ideas to such heterogeneous problems*. We use a reasoning that has been introduced in the study of variable-range hopping[31] in disordered semiconductors. Let us assume a distribution of uniform-size fractures of apertures δ such that $\log \delta$ is uniformly distributed between a maximum value $\log \delta_M$ (macrocracks) to $\log \delta_m$ (micro-fissures) ($\delta_M/\delta_m \ggg 1$). We do a partial reconstruction of the network by establishing the fractures in decreasing order of δ. There is a critical value δ_c such that the incomplete lattice first percolates. If we continue to add fractures the conductance (or the permeability) increases continuously from zero, as in ordinary percolation. As the resistance of the individual elements becomes larger, however, there is a saturation of this increase that takes place not far from term value δ_c if the distribution of $\log \delta$'s is very large. We call δ_e an estimate of the value of δ around which the saturation effect takes place.

As the fractures are continuously distributed, the critical value δ_c is obtained from the distribution by expressing that the total number of fractures integrated from δ_M to δ_c gives the criterion for percolation (Eq. (10)). The distance δ_e to δ_c corresponds to the distance to p_c, Δ_p, in ordinary percolation and thus can be associated to a correlation length value ξ (Δ_p). The R.E.V. constructed over this length ξ should be large enough to probe the statistical distribution of pore sizes around the value δ_c: having a broader distribution means larger ξ, a smaller "distance" between δ_c and δ_e and a weaker *sublattice* (that made with bonds between δ_c and δ_e values). This scaling argument can be expressed more precisely[32] and permits an estimate of the conductance of such a lattice.

We also stress that a similar argument should apply to other mechanical and electrical elements discussed above: (a) the elasticity and electrical transport of weak unconsolidated granular systems (in particular if tunneling enters in the electrical problem) and (b) the permeability and also the determination of the fast-exiting fluid in dispersion experiments in heterogeneous porous media. We will focus on this last property.

DISPERSION IN POORLY-CONNECTED POROUS MEDIA

We have already mentioned that dispersion was very sensitive to local heterogeneities. We clearly expect it to apply in an even more sensitive fashion to percolation in porous media near threshold.

There exists a fractal structure up to a length ξ. The multiplicity of scales shows itself in a dependence of $D_\parallel$ with distance of spread.

There is an extreme heterogeneity of the transit times across the percolation lattice associated with this distribution. deArcangelis et al[33] found a nearly log-normal distribution of currents on a finite percolation lattice at threshold (A. Coniglio lecture). The "shortest path" (sometimes called the mechanical backbone) of the percolation lattice which has received a fractal description uses the path used by the first exiting fluid, but the multiplicity of channels leads to a broad distribution of exit times in response to a stepwise injection of dye.

The problem is presently studied numerically as a biased "ant" problem.[34] The transition probability is a function of the local velocity (of the local Peclet number). The problem for $Pe = 0$ is the usual "ant in a labyrinth" problem. In the opposite limit $Pe \to \infty$, the problem is a purely *geometrical* dispersion problem. However it neglects the effect of velocity gradients across and along pores and the complex effects of mixing at the nodes (theoretical models mostly assume perfect mixing at nodes). We are performing experiments on hexagonal lattices etched on a $2D$ plexiglass model with the same geometry as in the numerical simulations to clarify these effects.

The percolation lattice has dead arms of scales extending to ξ that act as traps of the dye. Quite generally, such *"hold up"* effects associated with trapping times τ lead to a *hydrodynamical* dispersion term $D_\parallel \sim U^2 \tau$. In the case of percolation, $\tau = \tau_\xi \sim \xi^2/D_a$ is the crossover time for the diffusive ant walk on a distance ξ; the diffusion coefficient for an ant D_a goes to zero with the conductivity at threshold as $\Delta p^{\tau-\beta}$ (the correction exponent β in this Einstein-like relation comes from the small volume fraction of the infinite cluster near p_c). On the other hand, the flow rate U for a given applied pressure gets to zero with the permeability (or conductance) as Δp^t.

Developing this argument, de Gennes[35] found a critical dispersion $D_\parallel \propto \Delta p^{\mu-\beta-2\nu}$ (with $\mu - \beta - 2\nu$ negative in $2D$ and $3D$). This result is yet to be checked experimentally and numerically. A difficult problem comes from the coexistence of very different time scales for convective (L/U) and diffusive mechanisms at Peclet numbers $Pe \gg 1$ at which most experiments are made. In order to get a gaussian dispersion we need to have enough mixing between the stagnant fluid and the flowing one for the central limit theorem to apply; this means that one has to wait several τ_ξ and must have a large sample size given by a convection length $L \sim U\tau_\xi$. It may very well be that such a limit cannot be reached in simulations of the problem for the large Pe corresponding to realistic experiments. Asymptotic matching techniques where a problem is solved using two very different scales (time scales here) applicable to separate elements and by matching them may provide a solution to the numerical problem. The distribution of dye in response to a step injection should give very asymmetrical curves already obtained with weak heterogeneities (Fig. 1b). A direct geometrical study of the distribution of dye should complement the study of the moments of the response function. This is the object of a present program in our laboratory on $2D$ and $3D$ transparent models.

ON SPICULES AND SPICULAR SKELETONS

We have considered macroscopically-heterogeneous structures as can be obtained by various growth processes. In the two following examples—immiscible two-phase flows in porous structures and 2D-sheared concentrated suspensions—the hydrodynamics control the shape of the structures.

[1] A discussion of immiscible flow in porous media in the spirit of percolation processes is presented in this book by J. Willemsen. Gravity as well as finite flow can modify dramatically the structure of one phase pushing another in drainage or inhibition processes, which display strong analogies with growth processes (as discussed in R. Ball, this book). A detailed discussion of this chapter can be found in Ref. 36. In a permanent regime, the coexistence of two immiscible phases within the medium makes this system a candidate for the study of permeability and dispersion (as discussed in III) for both phases near relative percolation threshold.[28]

[2] Another example of growth concerns 2D simulations of aggregation using spheres deposited at the surface of a liquid. Attraction between particles can be controlled by surface capillary action. The effect of a 2D shear field has been studied in detail; three regimes can be produced:

No shear. The attraction between particles (irreversible aggregation) leads to fractal structures with $d_F = 1.7$. The value is larger than the cluster value $d_F \sim 1.4$ because of the local reorganization of grains.[37]

No attraction—shear applied. This case of *reversible* aggregation had been considered by de Gennes,[38] who suggested that it could lead a percolation-like transition for a critical surface concentration. The cluster structures and rheological behavior have been studied in detail from experiments and molecular dynamic simulation and indicate the role of the growing clusters in the increase of viscosity.[39]

Capillary attraction plus shear. The intermediate case with capillary attraction plus shear leads, quite surprisingly, to compact structures. The hydrodynamics and numerical simulations are also discussed in Ref. 39.

Such behavior has been described in classical colloidal science work. In addition to deserving much more experimental effort, in particular dealing with rheology, the field of growth and form discussed in this book will benefit from a careful analysis of the vast literature dealing with the subject (e.g., *Journal of Colloid and Interface Science*.

[1] [Sir] Darcy Thompson, *On Growth and Form* (Cambridge Univ. Press, 1982), ed. J. P. Bonner. We have labelled the paragraphs from chapters of this work.
[2] By using a small percentage of spheres of larger diameter it is possible to generate 2D disrodered structures which can simulate 3D contact disorder effects.
[3] T. J. Plona and D. L. Johnson, in *Physics and Chemistry of Porous Media*, eds. D. J. Johnson and P. N. Sen (American Institute of Physics **107**, New York, 1984), p. 89.

[4] Groupe des systèmes desordonnés (Marseille), Annales de Physique **8**, 1 (1983).
[5] L. Oger, Thèse Rennes (1983).
[6] J. P. Clerc, G. Giraud and E. Guyon, Pow. Tech. **35**, 107 (1983).
[7] L. Schwartz, D. L. Johnson and S. Feng, Phys. Rev. Lett. **52**, 831 (1984).
[8] A. Drescher and G. Josselin de Jong, J. Mech. Phys. Sol. **20**, 337 (1972).
[9] R. Ben Aim, Thèse Université de Nancy (1970).
[10] The effect of multiple potential scattering by the solid object, α, is directly related to the index of refraction of acoustic modes of the fluid (4th sound of superfluid He) $n^2 = \alpha$. D. L. Johnson, T. J. Plona, C. Scala, F. Pasierb and H. Kojima, Phys. Rev. Lett. **49**, 1840 (1982).
[11] P. Z. Wong, J. Koplik and J. P. Tonamic, Phys. Rev. B **30**, 6606 (1984).
[12] P. N. Sen, in *Macroscopic Properties of Disordered Media: Lectures Notes in Physics* (Springer-Verlag, 1982), Vol. 154.
[13] J. Lemaitre, Thèse Université de Rennes (1985).
[14] P. G. Saffman, J. Fl. Mech. **6**, 321 (1959).
[15] C. Baudet, E. Guyon and Y. Pomeau, J. Phys. Lett. (submitted).
[16] S. Grossman and I. Proccacia, Phys. Rev. A **29**, 1358 (1984).
[17] M. Fermigier, M. Cloitre, E. Guyon and P. Jenffer, Jour. de Mech. *1*, 123 (1982).
[18] A. J. Katz and A. H. Thompson, Phys. Rev. Lett. *54*, 1325 (1985).
[19] J. N. Roberts, comment to appear.
[20] S. Roux and E. Guyon, this book.
[21] B. Jouhier, C. Allain, B. Gauthier-Manuel and E. Guyon, in *Percolation, Structures and Processes*, eds. G. Deutscher, R. Zallen and J. Adler (Ist. Phys. Soc., 1983), p. 167.
[22] P. G. de Gennes, J. de Phys. Lett. **37**, L1 (1976).
[23] J. de Physique **45**, 1939 (1984).
[24] B. Gauthier-Manuel, Thèse Paris (1985).
[25] S. Roux and E. Guyon, Jour. de Phys. Lett. (submitted).
[26] D. Deptuck, J. P. Harrison and P. Zawadski, Phys. Rev. Lett. **54**, 913 (1985).
[27] J. Hammersley and D. J. A. Welsh, Cont. Phys. **21**, 593 (1980).
[28] Groupe poreux P.C., in *Scaling Concepts in Disordered Matter* (Plenum Press, 1985). This article contains many references on one- and two-phase flows in porous media.
[29] E. Charlaix, E. Guyon and N. River, Geol. Mag. **122**, 157 (1985), and E. Charlaix, à paraître.
[30] P. C. Robinson, J. Phys. A **16**, 605 (1983).
[31] B. Shklovskii and A. L. Efros, *Electronic Properties of Doped Semiconductors* (Springer-Verlag, Berlin, 1984), p. 130.
[32] E. Charlaix and E. Guyon, to appear and Ref. 28, Chapt. 2. We can also stress the correspondance with the problem of change of class of universality due to large bond disorder around a percolation threshold discussed in these

Proceedings by Sen. In the application of his analysis to fractures we should have δ_m go to zero and all the bonds down to this lowest limit are needed at percolation threshold.

[33] A. Coniglio, these Proceedings.

[34] C. D. Mitescu and J. Roussenq, private communication. In this problem the probabilities are biased by the local currents flowing across the lattice. This is quite different from biased probability problems with an average external field as studied by M. Barma and G. D. Dhar, J. Phys. C **16**, 1451 (1983).

[35] P. G. de Gennes, J. of Fl. Mech. **136**, 189 (1985).

[36] "Groupe poreux P.C.," in *LES HOUCHES SCHOOL ON FINELY DIVIDED MATTER*, ed. M. Daoud (Springer-Verlag, Berlin, 1985).

[37] C. Allain and B. Jouhier, J. Phys. Lett. (Paris) **44**, L421 (1983).

[38] P. G. de Gennes, J. Phys. Lett. (Paris) **40**, L783 (1979).

[39] C. Camoin, G. Bossis, E. Guyon, R. Blanc and J. F. Brady, Jour. de méca. (Paris), in press.

Etienne Guyon

DEVELOPMENT, GROWTH, AND FORM IN LIVING SYSTEMS

Jean-Pierre Boon and Alain Noullez

Faculté des Sciences CP231
Université Libre de Bruxelles
B-1050-Bruxelles, Belgium

INTRODUCTION

Just about everything in the living world grows and all multicellular organisms develop forms. In one way or another development inducing structural organization appears as a general rule of living systems. So the genesis of spatial patterns constitutes a central problem in theoretical biology.[1]

Biological patterns ultimately result from the selective action of genes: (i) genes are activated in a specific spatial and temporal order, (ii) the activation results in a specific structure, (iii) the structure induces cellular organization, (iv) organization manifests itself by the formation and growth of shapes.

In his proposal for a theory of morphogenesis, Turing[2] put forward the idea that, in the developmental scheme, patterns or structures (e.g., cell differentiation) result from prepatterns of non-homogeneous chemical concentration distributions, which arise as a consequence of instabilities in homogeneous steady states. A system of chemical substances (morphogens), reacting together and diffusing through cells, may become inhomogeneous because the original homogeneous state is unstable against random disturbances; then a non-homogeneous chemical distribution forms a prepattern wherefrom the ultimate struture originates. Turing's *chemical prepattern concept* constitutes the basic philosophy of the *reaction-diffusion* (R-D) models. A logical generalization follows from the idea that the developmental scheme is governed by successive bifurcations of one morphogen prepattern into another resulting ultimately into the final space- and time-ordering structure.[3]

Such systems which are modeled by highly non-linear coupled differential equations, must be solved by numerical computation. These models describe the development of interconnected cell systems. It is thus suggested that the pattern simulation of cellular growth and differentiation by automata nets would offer adequate representations when interaction rules for the composing automata are incorporated in compartmentalized R-D systems.

PATTERN FORMATION

Following the R-D system approach, based on the Turing prepattern concept, Gierer and Meinhardt[4] proposed, as a solution to the problem of generating a structure in an initially featureless domain, a model system of pattern generating chemical reactions, including diffusion, with three essential key features: (i) autocatalytic activation (positive feedback of an activator on its own production) and (ii) genesis and spread of an inhibitor with (iii) short-range activation and long-range inhibition.

The Turing-Gierer-Meinhardt basic scheme for activator (X)/inhibitor (Y) interaction (TGM) is most easily represented by the graph given in Fig. 1. When decay rates and diffusion are included, the full model reads (in dimensionless form)

$$\begin{aligned}\dot{X} &= 1 + rX^2Y^{-1} & -X & & -k^2X \\ \dot{Y} &= \underbrace{qX^2}_{\text{production}} & \underbrace{-qY}_{\text{decay}} & & \underbrace{-pk^2Y}_{\text{diffusion}}\end{aligned} \qquad (1)$$

where k^2 follows from Fourier transformations of the diffusion terms, and the parameters r, q and p have the following meaning: r is the dimensionless measure of the activator autocatalysis rate, $q = h/a$ is the ratio of inhibitor decay rate to activator decay rate, and $p = D/D^*$ is the ratio of inhibitor diffusion coefficient to activator diffusion coefficient.

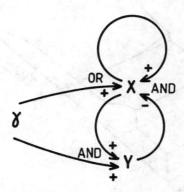

FIG. 1: Schematic representation of the TGM model; X = activator; Y = inhibitor; γ = concentration of undifferentiated cells.

The underlying idea of the pattern formation model is that the uniform steady state (X_s, Y_s) is stable against long and short wavelength perturbations, but unstable against perturbations of intermediate wavelength, i.e., those matching the observed repeating features in living systems. For the spatially homogeneous case, stability of the solution (X_s, Y_s) is guaranteed when the condition $q > (r-1)/(r+1)$ is satisfied. Furthermore linear stability analysis of the spatially inhomogeous system yields the stability condition: $q/p > (r-1)/(r+1)$, i.e., $p \leq 1$. Consequently if $p > 1$, that is when $D > D^*$, spatial non-uniformities can develop. Now the additional necessary condition for pattern formation is found to be: $(r-1)/(r+1) > (q/p) + 2\sqrt{q/p}$, which is compatible with the previous condition when $r \gg 1$ and $\sqrt{q/p} < \sqrt{2} - 1$. The latter results are easily interpreted in terms of the physical meaning of r, the measure of autocatalytic rate, and of $\sqrt{q/p} = (D^*/a)^{1/2}(D/h)^{-1/2} = \ell^*/\ell$, the ratio of distances over which activator and inhibitor diffuse during their respective decay times. So the two basic conditions for instability of the homogeneous state, i.e., for pattern formation, are (i) strong activator autocatalysis ($r \gg 1$), and (ii) excess of inhibition range with respect to activation range ($\ell^* < \ell < 0.4...$). Under such conditions, perturbations would grow exponentially were it not for the regulatory effect of the non-linear terms which act to limit that growth. Figure 2 illustrates the basic features of the TGM model for a system with finite size ($\sim 10\ell^*$). At initial

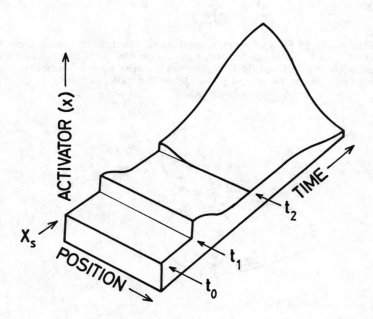

FIG. 2: Illustration of the TGM model, Eqs. (1) (after Ref. 5).

time t_0, the system is in the homogeneous steady state; when a homogeneous perturbation is introduced at time t_1, the system returns to its homogeneous state X_s; if at time t_2 a local perturbation is injected, it grows and induces the formation of a new non-homogeneous state which is stable.

The major feature of the TGM theory, as illustrated by the above example, is that suitable concentrations of morphogens can trigger developmental evolution. Various versions of the basic R-D model have been developed and show how specific patterns of morphogens emerge from the interplay of chemical reactions and diffusion. A particular model for the generation of netlike patterns offers an interesting scheme to describe the formation of leaf veins, capillary beds, or bronchial tubes.[5] Differentiation into net cells is induced by the irreversible switching on of a morphogen Z whose production is generated by an activator X; production of X in turn requires a substrate S, so that here the inhibition effect comes from the depletion of S. Both X and Z have autocatalytic production and C is the resulting differentiation product of the switching on of Z. The basic scheme is given in Fig. 3.

The model equations including decay and diffusion (but omitting numerical parameters) read*

$$\begin{aligned}
\dot{S} &= 1 & -(X^2 + Z)S & \quad +\nabla^2 S \\
\dot{X} &= SX^2 & -X & \quad +\nabla^2 X \\
\dot{Z} &= \underbrace{X + Z^2(1+Z^2)^{-1}}_{\text{production}} & \underbrace{-Z}_{\text{decay}} & \quad \underbrace{}_{\text{diffusion}}
\end{aligned} \qquad (2)$$

This set of equations generates an inhomogeneous distribution of morphogen Z in the form of a dichotomously branching leaf pattern. When an additional inhibitor is involved, further complexity is obtained in differentiation patterns (e.g., network of tracheae[5]).

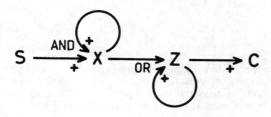

FIG. 3: Schematic representation of the Meinhardt model for netlike pattern formation.

---

* Note analogy with branching model discussed by T. Witten (this book).

DIFFERENTIATION AND GROWTH

What governs cell replication in a compartmentalized R-D system composed of N cells with n reacting and diffusing agents (X_i = activators, Y_i = inhibitors)? At any time the state of a cell is specified by the n concentrations, and an event will be characterized by inequalities between their values. Consider the simplest case with one activator and one inhibitor ($n = 2$) and $X > Y$ being the necessary and sufficient condition for occurence of the event responsible for cell replication. However (i) the event should not be the result of a mere chance pulse in concentration variation; so $X > Y$ should hold for a time large compared to the duration of a random fluctuation, so that the system can reach a steady state in concentration distribution; (ii) in a developing organism, there is differential replication; the steady state concentration distribution should thus be inhomogeneous (i.e., $X > Y$ is only local); (iii) growth patterns change during development as a result of changes in the underlying chemical prepattern; the latter changes come about because the inducing pattern (i.e., the inhomogeneous concentration steady state) becomes unstable after the growth it induced occurs; then the pattern becomes "unstable to itself" and the system switches to a new distribution that is stable under the new configuration.

We consider a hypothetical organism[3] specified as a linear array of cells (row, e.g., filamentous algae, or ring, e.g., 2-dimensional blastula, depending on boundary conditions). Replication occurs by cell division and is governed by the concentration of substance Z which depends on the relative values of morphogen concentrations X and Y. The activator and inhibitor concentrations control the rate of change of Z in such a way that Z reaches threshold value for replication if $X > Y$. A potential state of division and replication actually occurs for a cell when $X > Y$ has been maintained for a sufficiently long period of time. Since X and Y in turn are governed by intracellular reactions and intercellular diffusion, the delay for cell division reflects the fact that replication is a slow process compared to chemical changes and diffusion.

As a model reaction scheme, consider the trimolecular model[6]—often referred to as the "Brusselator"[7]—

$$\dot{X} = A + X^2Y - BX - X$$
$$\dot{Y} = BX - X^2Y, \qquad (3)$$

which applies to a single cell with perfect mixing and has steady state values $X_s = A$, $Y_s = B/A$ with stability condition $B < A^2 + 1$. Extension to a R-D system composed of a collection of adjacent cells is performed by insertion of diffusion terms in (3) which is generalized (in computational form) to

$$\dot{X} = f(X_j, Y_j) + \mu(X_{j+1} - 2X_j + X_{j-1})$$
$$\dot{Y} = g(X_j, Y_j) + \nu(Y_{j+1} - 2Y_j + Y_{j-1}), \qquad (4)$$

where f and g are the reaction functions of Eq. (3), μ and ν are the activator and inhibitor diffusion coefficients respectively, and $j = 1, \ldots, N$.[3] In the ring configuration (boundary conditions: $1 = N+1$, $0 = N$) the homogeneous steady state is unstable for large ν if $B > 4\mu + 1$, i.e., when the inhibitor diffusion is stronger than the activator diffusion. So starting with initial conditions corresponding to the homogeneous steady state, a local random perturbation will destabilize the system which will develop a stable non-homogeneous "pattern" of differentiated cells, as illustrated in Fig. 4a. An alternative example is shown in Fig. 4b where the R-D scheme is the TGM activator-inhibitor model, cf. Eq. (1). One observes that a stronger X versus Y competition model (TGM as compared to Brusselator) induces a more regular periodicity in the final structure (compare Figs. 4a and 4b).

To illustrate the growth process, let us consider an initial system with two cells (Fig. 5). Morphogens can be exchanged by intercellular diffusion making it possible for a non-homogeneous distribution to establish. There is then a strong probability that $X > Y$ occurs in one of the cells, for which z reaches replication threshold and division takes place. The new stage is a three cell system where the two daughter cells inherit the parent cell morphogen concentrations and therefore are in a position to divide. Moreover, since the configuration has changed, there will be a redistribution of morphogens, which

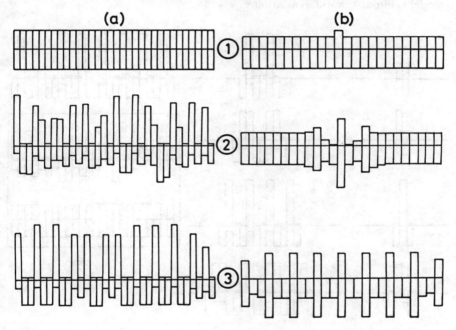

FIG. 4: Cellular differentiation (a) Brusselator model, Eq. (4), $N = 32$; (b) TGM model, Eq. (1), $N = 25$. Upper and lower boxes indicate the concentration values of activator and inhibitor respectively. **1**= initial state; **2**= intermediate state; **3**= final state.

is a fast process as compared to replication. As a result only one cell stays in the replication state, and divides, forming a transient four-cell state with two adjacent activated cells. Again morphogen redistribution takes place before further division occurs and growth proceeds. Successive stages of a growing "organism" are shown (up to 24 cells) in Fig. 5. Also shown, a 24-cell differentiating (non-growing) system to illustrate the symmetry difference in both systems.

A LEOPARD WHICH CAN CHANGE HIS SPOTS

As the survival value of skin patterns is of considerable importance in the animal world, a theoretical analysis of the morphogenesis of such patterns poses a valuable problem which is also of mathematical interest per se. The formations of spots and stripes result from the activation of pigment cells (melanocytes); so skin patterns arise from the distribution of activated cells whose differentiation can be conjectured to result from a Turing-type prepattern of morphogens with local cellular interaction by short range diffusion. On the basis of this philosophy, an activator-inhibitor theory[8] can be extended to model skin pattern differentiation.[9]

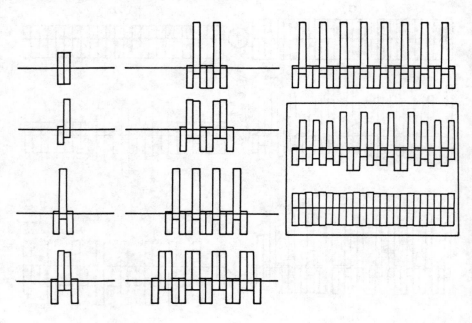

FIG. 5: Cellular growth according to model Eqs. (4); cells with $X > Y$ (upper box larger than lower box) are in replication state. In frame, initial and final states of differentiating system (same model).

The developmental scheme starts from an early stage skin with uniform distribution of pigment cells, some of which are differentiated (active cells, A), the others being undifferentiated (U). "A" cells produce an activator (morphogen X) which stimulates nearby U cells to become colored, and an inhibitor (morphogen Y) which deactivates nearby A cells. The two morphogens are governed by R-D equations

$$\dot{X} = f(X,Y) + \nabla^2 X$$
$$\dot{Y} = g(X,Y) + \nabla^2 Y, \tag{5}$$

where f and g contain production and decay terms, and the inhibitor diffusion range exceeds the activator diffusion range (see analysis in the pattern formation section). Each A cell produces X and Y at constant rate; X and Y diffuse to neighboring cells and are degrated according to the model equations. As a result, there is a steady state distribution forming a "morphogenic field" around each A cell (Fig. 6), with short range activation of nearby U cells and long range inhibition of A cells. The field values due to all cells are added up to determine the state of a given cell; each cell takes up the resulting field value instantly, i.e., differentiation is a slow process compared to the R-D mechanism.

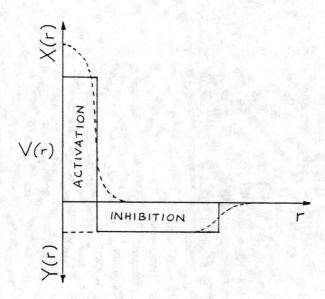

FIG. 6: Effective morphogenic field, $V(r)$ (solid line) represented as the schematically combined effects of activation and inhibition (dashed lines); r = radial distance from A cell.

The model is simulated on a cellular automaton starting from an initial random distribution of A and U cells on a 128×128 grid, with on-off switching according to the rule

FIG. 7: Automaton simulation of skin-pattern formation.

cell j	if	$\sum_{k(\neq j)} V(|\underline{r}_k - \underline{r}_j|)$	
becomes or remains A		> 0	
becomes or remains U		< 0	(6)
remains A or U		= 0	

The proceedure is repeated until the resulting pattern is stable. (Note that the final configuration is insensitive to initial A-cell distribution.)

The results of the simulations are presented in Fig. 7. The upper left configuration shows the initial distribution of A cells (black dots). The spot-type pattern (upper right) is obtained with a radially-symmetric morphogenic field. When the inhibition amplitude factor is lowered, one obtains the lower left configuration. The change from a leopard-type pattern to a tiger-type (lower right) is obtained by making the field anisotropic, i.e., by changing the symmetry of the field from circular to elliptic.

Since the underlying mechanism for skin pattern formation should be basically the same for all species belonging to a given class, different patterns should be generated according to a general sequence of processes with particular specifications (relative values of morphogen properties). Within the limits of the Turing philosophy, models presented here have the virtue of those which can be put to work with simple logics. Whether they actually contain the essentials to describe natural systems remains an open question.

We thank R. J. Silbey for his helpful suggestion. This work was supported by the FNRS, Belgium. One of us (AN) has benefited from a grant by the IRSIA.

[1] L. A. Segel, *Modeling Dynamic Phenomena in Molecular and Cellular Biology* (Cambridge University Press, Cambridge, 1984), Chapter 8.
[2] A. M. Turing, Philos. Trans. R. Soc. Lond. B **237**, 37-72 (1952).
[3] H. M. Martinez, J. Theor. Biol. **36**, 479-501 (1972).
[4] A. Gierer and H. Meinhardt, Kybernetik **12**, 30-39 (1972).
[5] H. Meinhardt, *Models of Biological Pattern Formation* (Academic Press, London, 1982).
[6] I. Prigogine and R. Lefever, J. Chem. Phys. **48**, 1695-1701 (1968).
[7] J. J. Tyson, J. Chem. Phys. **58**, 3919-3925 (1973).
[8] N. V. Swindale, Proc. R. Soc. Lond. B **208**, 243-264 (1980).
[9] D. A. Young, Math. Biosc. **72**, 51-58 (1984).

Cettina Amitrano

Max Kolb

PART B

. .

"THE SEMINARS"

Didier Roux and Didier Sornette

Francois Leyvraz

Aggregation of Colloidal Silica

David S. Cannell and Claude Aubert

Department of Physics
University of California
Santa Barbara, CA U.S.A. 93106

ABSTRACT

Spherical colloidal silica particles of $\sim 7nm$ diameter may easily be induced to aggregate into "large" ($\gtrsim 5\mu m$) structures even in dilute solution, by the addition of sufficient salt. The resulting aggregates appear to be fractal, scale invariant objects, as determined by classical light scattering. Studies carried out in $0.5M$ NaCl show that the fractal exponent characterizing the clusters is 2.08 ± 0.04 over a wide range of pH, and that the aggregation kinetics are slow and exponential in time. Studies in $1M$ NaCl reveal a transition to very rapid aggregation at high pH or very low particle concentration. This rapid aggregation regime results in a fractal dimension of either 2.08 or 1.77 depending upon conditions.

INTRODUCTION

Light scattering, either alone or in combination with small angle X-ray scattering or neutron scattering has proved to be an extremely powerful tool in the study of aggregation, permitting the aggregates to be studied while they are growing, without disturbing them in any way[1-4]. Aggregates formed from small ($\sim 7nm$) spherical colloidal silica particles were the first aggregates to be thus studied *in situ*[1], answering many questions raised by studies employing electron microscopy or other techniques which might seriously perturb the rather delicate aggregates. These studies were carried out[5] at pH 8.5, in $1M$ NaCl at a temperature of 25^0C. The structure factor of the fully formed aggregates, as determined by combined classical light scattering and X-ray scattering, was found to show a simple power law decay

$$S(q) \propto q^{-D} \tag{1}$$

over a range in q, the scattering wavevector, extending from $q = 0.0002$ Å^{-1} to $q = 0.02$ Å^{-1}. The exponent D was found to be 2.12 ± 0.05. This indicates that over length scales from a few particle diameters to several μm, the aggregates were described internally by a density-density correlation function

$$g(r) \propto r^{-A} \tag{2}$$

where $A = 3 - D$ in 3 dimensions. Such an object apparently has no intrinsic internal length scale, although to show this experimentally would require measuring higher order correlation functions as well.

It is noteworthy that the simplest realistic model of such an aggregation process, in which small clusters form and join to form larger clusters, the so-called cluster aggregation model[6,7], generates clusters having a fractal dimensionality of ~ 1.75, clearly different from the experimental result. The most obvious difference between the numerical simulations and the experimental system is that the sticking probabilities explored numerically are all rather high, while that for the real system must be many orders of magnitude less, considering that periods of days may be required for a system containing $0.5wt.\%$ colloidal silica to form large aggregates. This immediately raises two rather fundamental issues. First, is the result $D = 2.08$ in any way fundamental, in the sense of applying to at least some class of aggregates or aggregation processes; and second, are there conditions under which silica aggregation yields aggregates with the same fractal dimension observed in the numerical simulations ? It is the purpose of this presentation to provide at least partial answers to these questions.

RESULTS AND DISCUSSION

The experimental system studied was Ludox SM, graciously provided by Dupont. The original sample consisted of $\sim 7nm$ spherical particles of SiO_2 at $30wt.\%$ silica, and was diluted as required. Aggregation was initiated by introducing a small amount of diluted Ludox into a salt solution, whose pH was adjusted as desired by adding dilute NaOH or HCl. The aggregates were studied by means of classical light scattering, *in situ*, as they formed, using instrumentation which has been described previously[8]. These studies yielded the weight average molecular weight of the aggregates M_w, as a function of time, as well as the z-average radius of gyration R_G. Eventually the aggregates became so large that even small angle light scattering was unable to resolve R_G, and then only their internal structure could be probed through $S(q)$, which under such conditions was always found to be of the form given in Eq. 1.

Figure 1 shows the time evolution of $S(q)$ vs. the scattering wavevector q, during the aggregation process, for a sample of $0.10wt.\%$ Ludox in $0.5M$ NaCl at pH 8.70. We have always observed an initial aggregation stage which

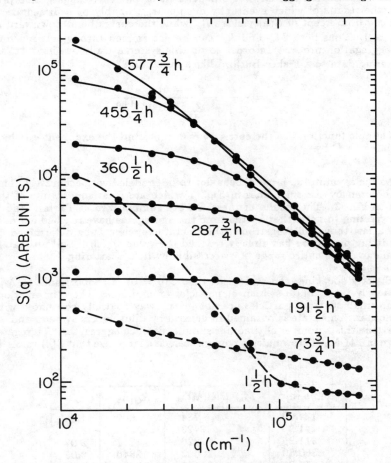

Figure 1. The static structure factor $S(q)$ vs. the scattering wavevector q at various times after initiating the aggregation of a solution of $\sim 7nm$ diameter colloidal silica spheres at $0.1wt.\%$ concentration, suspended in $0.5M$ NaCl solution at pH 8.7. Data at the two earliest times show the result of an initial relatively rapid aggregation of a small fraction of the silica, and are shown together with dashed lines. These aggregates eventually disappear, probably by sedimenting. The solid curves are fits as discussed in the text.

occured within a short time (dependent on salt concentration), and which generated a few very large ($\gtrsim 3\mu m$) aggregates. The longer term aggregation process represented in Fig. 1 is independent of this early aggregation process, the results of which eventually disappear, probably by sedimentation. The $q = 0$ intercept of $S(q)$ yields M_w, while the curves bend downward at $qR_G \simeq 1$. Thus both M_w and R_G can be determined throughout much of the aggregation process. In order to do this systematically, we have fit the scattering data to a Fisher-Burford-like approximate[9],

$$S(q) = \frac{S(q=0)}{\left[1 + q^2 R_G^2/3(1+\alpha)\right]^{1+\alpha}} \quad (3)$$

This simple function has the correct $q = 0$ limit, and the exponent α, which is related to D by

$$D = 2(1+\alpha) \quad (4)$$

allows the asymptotic, large q, behavior to be modeled very accurately. This can be seen by the solid lines in Fig. 1 which are least squares fits of the data to Eq. 3. The results of these fits are summarized in Table I, and it is interesting to note, that by the time the aggregates have reached a size of even a few thousand Angstroms, the fractal exponent D as determined by this fitting procedure has already reached the value which it will ultimately assume over the entire range of q accessible by light scattering.

Strictly speaking, the statement that $S(q = 0)$ is proportional to M_w, holds only for a dilute system of particles or molecules where interactions are so small that the particles are not in any way spatially correlated with each other. This state is obtained in general by diluting the system, and we have done this a number of times for samples of silica aggregates. Aggregates grown in $1M$ NaCl grow quickly enough to such a large size that $S(q)$ varies as

Time [h]	M_w[arb.units]	R_G[Å]	D
174.00	734.	609.	1.94
241.25	2423.	1045.	2.00
311.25	7826.	1771.	2.07
360.50	19662.	2846.	2.05
455.25	90261.	5971.	2.05
509.00	174531.	8478.	2.06
624.00	233705.	11844.	2.05

Table I. Values for the weight average molecular weight, z-average radius of gyration and fractal dimensionality D of growing silica aggregates for various times after the aggregation process was initated.

q^{-D} over the entire range of q that is experimentally accessible. Upon dilution of such a fully aggregated sample, $S(q)$ simply decreases by the expected factor at each q-value. Aggregates grown in $0.5M$ NaCl form sufficiently slowly as to allow the very largest clusters to sediment, at least partially, during the course of the experiments. Therefore we never observe $S(q)$ to vary simply as q^{-D} over the entire q-range, but instead $log\ S(q)$ vs. $log\ q$ always bends downward at small q as shown in Fig. 1. This effect however, vanishes when such a sample is diluted, as is shown by the data of Fig. 2. These data were obtained by diluting the sample of Fig. 1, by a factor of ~ 10 into pure water $624h$ after the initiation of aggregation. As shown by the figure this immediately resulted in $S(q)$ becoming proportional to q^{-D} at all q. Since the dilution sharply reduced the salt concentration it should have completely halted any further aggregation. Thus it seems most likely that all that happened, was to resuspend some of the larger aggregates which had sedimented. This effect has only been observed very late in the aggregation process, but deserves further study. In any event, the fractal exponent D, as determined directly from $S(q)$ by fitting, showed no significant change upon dilution, as indicated by the two values shown in Fig. 2. Thus we believe that the results shown in Fig. 1 are indicative of the properties of the individual clusters themselves. The growing clusters must, however, be present in a variety of sizes, and as usual[10] the light scattering results represent averages over that cluster size distribution. The averages involved weigh the largest clusters heavily, and so the results should be close to those which would be observed in studying only the largest clusters, but this question certainly deserves further investigation.

With the above caveats in mind we may extract M_w and R_G from the data for $S(q)$ at various times, and thus examine the kinetics of the aggregation process. These results are presented in Figs. 3 and 4 which show $log\ M_w$ and $log\ R_G$ vs. time in seconds after the initiation of aggregation. As may be seen, both M_w and R_G increase exponentially with time, and the solid lines are fits of the form

$$M_w = M_o e^{\Gamma_M t} \qquad (5a)$$

and

$$R_G = R_o e^{\Gamma_R t} \qquad (5b)$$

An alternative method of determining the fractal exponent D is to plot $log\ M_w$ vs. $log\ R_G$, since $M \propto \int_0^R g(r) r^2\ dr \propto R^{3-A} = R^D$. Figure 5 shows the typical result of such an analysis. The straight line is given by $M_w = R_G^{2.07}$. We usually find that the value of D obtained in this way is about 3% higher than the result obtained by fitting $S(q)$ directly, but basically both methods agree quite well.

We have carried out measurements similar to those described above ($S(q)$, $M_w(t)$, $R_G(t)$ and $M_w(R_G)$) under a variety of conditions, and some of the

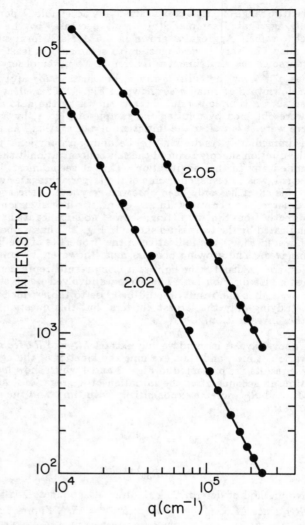

Figure 2. Results of diluting the sample of Fig. 1 by a factor of ~ 10 into pure water, $624h$ after initiating aggregation. Before dilution $S(q)$ bent downward at small q indicating aggregates with a radius of gyration of $\sim 1.2 \mu m$, but we believe the dilution process resuspended enough very large aggregates to yield a linear $S(q)$ afterwards. Fits, as discussed in the text, gave the same fractal exponent D before and after dilution, as indicated by the solid lines and numbers in the figure.

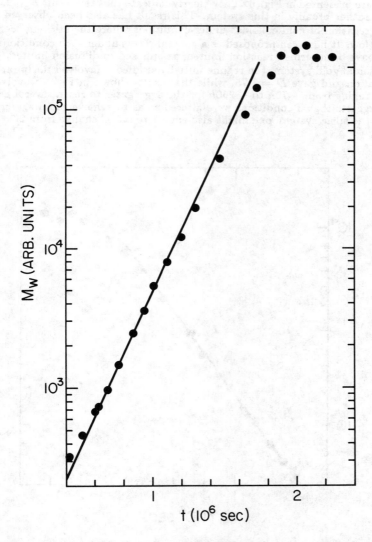

Figure 3. The weight average molecular weight M_w (in arbitrary units) vs. time after initiation of aggregation, for the sample of Fig. 1. The solid line is a fit of the data over the indicated range and yielded $M_w = 27.6 exp(5.15 \times 10^{-6} t)$.

results will now be summarized. First, at a fixed silica concentration of 0.5$wt.$% in 0.5M NaCl we have studied the pH dependence of D, and the results are presented in Fig. 6. They clearly indicate that the result $D = 2.08$ applies rather broadly in this system. This result has also been observed in the aggregation of gold colloids[4] under conditions where the sticking coefficient is low. It has been proposed as a general result under such conditions[4], which have been termed reaction limited, as opposed to diffusion limited. In the colloidal gold system, the regime initially studied[11] involved fairly rapid aggregation and gave $D = 1.75$, while the results shown in Fig. 6 involved times ranging from $\sim 70h$ to $\sim 260h$ for the aggregates to reach several μm. This suggests that if conditions were altered so as to achieve rapid aggregation in the silica system, one might also expect to see D change from 2.08 to ~ 1.75.

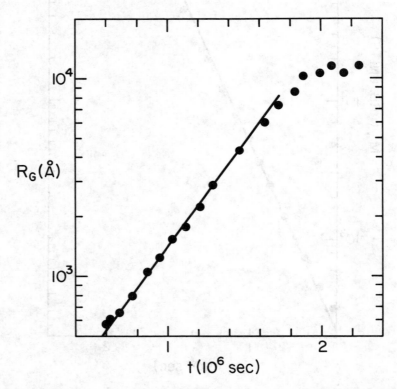

Figure 4. The z-average radius of gyration $vs.$ time after initiation of aggregation, for the sample of Fig.1. The solid line is a fit of the data over the indicated range and yielded $R_G = 125.2 exp(2.42 \times 10^{-6} t)$ Å.

In attempting to observe rapid aggregation, we have used NaCl concentrations as high as $1.25 M$, with silica concentrations ranging from $0.02 wt.\%$ to $0.5 wt.\%$ at pH $\simeq 8.5$, but D remains 2.08 under these conditions, and aggregation is fairly slow, requiring at least 2 hours. On the other hand,

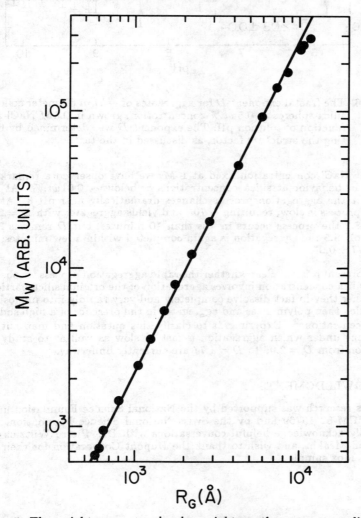

Figure 5. The weight average molecular weight *vs.* the z-average radius of gyration for the growing aggregates of Fig. 1. The solid line is a fit over the indicated range which yielded $M_w \propto R_G^{2.07 \pm 0.04}$.

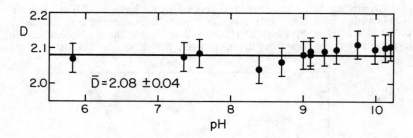

Figure 6. The fractal exponent D for aggregates of $\sim 7nm$ diameter colloidal silica spheres at $0.5wt.\%$ concentration, grown in $0.5M$ NaCl as a function of solution pH. The exponent D was determined by fitting the structure factor, as discussed in the text.

with the NaCl concentration fixed at $1\,M$, we have observed a remarkable change in behavior at silica concentrations c, below $\sim 0.01wt.\%$. At $c = 0.01wt.\%$ the aggregation process changes dramatically near pH 8. At pH 7.7 the process is slow, requiring $\sim 70h$ and yields aggregates with $D = 2.08$. At pH 8.2 the process occurs in less than 10 minutes, but D remains 2.08, and at pH 8.5 the aggregation is again completed within a few minutes, but D is 1.77 ± 0.05.

At present it is not clear whether the rapid aggregation process we observe at low silica concentration involves aggregation of the original silica particles, or whether they in fact dissolve completely and very rapidly into monosilicic acid which then polymerizes and aggregates in the presence of a high sodium ion concentration[12]. Experiments to clarify this question and map out the conditions under which aggregation is fast or slow as well as to study the transition from $D = 2.08$ to $D = 1.77$ are currently underway.

ACKNOWLEDGMENTS

This research was supported by the National Science Foundation under grant PCM-82 15769 and by the Swiss National Science Foundation. We gratefully acknowledge helpful conversations with Dr. Pierre Wiltzius and Dr. Paul Meakin, and wish to thank the Dupont Corporation for their gift of the Ludox sample.

REFERENCES

1. D.W. Schaefer, J.E. Martin, P. Wiltzius and D.S. Cannell, Phys. Rev. Lett. 52, 2371 (1984).
2. D.A. Weitz, J.S. Huang, M.Y. Lin and J. Sung, Phys. Rev. Lett. 53, 1657 (1984).
3. J.E. Martin and D.W. Schaefer, Phys. Rev. Lett. 53, 2457 (1984).
4. D.A. Weitz, J.S. Huang, M.Y. Lin and J. Sung, Phys. Rev. Lett. 54, 1416 (1985).
5. For some details not presented in reference 1 see D.W. Schaefer, J.E. Martin, P. Wiltzius and D.S. Cannell in *Kinetics of Aggregation and Gelation*, ed. by F. Family and D.P. Landau, p. 71 (Elsevier, 1984).
6. P. Meakin, Phys. Rev. Lett. 51, 1119 (1983).
7. M. Kolb, R. Botet and J. Jullien, Phys. Rev. Lett. 51, 1123 (1983).
8. H.R. Haller, C. Destor and D.S. Cannell, Rev. Sci. Instrum. 54, 973 (1983).
9. M.E. Fisher and R.J. Burford, Phys. Rev. B10, 2818 (1974).
10. C. Tanford, *Physical Chemistry of Macromolecules* (Wiley, N.Y., 1961)
11. D.A. Weitz and M. Olivera, Phys. Rev. Lett. 52, 1433 (1984).
12. See, for example R.K. Iler, *The Chemistry of Silica* (Wiley, N.Y., 1979).

Dietrich Stauffer

DYNAMICS OF FRACTALS

Dale W. Schaefer, James E. Martin and Alan J. Hurd

Sandia National Laboratories
Albuquerque, NM 87185

INTRODUCTION

Apart from linear polymers, the dynamic properties of fractal structures are essentially unexplored. This paucity of data is due to the fact that dynamic light scattering, the technique of choice for such studies, is expected to display unique features only for objects which are fractal over dimensional scales comparable to the wavelength of light (5000Å). Recently, however, colloidal aggregates were shown 1 both to be fractal over lengths up to 1μ and to display interesting dynamics. (2) Our purpose here is not only to discuss the dynamics of such aggregates, but also to review the results for polymers and colloidal particles. In all cases, these systems are studied in a dilute suspension by quasi-elastic light scattering.

Dynamic information is extracted from the initial slope of the electric field correlation function of light scattered from particles in suspension. The slope is the mean relaxation rate $\langle \Gamma \rangle$ of fluctuations of spatial frequency $K = 4\pi\lambda\sin(\theta/2)$, where λ is the wavelength in the medium and θ is the scattering angle. Depending on KR, R being a mean radius of the scatterers (i.e., clusters, polymers, particles, etc.), $\langle \Gamma \rangle$ may depend on either center-of-mass (CM) motion or the internal breathing modes of flexible objects. When $KR \ll 1$, for example, fluctuations in concentration relax by CM motion so $\langle \Gamma \rangle = D_0 K^2$, D_0 being the cluster diffusion constant. In general $D_0 \sim 1/R_h$ where R_h is defined as the hydrodynamic radius.

For $KR \gg 1$, a scaling argument can be used to predict the K dependence of $\langle \Gamma \rangle$,

$$\langle \Gamma \rangle \sim D_0 K^2 (KR)^x; KR \gg 1, \qquad (1)$$

where the exponent x is determined by the requirement that $\langle \Gamma \rangle$ be independent of R for $KR \gg 1$. That is, $x = 1$ and $\langle \Gamma \rangle \sim K^3$. This result is well known for polymers but it should be true for all flexible self-similar objects. Note that when $KR \sim 1$, $\langle \Gamma \rangle_0 \sim 1/R^3$ where $\langle \Gamma \rangle$ is called the fundamental relaxation rate. At $KR = 1$, all processes (CM diffusion, rotation and configurational relaxation) have the same characteristic rate.[3]

Equation (1) is the dynamic equivalent of self-similarity. At a given value of K one probes the motions of sub-clusters of size $R \sim K^{-1}$. The fundamental rate of such a structure, however, is $\langle \Gamma(R) \rangle_0 \sim R^{-3} \sim K^3$, consistent with Eq. (1).

Equation (1) is expected to fail when $Ka \sim 1$, where "a" is a chemical length that characterizes the monomer size or persistence length of the structure. Spin-echo neutron scattering is the only technique that can probe the regime $Ka = 1$ at the present time.

POLYMERS

The first figure shows the normalized relaxation rate for linear polymers (polystyrene) in several solvents. As expected, these data show $\langle \Gamma \rangle \sim K^2$ for $KR_h \ll 1$ and $\langle \Gamma \rangle \sim K^3$ for $KR_h \gg 1$. Note that $D_0 = kT\, 6\pi\eta R_h$ is the diffusion constant at infinite dilution and η is the solvent viscosity. Although the issue has not been studied extensively, it is believed that $\langle \Gamma \rangle$ is independent of R in the regime $KR \gg 1$. Later we show this is not the case for solution aggregates of silica.

SOLUTION AGGREGATES

Colloidal aggregates are the simplest branched structures which display fractal geometry. Dynamic data, similar to Fig. 1, are shown in Fig. 2 (and Curve D of Fig. 4) for aggregates of colloidal silica grown in solution 2 Although these data shown the general form of polymer data, the slope for $KR > 1$ is 2.7, somewhat less than that found for polymers. Several factors might explain the

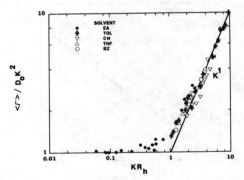

FIG. 1: Normalized relaxation rate for polymers in several solvents: EA = ethyl acetate, CH = cyclohexane, THR = tetrahydrofuran, BZ = benzene. From Ref. 3.

slope of Fig. 2. The first possibility is suggested by Fig. 3, which shows the unscaled data of Fig. 2. This plot demonstrates that $\langle \Gamma \rangle$ is not independent of R for $KR > 1$ as assumed in the derivation $x = 3$ in Eq. (1), implying that the asymptotic $KR \gg 1$ has not been reached. A problem arises because R_h is very sensitive to short length scales, so enormous structures are necessary to reach an asymptote.[3] Non-asymptotic hydrodynamics can be treated[2] crudely by using Eq. (1) with $D^{-1} = R_h \sim M^{\nu_h}$ and $R \sim M^\nu$, ν being the asymptotic exponent $(\nu > \nu_h)$. The result is $x = 2 + \nu_h/\nu$ or $\nu_h/\nu = 0.7$ for the data in Fig. 2. This explanation has some support from polymers since asymptotic exponents $\nu_h = \nu$, have never been observed.

It should also be noted that Eq. (1) assumes that the clusters are completely flexible. If the clusters are stiff, elastic forces will affect the dynamics. No treatment is available for the elastic properties of overdamped fractals because different modes shown different scaling behavior. If the arms of a cluster are represented by stiff rods of length L, for example, the bending modulus goes as L^{-3} whereas the extensional modulus goes as L^{-1}. Clearly no simple behavior can be expected.

In the absence of a theory for elastic processes, we can still conclude that elasticity does not account for the data in Fig. 2. In the regime $KR > 1$, the aggregate data in Fig. 2 fall below the polymer data in Fig. 1. If elasticity were present, however, the relaxation rates should be enhanced with respect to polymers. Since the opposite is actually observed, we conclude that elasticity is not the issue.

Polydispersity can also explain the slope in Fig. 2. For highly polydisperse systems (percolation) Martin and Ackerson[5] show that $\langle \Gamma \rangle \sim K^3$. For systems with weaker polydispersity $\langle \Gamma \rangle$ is expected to yield expoanents less than three. This issue is discussed in detail by Martin.[6] Dynamics of diffusion limited cluster-cluster aggregates (CCA) is discussed in Ref. 2. Note that—in contrast to the above results—Weitz et al find $\langle \Gamma \rangle \sim K^2$ for CAA.

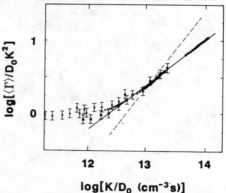

FIG. 2: Normalized relaxation rate for colloidal silica. The dashed line is the asymptotic result in Fig. 1. Note that the axes are mislabeled in Ref. 2, the source of these data.

VAPOR AGGREGATES

Further information on dynamics of fractals is available from aggregates grown in the gas phase.[8] In this case, static scattering data show a fractal dimensionm of 1.8 ± 0.1 which is consistent with simple CCA. Whereas CCA produces only weak polydispersity, however, the dynamic data (curve C, Fig. 4) for vapor aggregates (Cab-O-Sil) look identical to the solution case. We conclude that either CCA is not a suitable model for this system or the dynamic exponent is not related to polydispersity.

COLLOIDAL POWDERS

The final systems discussed here are high surface area powders (carbon black and alumina). Alumina powders are polydisperse dense colloidal par carbon black is ramified. Static scattering[8,9] from these systems is very similar with a small "apparent" power-law regime between $K = 0.002$ and $0.004 Å^{-1}$. The slope of the scattered intensity in this regime gives an apparent fractal dimension of $D = 2.1$ for carbon black and $D = 1.1$ for alumina. For alumina the exponent is sensitive to preparationm procedures. The dynamics for these systems is shown as curve A and B in Fig. 4. Both curves give maximum slopes near 0.4 so $\langle \Gamma \rangle \sim K^{2.4}$. If we assume that the particles are uniform, the static data yield the exponent τ which characterizes the power-law size distribution.[10] In both cases, τ is greater than 2 indicting strong polydispersity, a situation which should yeild $\langle \Gamma \rangle \sim K^3$. The fact that $\langle \Gamma \rangle$ is not cubic in K is therefore evidence that these systems are not fractal (i.e., do not have a power-law size distribution).

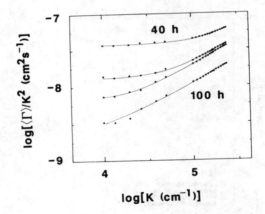

FIG. 3: Unscaled data from Fig. 1. Although static data are fully asymptotic,[2] the relaxation rate does not approach a size independent value at large K.

SUMMARY

We have shown that simple power-law dynamics is expected for flexible fractal objects. Although the predicted behavior is well established for linear polymers, the situationm is considerably more complex for colloidal aggregates. In the latter case, the observed K-dependence of $\langle \Gamma \rangle$ can be explained either in terms of non-asymptotic hydrodynamics or in terms of weak power-law polydispersity. In the case of powders (alumina, in particular) apparent fractal behavior seen in static scattering is not found in the dynamics.

[1] D. W. Schaefer, J. E. Martin, P. Wiltzius, and D. S. Cannell, Phys. Rev. Lett. **52**, 2371 (1984).

[2] J. E. Martin and D. W. Schaefer, Phys. Rev. Lett. **53**, 2457 (1984).

[3] D. W. Schaefer and C. C. Han in *Dynamic Light Scattering*, R. Pecora ed, Plenum, NY, 1985) p. 181.

[4] P. Sen, this book.

[5] J. E. Martin and B. J. Ackerson, Phys. Rev. A **31**, 1180 (1985).

[6] J. E. Martin, to be published.

[7] D. A. Weitz, J. S. Huang, M. Y. Lin and J. Sung, Phys. Rev. Lett. **53**, 1657 (1984).

[8] J. E. Martin, D. W. Schaefer and A. J. Hurd, to be published; D. W. Schaefer, K. D. Keefer, J. E. Martin, and A. J. Hurd, in *Physics of Finely Divided Matter*, M. Daoud, Ed., Springer Verlag, NY, 1985.

[9] D. W. Schaefer and A. J. Hurd, to be published.

[10] J. E. Martin, J. Appl. Cryst. (to be published).

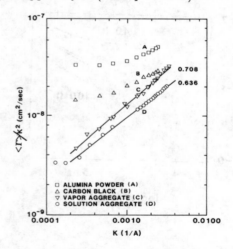

FIG. 4: K-dependence of thef mean relaxation rate, $\langle \Gamma \rangle$, for (A) Alumina (surface area $265 m^2/gm$, (B) carbon black (Spheron 6), (C) fumed silica (Cab-O-Sil), (D) solution-grown aggregates of silica.[2]

FRACTAL VISCOUS FINGERS:
EXPERIMENTAL RESULTS

G. Daccord, J. Nittmann
Etudes et fabrication Dowell–Schlumberger
ZI Molina la Chazotte - BP 90
42003 Saint Etienne - FRANCE

and

H. E. Stanley
Center for Polymer Studies and Department of Physics
Boston University, Boston, MA 02215 - USA

1 INTRODUCTION

Viscous fingering has been actively studied for some time, not only because of its economic importance, but also because it is an area in which both experimental and theoretical studies can be performed in parallel. Moreover, recently developed statistical models of aggregation and growth have been found to be a useful means of representing fingering.

It has been known that ramified patterns can be obtained in viscous fingering if the capillary number is sufficiently high (2). However the ramification is not very important and if the fluids are miscible the low viscosity fingers do not displace the viscous fluid as a plug and most of time they are not steady with respect to time (3). These major drawbacks can be overcome by the selection of appropriate experimental conditions. The displacement patterns so obtained can be more closely studied due to the simplifications these conditions bring to the flow equations.

Hele–Shaw cells were used in this study. Unlike packed beads beds or etched network models such models are perfectly isotropic. The viscous fluids used were non-Newtonian (shear thinning) which causes a plug-like motion of the interface. This paper will focus on the experiments side, numerical simulations being described in other articles (1,4).

2 APPARATUS AND EXPERIMENTAL PROCEDURE

A schematic diagram of the experiment is shown on figure 1. The fluids are injected into the cell, the displacement patterns are recorded onto a video tape. The pictures are then digitized and sent to the main computer for numerical computations.

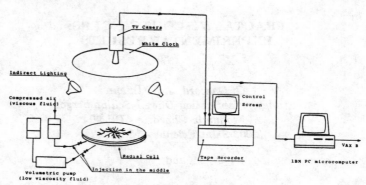

Figure 1: schematic representation of the experimental set up.

A linear cell (length 90 cm, breadth W_{HS} adjustable between 5 and 11 cm and thickness b adjustable between 0.1 and 1 mm) and a radial cell (diameter 95 cm, thickness 0.5 mm) were used. A black and white video camera was used for recording the images. A few frames are later selected for digitalization. Each image is digitalized into a 256 × 256 matrix with 64 levels of intensity. A threshold grey level is selected to distinguish the two fluids and locate exactly the interface. The matrix is then compacted and sent to a VAX 11/785 for numerical treatment.

The viscous fluids were of several kinds: polysaccharide solutions (shear thinning), oligosaccharide solutions (Newtonian) and a latex (a latex is a suspension of polymer spheres. It was also shear thinning and will be designated by its trade name CEF). The different polysaccharides used were: scleroglucan (SC), a guar gum derivative (GG) and an hydroxyethyl cellulose (HEC) at concentrations ranging from 2.4 to 12 g/l. Figure 4b shows a typical rheological behaviour of a shear thinning fluid. The slope of the log-log plot of the viscosity versus shear rate represents the power law index n and the viscosity at an abcissa of 1 s^{-1} is called the consistency k. $n = 0.1$ for SC (for the two concentrations used) while for GG and HEC it is higher and depends on their concentration. For CEF, the rheological measurements yield an n of around 0.2.

The displacing fluids were always Newtonian with a viscosity k_o from 1 cP up to 90 cP. Water, glycerine, corn syrup or low molecular weight polyvinylalcohol (PVA) dyed with methylene blue were used.

In all the experiments the maximum shear rate at the wall in the viscous fluid (at the tip of the fingers) was between 100 and 1000 s^{-1}, while the Reynolds number for the low viscosity fluid was around 100.

3 RESULTS

All the numerical treatment was performed on the digitized figures. Each pixel represents the average intensity on a rectangle the aspect ratio of which is 1.43.

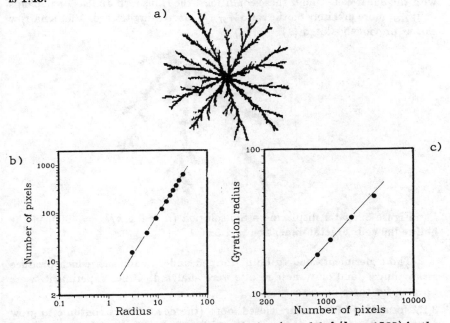

Figure 2: a) water displacing a SC solution ($n = 0.1, k/k_o = 1500$) in the radial cell. Flow rate 20 ml/min. b) Hausdorff plot yielding a D_f of 1.7±0.05. c) gyration radius versus time, yielding an initial D_f of 1.6.

3.1 Fractal Dimension

When using the Hausdorff method, the structures appear to have a self similarity property, the lower cuttoff corresponding to the finger width (a few pixels) and the upper one being higher than the maximum size of the radius used. The corresponding fractal dimension D_f, when using the radial cell, is 1.7±0.05 (figure 2).

For the radial fingers, the log–log plot of the gyration radius versus time does not yield a straight line (figure 2c). The extend of the bending depends on the conditions of the experiment: it is more important at high flow rates and low polymer concentrations. However, the tangent at its beginning yields a fractal dimension between 1.7 and 1.6, averaged on several runs.

The coastline method has not been used because of the rectangular pixel.

Experiments in a semi–radial cell (figure 3) result in a lowering of D_f (1.5±0.1 on only a few experiments). The geometry is still further restricted with the linear cell. Under these conditions, the Hausdorff dimension depends on W_{HS}, more precisely on the ratio W_{HS}/b, decreasing towards 1 for a narrow slot as previously stated (1).

Figure 3: water displacing a SC solution ($n = 0.1, k/k_o = 4000$) in the half radial cell. Fractal dimension of 1.5±0.1.

The experiments had to be performed under conditions which produce steady fingers and only these results were analysed. Other experiments were rejected for three main reasons:

3.1.1 formation of too many closed loops (the dead fingers continue to grow from their tip towards the main branch). Their conditions of appearance depend on the rheologies of both fluids and they are observed mainly for high flow rates.

3.1.2 non steady fingers i.e. the breadth of all the fingers increases continuously. This appears when the power law index is not sufficiently low.

3.1.3 seepage of the water between the wall of the cell and the mass of the viscous fluid (figure 4). It depends mainly on the rheology of the viscous fluid and occurs at low flow rates or at the tips of the dead fingers.

3.2 Influence of the Thickness of the Gap

The thickness of the linear cell was decreased from 1 to 0.1 mm. This did not affect the nature of the pattern which was always self similar but the average width of the fingers was found to be roughly proportional to the thickness, between 5 and 10 times the gap. Other experiments (5) with a gap of 3 mm fit this conjecture.

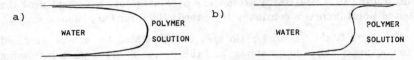

Figure 4: a) cross section of the slot showing the front in normal conditions and b) when seepage occurs between the wall and the viscous solution.

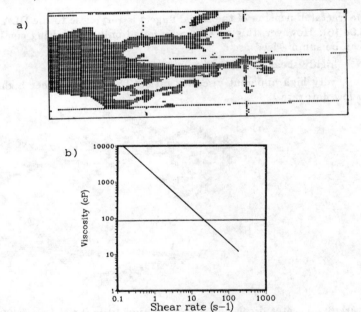

Figure 5: a) glycerol solution displacing a SC solution in the linear cell. Flow rate 10 ml/min. b) rheological curves of the two fluids ($n = 0.1$, $k/k_o = 17$)

Decreasing the gap narrows the range of conditions for obtaining steady fingers. Even when n is low, seepage and closed loops can be observed simultaneously for 0.1 mm thin gaps.

3.3 Influence of the Rheology of the Fluids

The rheology of the fluids can be adjusted by changing the polymer concentration or by using a mixture of two different fluids. Figure 5 shows the effect of lowering the ratio k/k_o while keeping a power law index $n = 0.1$. The displacement is similar to that observed when displacing one Newtonian fluid by another: the finger width increases continuously after one has overtaken them, leading to a non steady state. The percentage of area swept near the injection point increases continuously towards 100% for long enough times.

On figure 6, the viscosity ratio k/k_o was kept high but n was increased up to 0.4: many closed loops appear and this is also observed when increasing the flow rate for experiments with fluids with very low power law index.

4 DISCUSSION

The fractal dimension of the radial fingers is very close to the DLA value 1.68±0.04 (6). However, this agreement is found only for limiting conditions which can be summarized as:

- highly shear thinning viscous fluid ($n < 0.3$).

- very high apparent viscosity ratio. This means either high k/k_o ratio or low flow rate.

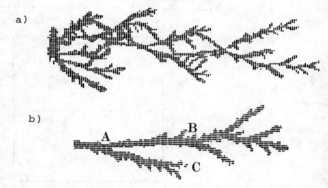

Figure 6: a) water displacing a GG solution ($n = 0.4$, $k/k_o = 400$). b) schematic representation of the closure of a loop.

Under these conditions, at each time the average Hausdorff dimension is 1.7 but the gyration radius method yields this value only at the start of the curve. This may be due to the fattening of the fingers close to the center as time goes on, resulting either from the actual displacement of some viscous fluid (mechanical erosion of the finger sides at high flow rates) or from the finite resolution of the digitalizing system. This strongly affects the gyration radius which increases more slowly than it would if the growth took place only at the tip of the fingers.

The lower D_f obtained when using a restricted geometry appears to be due to these boundary conditions but no convincing physical explanation can be made. No comparison with simulations have been made for semi-radial conditions, but for the linear cell such simulations were performed and agree well with the experiments.

4.1 Finger Width

The finger width was studied by Paterson (7) in the case of Newtonian fluids (studying the dissipation of energy), for which he predicted and verified that the wavelength should be around 4 times the thickness b. Our results show that for shear thinning fluids, it is between $5b$ and $10b$. This agreement is in fact better if we consider that for Newtonian fluids, the low viscosity fluid displaces only partially (50%) the fluid initially in the cell (figure 4 in ref. 7), while for shear thinning fluids the displacement is almost total (This fact was checked in two ways: by looking at the interface in a tube (figure 4) and by comparing the actual flow rate with that calculated from the swept area multiplied by the cell thickness). So, in fact, considering the ratio finger width/finger thickness (and not slot thickness), our results are in agreement with those of Paterson although the latter were obtained for Newtonian fluids for which the mechanism of energy dissipation is fundamentally different from that of polymer solutions.

4.2 Other Phenomena

As previously noted, several phenomena restrict the range of experimental conditions.

Closed loops appear when the pressure drop in the fingers cannot be neglected. Considering one growing finger and a dead one (see figure 6b), at high enough flow rates there is a non-negligible pressure drop between the tip of the dead finger and the corresponding point in the growing finger. If the time necessary to close this loop (which depends on the ratio $P_C - P_B$/viscosity of the viscous solution) is less than the time of the experiment, such loops will be observed. One of the effects of n comes in here: since it is the static viscosity which has to be considered, high values of n which are correlated with a newtonian plateau at low shear rates will present a lower viscosity than SC for example (figure 5b).

This result is in agreement with numerical simulations of Gradient Governed Growth (8) which produce closed loops only when the viscosity ratio is not too high.

The seepage of water between the wall and the mass of the viscous fluid is an instability which arises at the tip of a finger when the polymer solution viscosity is very high (quasi static or concentrated solution) and its n low. This phenomenon is the rule when using cross-linked solutions, and on the other hand is observed less and less as the power law index increases to 1.

The continuous growth of the finger breadth (fattening) is directly related to the Newtonian character of the viscous fluid. It is never observed for n less than 0.3–0.4.

Some mechanical erosion of the finger sides is also observed at high flow rates.

5 CONCLUSIONS

Fractal structures can be obtained with viscous fingers in a wide range of conditions. The main parameters appears to be the rheological behaviours of the two fluids, the flow rate and the gap thickness.

Most of the various phenomena taking place can be at least qualitatively understood, except the basic mechanism of tip bifurcation which is still obscure.

6 REFERENCES

1– J. Nittmann, G. Daccord, H. E. Stanley. Fractal growth of viscous fingers: quantitative characterization of a fluid instability phenomenon. *Nature*, 314, 141–144, 14 march 1985.
2– P. G. Saffmann, G. I. Taylor. The penetration of a fluid into a porous medium or Hele-Shaw cell containing a more viscous liquid. *Proc. Roy. Soc.*, A242, 312–320, 1958.
3– B. Habermann. The efficiency of miscible displacement as a function of mobility ratio. *Pet. Trans. AIME*, 219, 264–272, 1960.
4– G. Daccord, J. Nittmann, H. E. Stanley. In preparation.
5– Unpublished experiments performed jointly by Shell and Dowell-Schlumberger with a 1.2 × 3.6 m cell and a gap of 3.17 mm.
6– T. A. Witten, L. M. Sander, *Phys. Rev. Lett.* 47, 1499–1501 (1981).
7– L. Paterson. Fingering with miscible fluids in a Hele Shaw cell. Phys. Fluids 28 (1), 26–30, Jan. 1985.
8– J. D. Sherwood, J. Nittmann, Submitted to *J. Phys.*

WETTING INDUCED AGGREGATION

Daniel Beysens, Coggio Houessou and Françoise Perrot

SPSRM—CEN Saclay
F. 91191—Gif/Yvette Cedex, France

We present experimental results which show that colloids immersed in an homogeneous binary fluid can aggregate because of the preferential adsorption of one component. This aggregation phenomenon seems to be connected to the general wetting properties related to the existence of a wetting transition near a liquid-liquid critical point.

INTRODUCTION

The basic experiment is as follows. Silica spheres of diameter 1600Å or 4000Å immersed in the binary fluid $2-6$ lutidine + water (L-W) aggregate when the temperature (T) is increased above a given temperature T_A. The aggregation temperature is a function of concentration (C) so that a line forms in the plane (T,C) whose shape is related to the coexistence curve of the mixture. This suggests a connection with the wetting properties of binary mixtures and especially with the prewetting line.[1] Let us now examine in greater detail the different aspects of this experiment.

SILICA SPHERES PROPERTIES

Most of the experiments have been performed with 1600Å diameter silica spheres. The 4000Å diameter spheres gave nearly the same results. The spheres were obtained using the Störber method.[2] The measurements of the size of the spheres were made by electron microscopy, static light scattering and photon beating experiments. All methods are in agreement; a $\pm 30\%$ polydispersity for the 1600Å diameter spheres has been found.

The surface of silica is somewhat complicated and depends on the fabrication process.[3] Some groups (siloxane bridges) are hydrophobic, others are hydrophylic, so that preferential wetting by water cannot be determined without experiment on the sample itself. Also some ions can be adsorbed (Na^+, etc.) which may change the surface properties.

In solution, silica spheres acquire a surface charge (σ) which is a function of the density of charges O^- and OH^+, and therefore is a function of the pH and of the ionic strength of the solution (Fig. 1). The isoelectric point $\sigma = 0$ corresponds to the aggregation of the colloids, and occurs generally for an acid pH. Large negative surface charges are found for basic pH, which is a region of high stability.

THE LUTIDINE-WATER MIXTURE

2-6 Lutidine (2-6) dimethyl piridine) is dissociated in solution with water. The pH lies between 9 and 10 and decreases with increasing temperature ($T = 20°C$: pH = 9.8; $T = 30°C$: pH = 9.5 for an L-mass fraction $c_L = 0.30$). Miscibility with water is governed by a coexistence curve which exhibits upper and lower critical end points.[4] We shall be concerned here only with the lowest part of the coexistence curve which exhibits a critical point at $T_c = 34°C$ and $c_c = 0.290$.

According to the wetting properties of such systems in the 2-phases region (i.e., on the coexistence curve), one phase should completely wet any extra phase for temperatures close enough to T_c. This extra phase can be either the walls of the container, the meniscus with vapor or any immersed colloids.

Pohl and Goldburg[5] have observed in the L-W system that W preferentially wets a glass wall. However L preferentially wets the silica spheres, as is shown by the L-rich drops which form at the surface of the aggregates when the fluid phase-separates.

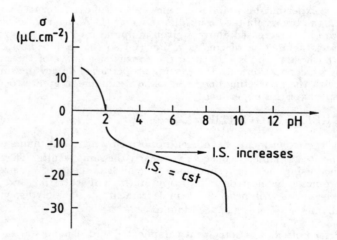

FIG. 1: Schematic behavior of the charge per unit surface of silica colloids in solution vs. pH at constant ionic strength (I.S.). When I.S. is increased, the curve is displaced toward large pH.

In the one-phase region a L-preferential adsorption layer should therefore exist at the surface of the spheres. A first-order transition between a "thin" layer, of order ξ (the correlation length) to a "thick" layer, of order $\xi + L$, is expected. This is the so-called prewetting transition, which should end by a 2nd-order transition point.[1] This transition has not been observed up to now.

EXPERIMENTAL METHODS

We used mass fraction of silica in the range $10^{-2} - 10^{-4}$. The results did not show any obvious mass fraction dependence. The observations were performed in several ways (Fig. 3):

—visually, with photos

—using small-angle light scattering, thus providing the means size of the aggregates still present in the laser beam

—recording the transmitted light, which allows the turbidity (τ) of the sample to be obtained. When the spheres are not aggregated, τ is roughly proportional to the 6th power of the radius (a).[6] This is a very sensitive way to control any change of the surface properties of the spheres, and especially the appearance of any L-rich (or W-rich) layer.

—analysing the fluctuations of the 90° angle scattering by photon beating spectroscopy. The hydrodynamic radius a_H of the spheres before aggregation

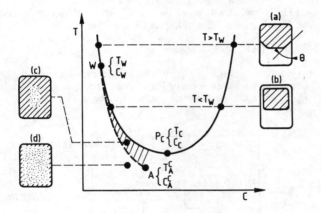

FIG. 2: Phase diagram of a binary fluid with a lower consolute point. Solid curve: coexistence curve with with $P_c(c_c, T_c)$ the liquid-liquid critical point. Dotted curve, first order prewetting line, which separates a region of large adsorption (hatched region) from a region of small absorption. The overall aspect of the mixture is represented by (c) and (d), with an excess concentration of one component near the sample walls. $A(c_A^c, T_A^c)$ is a critical absorption point, and W is the wetting point which separates, when 2 phases are in coexistence, a region of partial wetting (a), with a finite angle θ, from a region of complete wetting (b), where $\theta = 0$.

can be determined and compared to the radius obtained from the turbitidy measurements.

—inteferometry has also been used with the spheres sedimented $(T < T_A)$ or aggregated $(T > T_A)$ in the bottom of the cell. This enabled the change of the bulk concentration (δ_c) to be measured in order to deduce the nature and the amounts of the adsorbed layers.

AGGREGATION

At a given L-concentration, the spheres are moving classically according to their Brownian motion, as shown by photon beating measurements (Fig. 4). When T is increased, there exists a temperature T_A which is a function of c, as shown in Fig. 5, for which a sudden aggregation occurs. Two different processes can be qualitatively distinguished.

(1) A fast and well-defined aggregation process: for $c \lesssim c_c$. The temperature resolution of the phenomenon is lower than 5mk and the time scale is of the order of a few hours.

(2) A weaker and not-well-defined temperature aggregation process for $c \gtrsim c_c$. The time scale is of the order of a few days.

In any case macroscopic aggregates are formed at the liquid-vapor interface of the sample (Fig. 6), with a typical length of 5mm. When T is decreased, all aggregates are dissolved and the spheres recover their Brownian motion.

AGGREGATION PROCESS

All measurements, including a previous interferometric study by Gladden and Brener[7] show that an L-rich layer formed on the spheres. The layer thickness (e) can be simply deduced from the turbidity if the layer is assumed to be made of pure L.

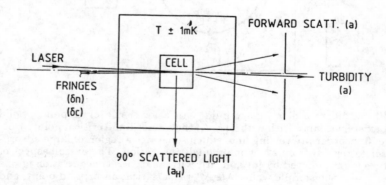

FIG. 3: Schematic set-up.

All measurements including bulk interferometry, show that the layer thickness increased with T, up to T_A, where it seems to diverge. That e increases when the coexistence curve is approached is in accordance with the fact that e should vary as the correlation length.[1] However the apparent divergence of e for $T \stackrel{<}{\sim} T_A$ can be understood in two different ways:

(i) A sudden change of layer: $e \to e + L$. The aggregation line should be the prewetting line for the fast aggregation process, i.e., for $c \stackrel{<}{\sim} c_c$.

(ii) Appearance of small clusters which are very strong scatterers and will give an apparent increase of e. The hydrodynamic radius of the spheres should be affected. This was not the case for $c \stackrel{<}{\sim} c_c$ (Fig. 4). However when $c \stackrel{>}{\sim} c_c$, such measurements became meaningless, the polydispersity being increased by a large amount, thus suggesting the presence of small clusters. In the strong aggregation regime ($c \stackrel{<}{\sim} c_c$) the aggregation occurs for temperature steps as small as $dT = 10 mk$. This corresponds to an increase of the correlation length which is 100 times lower than the increase of the L-layer thickness as deduced from bulk interferometry. It seems therefore that (i) is a more likely explanation of the strong aggregation process than (ii).

In any case the aggregation mechanism seems clearly due to the screening of the coulombian repulsive interaction by the temperature reversible adsorp-

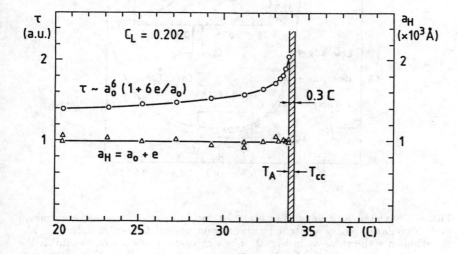

FIG. 4: Layer thickness assuming pure L wetting the surface of the spheres, from turbidity measurements (o) or photon beating measurements. T_c is the coexistence curve temperature, and T_A the aggregation temperature.

tions of L^+ ions and/or the attractive profile of the L-layer, as proposed by deGennes.[8]

CONCLUSION

We have observed in silica colloids immersed in a homogeneous binary fluid an aggregation due to the increase of an adsorbed layer. This layer was seen to obey the universal wetting properties connected to the existence of a bulk critical point.

Macroscopic aggregates can be formed at the meniscus between liquid and vapor. Simply varying the temperature allows the colloids to aggregate or de-aggregate, suggesting a way of obtaining reversible aggregation. Changing the binary fluid concentration modifies the interparticle potential so that a systematical study of the aggregation processes with a tunable interaction potential could be performed.

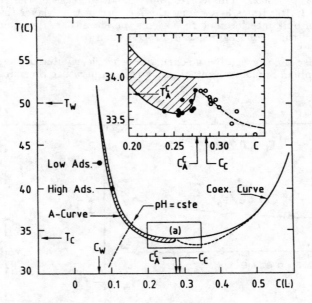

FIG. 5: Coexistence curve of the water+2-6 lutidine system (coexistence curve) and aggregation curve (hatched curve), upon addition of silica spheres. The notations are the same as in Fig. 2. The aggregation line seems to exhibit the same features as expected for the first order prewetting line. The dotted line corresonds to a much weaker aggregation process (see text). For clarity only the smooth curves are shown. The critial region (a) is magnified in the insert. Solid (open) circles correspond to a high(low) aggregation process. A line of constant pH has been reported for comparison.

[1]See, e.g., the review by M. R. Moldover and J. W. Schmidt, Physica **12D**, 351 (1984).
[2]W. Störber, A. Fink and E. Bohn, J. Colloid Interface Sci. **26**, 62 (1968).
[3]R. K. Iler, *The Chemistry of Silica* (J. Wiley, 1979).
[4]J. D. Cox and E. F. G. Herington, Trans. Faraday Soc. **52**, 926 (1956).
[5]D. W. Pohl and W. I. Goldburg, Phys. Rev. Lett. **48**, 1111 (1982).
[6]M. Kerker, *The Scattering of Light* (Academic Press, 1969).
[7]G. P. Gladden and M. M. Brener, J. Colloid Sci. **53**, 249 (1975).
[8]P. G. de Gennes, C. R. Acad. Sci. (Paris) **292**, 701 (1981).

FIG. 6. Macroscopic aggregates are formed from the liquid-vapor interface. Other aggregates lie in the bottom of the cell. A few remain in suspension.

LIGHT SCATTERING FROM AGGREGATING SYSTEMS: STATIC, DYNAMIC (QELS) AND NUMBER FLUCTUATIONS

J G RARITY AND P N PUSEY

Royal Signals & Radar Establishment, Malvern, Worcestershire, UK

We have studied the kinetics of salt induced aggregation of polystyrene microspheres (220 Å radius) using both static and dynamic light scattering techniques. At present the study has been limited to slow (reaction limited) aggregation (in 0.15M NaCl). Preliminary measurements show an exponential like growth of both the mean scattered intensity and the mean hydrodynamic radius R_h (determined from the initial decay of the intensity autocorrelation function) in the low Q region. In this region the typical aggregate radius is much smaller than the inverse modulus of the scattering vector $\underline{Q}$ ($QR_h \ll 1$) and aggregates scatter as the square of their mass(m). R_h probes non-integer moments of the aggregate distribution

$$R_h \sim \frac{1}{D} \sim \frac{\langle m^2 \rangle}{\langle m^{(2-1/d_h)} \rangle} \qquad (1)$$

where D is the intensity-weighted average diffusion coefficient (1) and d_h is the fractal exponent relating aggregate mass to its hydrodynamic radius ($m \sim R_h^{d_h}$). The mean scattered intensity probes the ratio of the second mass moment to the mean mass

$$\langle I \rangle \sim \frac{\langle m^2 \rangle}{\langle m \rangle} \qquad (2)$$

Under the assumption of a scale invariant aggregate distribution(2) equations 1 and 2 reduce to

$$R_h \sim \langle m \rangle^{1/d_h} \qquad (3a)$$

$$\langle I \rangle \sim \langle m \rangle \qquad (3b)$$

On plotting measured log <I> against log R_h (figure 1) we obtain a straight line of slope 2.02 ± .05 suggesting $d_h \approx 2$ and supporting the premise of scale invariance. The exponential like growth (in time) of these measurands further supports this premise and agrees with the results of other workers (3,4,5).

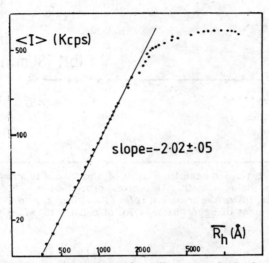

Fig 1. The mean scattered intensity <I> versus hydrodynamic radius Rh measured during aggregation of 220 Å polystyrene microspheres. Measurement made at low scattering angle (20°) with conventional Photon Correlation Spectroscopy apparatus (1), incident wavelength 6471Å.

When the aggregates grow larger than Q^{-1} individual particles add incoherently to the aggregate scattering and we observe a constant scattered intensity while the measured R_h continues to grow. Measuring the variation of scattered intensity with Q in this high Q region (by changing the scattering angle) has allowed estimation of d_f, the exponent relating aggregate mass to its radius of gyration Rg. For these samples d_f = 2.1 ± .05 close to the measured d_h and again agreeing with other measurements on slowly aggregating systems (3,4,5).

The scattered intensity cross-correlation function could also be measured during aggregation. Using two detectors at different scattering angles viewing the same small (30-300μm³) scattering volume inter-aggregate interference effects are suppressed and only fluctuations in number and scattering power of individual aggregates are correlated. Measurements of this type are often called Number Fluctuation Spectroscopy (6).

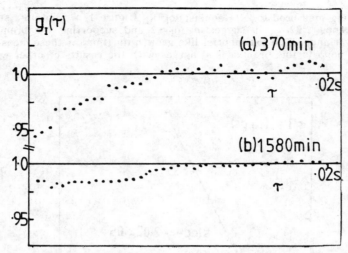

Fig 2. The short time modulation $G_I(\tau)$ of the intensity cross-correlation as a function of delay time τ. Detectors placed at +/- 90° in a cross-correlation apparatus as described in refs (8) and (9). Curve (a) 370 minutes after commencement of aggregation, QR_h of order 1 (b) after 1580 minutes when $QR_h > 1$.

The intercept of the cross-correlation function in the high Q region (when the mean scattered intensity has become constant) probes the same moment of the aggregate distribution as the low Q scattered intensity (eq. 2 and see (6)). Initial measurements taken during aggregation suggest an exponential like growth of this intercept over two decades before sedimentation effects set in.

The cross-correlation function itself decays on two timescales. The long time decay is due to slow fluctuations in the number of particles in the volume while the short time portion contains information about two, three and four point structure within the aggregate (7) decaying due to structural fluctuations or overall particle Brownian rotation (8, 9).

In the limit of large random aggregates and high Q this short time decay should only appear when the two detectors are closely spaced, its amplitude reducing in a fashion analagous to the low angle scattered intensity form factor as the detectors are separated. Measurements with nearly superimposed detectors (at 90 ± 2°) confirm a large amplitude effect which is almost neglible when detectors are placed at +/- 90° (opposite sides of the scattering volume). In the intermediate Q region when the aggregates have grown to sizes where $QRg \simeq 1$ the "opposed" detector technique probes correlations in aggregate dimension measured along orthogonal directions (the scattering vectors are orthogonal). Any aggregate asymmetry should produce a short time <u>anti</u>correlation effect in this size region (8,9). It is clear that aggregates formed in cluster-cluster simulations are asymmetric (10) and our results (fig 2) show a short time anticorrelation which becomes less pronounced as the aggregates grow. More quantitative measurements may allow estimation of this effective asymmetry for comparison with simulation results.

References

1. Brown J C, Pusey P N and Deitz R, J Chem Phys. 62, (1975) 1136

2. Leyvraz F, Growth Models: Rate Equation Approach, this volume.

3. Cannell D F, Scattering in Colloids and Gels, this volume.

4. Schaefer D W and Martin J E, Dynamics of Factals, this volume.

5. Weitz D A, Huang J S, Lin M Y and Sung J, Phys Rev Letts, 13 (1985) 1416

6. Rarity J G and Randle K J, J Chem Soc. Faraday Trans I, 81 (1985) 285.

7. Pusey P N, Macromolecules, in press.

8. Griffin W G and Pusey P N, Phys Rev Lett 43 (1979) 1100

9. Rarity J G and Randle K J, Opt Comm, 50 (1984) 106.

10. Kolb M, Flocculation and Gelation in Cluster-Cluster Aggregation, this volume.

FLOCCULATION AND GELATION IN CLUSTER AGGREGATION

Max Kolb,*† Robert Botet,* Rémi Jullien* and Hans J. Herrmann‡

*Laboratoire de Physique des Solides
Bat. 510, Université de Paris-sud
91405 Orsay, France

†Institut für Theorie der kondensierten Materie, FU Berlin
Arnimallee 14, 1000—Berlin 33, West Germany

‡Service de Physique Théorique—CEN Saclay
F. 91191—Gif/Yvette Cedex, France

Clustering of clusters is introduced as a model to describe irreversible diffusion-limited aggregation of clusters. The scaling properties are determined using numerical simulations and mean field arguments. The results are compared with the Smoluchowski equation approach. The scaling behavior is different for low cluster concentration (flocculation) and for high cluster concentration ("gelation"). Reversible cluster aggregation can be modelled by randomly breaking bonds. This leads to a different cluster structure, with scaling behavior like lattice animals. The clusters of irreversible cluster and particle (DLA) aggregation are investigated for their internal anisotropy: particle aggregates appear to scale differently parallel and perpendicular to the direction of growth.

Clustering of clusters has been introduced to describe irreversible diffusion-limited cluster aggregation as observed in colloidal growth processes. Here, we define the basic model and discuss its scaling properties.[1] Then, it is shown that the classical approach due to Smoluchowski gives an adequate description of this growth process, provided the fractal aspect of the clusters is taken into account. At high cluster concentration, the growth mechanism is different because of the entanglement of the clusters. This leads to (infinite time) gelation.

Both irreversibility and diffusion are relevant factors in determining the fractal properties of cluster aggregation. This is shown clearly for reversible cluster aggregation, where the bonds between the particles have a finite lifetime: the fractal dimension of the clusters becomes the same as the one for lattice animals. Finally, the question of the cluster anisotropy is addressed by calculating angular correlations for cluster aggregation (CA) and particle aggregation (PA or DLA). The PA growth process leads to different scaling along and orthogonal to the growth direction, whereas CA is isotropic.

THE CLUSTERING OF CLUSTERS MODEL

The model for irreversible, diffusion-limited cluster aggregation is defined as follows:[2] N_o particles are randomly placed on a hypercubic d-dimensional lattice of length L. Each particle moves independently and randomly (hopping to nearest neighbors). If two particles sit on neighboring sites, a bond is placed between them—they form a dimer. The dimer diffuses like the particles and larger clusters form in the same way. As the bonds are never broken, this leads to ever larger, rigid clusters. The growth ends when all particles belong to one big cluster. No rotation and no restructuring is allowed. Aggregates of many particles can be analyzed as fractals, i.e., $m = R^D$. R is the radius of gyration of the aggregate and m its mass (number of particles). The fractal dimension D is $D = 1.42 \pm 0.04$ $(d = 2)$ and $D = 1.78 \pm 0.05$ $(d = 3)$. Physically, the diffusivity of the clusters depends on the cluster size. This can be taken into account by fixing the diffusive velocity to be

$$V(m) = m^\alpha$$

for a cluster of mass m. The parameters of the model are then N_0, L and α. The quoted fractal dimensions describe aggregation at low cluster concentration and for $\alpha \lesssim 0$, when large clusters diffuse slower than small clusters. This result agrees with experimental observations in aerosols and colloids.[3]

The following scaling arguments can be used to describe the static and dynamic properties of this model. Let $N(t)$ be the number of clusters at time t $(N(t = 0) = N_0)$. The average mass then is $\overline{m}(t) = N_0/N(t)$ and the average radius $\overline{R}(t) = \overline{m}(t)^{1/D}$. Scaling implies that a system of N clusters of size $\overline{R}$ in a box of length L can be described by N particles of unit size in a box of length $L/\overline{R}$. The initial cluster concentration ρ_0 becomes at time t effectively

$$\rho = \frac{N}{(L/R)^d} = \rho_0 \overline{m}^{(d-D)/D}.$$

For fractal clusters $(D < d)$, the concentration ρ increases steadily. The intercluster distance eventually becomes comparable to the cluster radius $\overline{R}$ (if $L \to \infty$), no matter how small ρ_0. The scaling regime relevant for colloids is $\rho \ll 1$, and will be considered first. The situation where $\rho \gtrsim 1$ will be discussed further below. Supposing that mean field theory is valid, similar arguments can be used to describe the time dependent scaling of clustering of clusters.[4] Here, of course, the diffusivity $v(m)$ will enter the formulas through the exponent α. The average mass scales in time like $\overline{m} \sim t^\theta$ with

$$\theta = 1/(1 - \alpha - \frac{d-2}{D}).$$

Direct numerical simulations confirm this behavior qualitatively for monodisperse systems, and $\tilde{\theta} < 1$.

KINETIC EQUATION APPROACH TO CLUSTER AGGREGATION

The Smoluchowski equation has been used very successfully as a mean field description of irreversible cluster aggregation. The cluster size distribution, the number $N(m)$ of clusters of mass m, evolves in time according to

$$\frac{dN(m)}{dt} = \frac{1}{2} \sum_{m'+m''=m} K(m',m'')N(m')N(m'') - \sum_{m'} K(m,m')N(m)N(m'),$$

where the kernel $K(m,m')$ contains all the physics. In order to describe clustering of clusters in terms of the Smoluchowski equation, the fractal geometrical aspects have to determine the kernel. For a scaling analysis, one attempts to express the size distribution $N(m)$ in terms of a scaling function $p(x)$

$$N(m,t) \sim \bar{m}^{-2} p(m/\bar{m}),$$

which, for large times, is time-independent and universal.[4-6] The time dependence then enters only through $\bar{m}$. Using this scaling Ansatz for $N(m)$ and a homogenous kernel of degree 2ω,

$$K(\lambda m, \lambda m') = \lambda^{2\omega} K(m, m'),$$

i.e., $K(m,m') = (mm')^\omega$, the total number of clusters $N = \sum_m N(m) \sim t^{-\theta}$ or $\bar{m} \sim t^\theta$, with $\theta = 1 - 2\omega$. Comparing θ with the expression obtained above yields $2\omega = \alpha + (d-2)/D$. For monodisperse cluster size distribution ($p(x \to 0) \to 0$) there is agreement between $p(x)$ obtained from direct numerical simulations and $p(x)$ calculated from the Smoluchowski equation with the corresponding ω. For the polydisperse case, $\omega > 0$, care must be taken to distinguish between average and typical cluster size. The Smoluchowski equation distinguishes qualitatively between monodisperse ($\omega < 0$), polydisperse non-gelling ($0 < \omega < 1/2$) and gelling ($\omega > 1/2$). Varying α in the simulations lets one cover all three regions. In particular for $\omega > 1/2$ one finds that the fractal dimension is the same as DLA. The case where large clusters move faster than small ones corresponds to this situation. While the cluster-size distribution has the characteristics of gelation, this process does not actually describe gel formation, as we restrict ourselves to low concentration, $\rho \ll 1$. This condition fails before a gelling network appears.

KINETIC GEL FORMATION

As pointed out above, the density ρ increases monotonically, as long as $D < d$. This leads necessarily to the interpenetration of the clusters when $\rho \tilde{>} 1$. The growth then is not determined by the diffusive screening which causes two cluters to form bonds at the outermost tips at low concentration. Nevertheless, one may still attempt to use a scaling analysis.[7] The fractal dimension measured for $\rho \tilde{>} 1$ is $D = 1.75 \pm 0.07$ ($d = 2$), and the cluster size distribution is monodisperse when $\alpha \tilde{<} -0.40$. Consistent with the Smoluchowski picture, the time to form an infinite cluster is infinite. Experimentally, structures with a fractal dimension close to the one reported here have been observed in the aggregation of wax and latex spheres.[8] The model of cluster aggregation at high concentration effectively induces long range interactions across large, rigid clusters. This is probably valid experimentally only up to a certain cluster size, then the condition of total irreversibility must be relaxed.

REVERSIBLE CLUSTER AGGREGATION

In addition to the fact that, experimentally, growth can only be considered irreversible over some range of cluster sizes, it is also useful from a theoretical point of view to study the effect of relaxation of the clusters. The following model of reversible cluster aggregation has been investigated:[9] clusters aggregate in the same fashion as described above; they diffuse and form bonds whenever they touch each other. In addition, a fragmentation mechanism is introduced. Every bond now has a finite lifetime τ. If a bond breaks and a cluster breaks into two pieces, the two clusters diffuse independently of each other. In practice, every bond is cut with probability $1/\tau$ per unit time. Thus, aggregation is diffusion limited, fragmentation is random. Two situations have been considered, though the results do not depend on it. The clusters are either loopless (when in a collision, several bonds would form; one is picked at random) or with loops (when a bond is broken, a cluster only falls apart if no other bond connects the pieces).

Starting initially with single particles and τ large, the clusters grow until they reach a size where aggregation and fragmentation balance each other. The scaling analysis of the clusters has been performed in this regime of dynamic equilibrium. The fractal dimension of the clusters in $d = 2$ $(d = 3)$ is $D = 1.57 \pm 0.06$ (2.03 ± 0.05), both for clusters with and without loops. (Varying the lifetime τ varies $\overline{m}$ and hence $\overline{R}$). The mobility parameter α does not change this value appreciably, even for $\alpha > 0$. This indicates that there is no crossover analogous to the irreversible model. If one lets the clusters stick together with a sticking probability $p < 1$ (the actual value used is $p = 0.05$), the results do not change either. This independence of the exponent D with respect to the kinetics suggests that the fractal properties are those of static lattice-animals (which have the same D to within numerical errors). As the aggregation and fragmentation mechanisms are very different, there is no detailed balance in this process. A reduced cluster size distribution can be determined for this model as well. For every value of τ, a characteristic $\overline{m}(\tau)$ (and $\overline{R}(\tau)$) exist, from which $p(x)$ with $x = m/\overline{m}$ can be calculated. The shape of the size distribution is different from irreversible aggregation; notably the distribution is much broader for the monodisperse kinetics (i.e., $\alpha = -2$. It suggests that the Smoluchowski equation produces a different solution when a fragmentation term is added.[10]

INTERNAL STRUCTURE OF AGGREGATION CLUSTERS

The many simulations on growth processes indicate that the irreversible growth influences the structure in an important way. In order to quantify this statement in a specific example, both cluster (CA) and particle aggregation (PA) have been investigated for cluster anisotropy. This is motivated by the observation that PA grows out of a center, while CA does not. To measure the anisotropy, the correlation function $c(r)$ inside the cluster has been calculated as a function of the angle with respect to the center of the cluster.[11] In order to eliminate the effects of the underlying lattice, both parallel and diagonal correlations (with respect to the square lattice) were determined. In both CA and PA an amplitude anisotropy with respect to the lattice is observed. More importantly ,(for PA only), the scaling powers parallel and perpendicular to the direction of growth appear to differ:

$$c(r) \sim r^{-\eta}, \qquad \eta_\perp - \eta_\| = 0.16 \pm 0.05.$$

If this anisotropy persists asymptotically, the scaling region around a particle inside the cluster, where the correlation function $c(r)$ has scaling form, is narrower in the lateral direction. This suggests that DLA in different geometries could give a better idea of how the growth direction influences the resulting cluster.

[1] An introduction to random aggregation processes can be found in *Kinetics of Aggregation and Gelation*, eds. F. Family and D. P. Landau (North Holland, 1984).

[2] P. Meakin, Phys. Rev. Lett. **51**, 1119 (1983); M. Kolb, R. Botet and R. Jullien, Phys. Rev. Lett. **51**, 1123 (1983).

[3] S. R. Forrest and T. A. Witten, J. Phys. A **12**, 1109 (1979); D. A. Weitz and M. Oliveira, Phys. Rev. Lett. **52**, 1433 (1984).

[4] M. Kolb, Phys. Rev. Lett. **53**, 1653 (1984).

[5] R. Botet and R. Jullien, J. Phys. A **17**, 2517 (1984).

[6] T. Vicsec and F. Family, Phys. Rev. Lett. **52**, 1669 (1984); G. J. vanDongen and M. H. Ernst, CECAM Workshop on "Kinetic Models for Cluster Formation," J. Stat. Phys. **39**, 241 (1985).

[7] M. Kolb and H. Herrmann, J. Phys. A **18**, L435 (1985).

[8] C. Allain and B. Jouhier, J. de Physique Lett. **44**, L421 (1983); P. Ricetti, J. Prost and P. Barois, J. de Physique Lett. **45**, 1137 (1984).

[9] M. Kolb, preprint; R. Botet and R. Jullien (preprint) have considered restructuring of DLA clusters.

[10] E. M. Hendriks, CECAM Workshop on "Kinetic Models for Cluster Formation," J. Stat. Phys. **39**, 241 (1985).

[11] M. Kolb, J. de Physique LETT. **46**,L 631 (1985); similar anisotropy studies for DLA have been performed independently by P. Meakin, by T. Vicsek and by R. Voss (preprints).

BRANCHED POLYMERS

Mohamed Daoud

Laboratoire Léon Brillouin—CEN Saclay
F. 91191—Gif/Yvette Cedex, France

By polycondensing multifunctional units one may synthesize randomly branched polymers. Unlike aggregates that are discussed elsewhere,[1-3] these are not rigid but are very flexible. In the following we would like to discuss the conformation of these randomly branched polymers, or animals. One difficulty in this discussion resides in the synthesis itself, which takes place usually in a concentrated solution and leads to a wide distribution of molecular weights. Thus we will first discuss the conformation of a single animal, and later we will take into account the molecular weight distribution and show that it has important consequences on the results of any measurements. We will focus mainly on the exponents which appear in different power laws. All this discussion will be in the framework of the Flory theory which, although not correct, is usually a very good approximation.

THE SINGLE POLYMER

We first consider a single randomly-branched polymer made of N monomers in a good solvent. The only interaction is the excluded volume which prevents monomers from being on top of each other. One may write down a Flory free energy

$$F = \frac{R^2}{R_0^2} + v\frac{N^2}{N^d}, \tag{1}$$

where the first term is the elastic term and the second one the interaction term. v is the excluded volume parameter, R_0 the ideal radius when no interaction is present and R the actual radius. The mean field variation for the radius was calculated earlier by Zimm and Stockmayer[4]

$$R_0 \sim N^{1/4}\ell, \tag{2}$$

where ℓ is the length of the elementary unit. Minimizing F with respect to R leads to the following relation for D_A, the fractal dimension[5,6]

$$D_A = \frac{2}{5}(d+2), \qquad (3)$$

a result which is supposed to be exact for $d = 3$.[7]

DISTRIBUTION OF SIZES

The usual way to synthesize branched polymers is to start with a solution made of polyfunctional units, let them react for some time, and quench the reaction below the gelation threshold. This leads to a very wide distribution of sizes. The larger the molecular weight, the wider the distribution. This is easily understood in terms of percolation. The distribution of sizes corresponds to the cluster size distribution

$$P(N,\varepsilon) = N^{-\tau} f(\varepsilon N^\sigma), \qquad (4)$$

where τ and σ are the usual percolation exponents,[8] and $P(N,\varepsilon)$ the probability of finding a polymer with N monomers at a distance $p - p_c \equiv \varepsilon$ from the threshold. Usually it is very difficult to determine precisely the threshold and thus ε is not the best parameter for experimentalists. It is easier to determine the weight average molecular weight N_w, where

$$N_w \sim \varepsilon^{-\gamma}. \qquad (5)$$

At this point we describe the conventional experimental procedure. First the sol is synthesized as described above. The reaction is then quenched and the sol heavily diluted supposedly in a good solvent. Finally the measurement is made and leads to an average over the distribution of molecular weights $P(N,\varepsilon)$.

It is important to realize that after the reaction is quenched, this distribution is fixed and will not change anymore whatever is done to the solution. Because these are flexible polymers, the exact conditions of the solution are also crucial to the conformations. Whereas in the initial sol screening effects are important, they disappear upon dilution and each macromolecule swells. After the dilution is performed, the radius of each polymer is given by Eq. (3), whereas it is different before dilution.

We consider now two types of measurements that were actually made, namely those of the average radius of gyration and of the average diffusion coefficients. These were measured respectively by elastic and quasi-elastic light scattering. The same technique also gives N_w. What one gets then is the z-average:

$$\langle A \rangle_z = \frac{\int A(N) N^2 P(N,\varepsilon) dN}{\int N^2 P(N,\varepsilon) dN}, \qquad (6)$$

where $A(N)$ is either the radius $R(N)$ or the diffusion coefficient $D(N)$. For the latter we assume a Zimm behavior

$$D(N) \sim R^{-1}(N).$$

Performing the averages[9] leads to

$$\langle R^2 \rangle_z \sim N_w^{1.25} \tag{7}$$

$$\langle D \rangle_z \sim N_w^{-0.63}. \tag{8}$$

The experiments were performed by Leibler and Schosseler[10] on a polystyrene solution crosslinked by irradiation with γ rays and by Candau et al[11] on a polystyrene solution cross-linked with divinyl benzene. Their results are respectively

$$\langle R^2 \rangle_z \sim N_w^{1.16}$$

and

$$\langle D \rangle_z \sim N_w^{-0.58},$$

in good agreement with (7) and (8).

Thus we conclude that percolation seems to be a correct model for describing some gelation cases, and that the configuration of the macromolecules of the sol are very sensitive to the exact conditions of the solution.

SWELLING OF A GEL

The same arguments as above are also valid for the swelling of a gel. Suppose we start with a gel in its reaction bath, and then add solvent. It is known experimentally that the gel may swell in such a way that the ratio Q of its final to initial volumes may be on the order of 10 or more. This may be explained in the following way. In the initial state, the sol screens out the excluded volume interaction as above, and the fractal dimension of the gel is that of percolation. For three-dimensional systems in the Flory approximation the mass in a volume element with radius ξ_i is

$$M \sim \xi_i^{5/2}, \tag{9}$$

where ξ_i is the characteristic distance in the percolation problem and is related to the gel fraction G

$$\xi_i \sim G^{-\nu/\beta} \approx G^{-2}. \tag{10}$$

After an excess solvent has been added, we may reasonably assume that the sol fraction has been washed out of the gel. Then the screening effects are no longer present and the gel has the same fractal dimension as random animals. If we consider the same mass as above in relation (9) it is now in a volume with radius ξ_f, with, in d=3,

$$M \sim \xi_f^2. \tag{11}$$

From (9) and (11) we get

$$\xi_f \sim \xi_i^{5/4}$$

and

$$Q \sim \left(\frac{\xi_f}{\xi_i}\right)^3 \sim \xi_i^{3/4} \sim G^{-3/2}, \tag{12}$$

which shows that the swelling ratio Q may become very large when the gelation threshold is approached.

A final result we would like to mention here concerns the swelling of vulcanized linear polymer chains initially in a melt.[12] Here one starts with long linear polymer chains made of z monomers in a melt. These chains are crosslinked in order to get a gel. It has been shown that percolation describes this gelation process, but that the width of the critical reg should be observed. The main difference from the previous case is that scaling breaks down and that several networks are interpenetrating now, instead of a single network in the critical case. The number of interpenetrating networks is

$$I \sim (\varepsilon z^{1/3})^{3/2}, \tag{13}$$

where the distance ε to the threshold has to be larger than $\varepsilon^* \sim z^{-1/3}$ in order to be out of the critical region. As a consequence, the swelling ratio Q is

$$Q \simeq I\left(\frac{\xi_f}{\xi_i}\right)^3 \tag{14}$$

$$\sim (qG^{1/2})^8 \qquad (G > G^* \sim \varepsilon^*), \tag{15}$$

where G is the gel fraction and q the ratio of the distance

$$q = \frac{\xi_f}{\xi_i} \sim z^{1/10} G^{-1/2}. \tag{16}$$

For $G < G^*$ one crosses over to the critical behavior discussed above. It is important to note that because of interpenetration, affineness is no longer valid and Q has a very large dependence in q.

[1] T. A. Witten, this book.
[2] P. Meakin, this book.
[3] H. E. Stanley, this book.
[4] B. H. Zimm and W. H. Stockmayer, J. Chem. Phys. **17**, 1301 (1949).
[5] J. Isaacson and T. C. Lubensky, J. Physique Lett. **41**, 469 (1980); P. G. deGennes, Comptes Rendus Ac. Sci. (Paris) **291**, 17 (1980).
[6] M. Daoud and J. F. Joanny, J. Physique **42**, 1359 (1981).
[7] G. Parisi and N. Soulas, Phys. Rev. Lett. **46**, 871 (1981).
[8] D. Stauffer, *Introduction to Percolation Theory* (Taylor and Francis, 1985).
[9] M. Daoud, F. Family and G. Jannink, J. Physique Lett. **45**, 199 (1984).
[10] L. Leibler and F. Schosseler, J. Physique Lett. **45**, 501 (1984).
[11] J. Candau et al., Proc. Microsymposium on Macromolecules, Prague 1984.
[12] M. Daoud and G. Jannink, to be published.

DYNAMICS OF AGGREGATION PROCESSES

Fereydoon Family

Department of Physics, Emory University
Atlanta, Georgia 30322

The time dependent cluster size distribution function and its moments are used to describe the dynamics of aggregation processes. In particular, the scaling theory for the cluster size distribution in both irreversible and steady-state aggregation phenomena is reviewed.

The formation of clusters by aggregation of particles or the clusters themselves is the main process in many phenomena of practical importance in physics, chemistry, biology, medicine and engineering.[1] A widely-used method in describing the dynamics of such systems is the determination of the cluster size distribution function.[2,3] This quantity describes the time-dependence of the number of clusters of a particular mass and can be measured experimentally. In this talk I will describe the time-dependent properties of aggregation phenomena through the scaling theory for the cluster size distribution function and its moments.

In general, the kinetic aggregation processes can be divided into two types depending on their temporal evolution. The first type is irreversible aggregation with a permanent evolution in time in which the total number of clusters is always decreasing. Flocculation processes leading to the formation of low density colloidal aggregates are the best example of irreversible aggregation. The second class of kinetic aggregation processes are those which reach a steady state with a constant number of clusters after a long time. This type of aggregation occurs in stirred tank reactors for aerosols in which single particles are constantly fed into the system and the large clusters are removed. In this review I will describe the dynamics of these two types of aggregation through the scaling theory for the cluster size distribution function and its moments.

IRREVERSIBLE AGGREGATION

A model which embodies many of the physical features of an irreversible aggregation process and is well suited for the investigation of the dynamic cluster size distribution is the cluster-cluster aggregation (CCA) model, recently introduced by Meakin (4) and independently by Kolb et al (5). I will not discuss the CCA model in detail here because it is discussed elsewhere in this volume. Briefly, in CCA, initially N_0 particles are randomly distributed on a periodic d-dimensional lattice. Particles are then picked at random and moved by one lattice spacing in one of the q equally probable directions, where q is the coordination number of the lattice. If a particle becomes the nearest-neighbor of another particle, they are permanently bound together and form a two-particle cluster. These clusters and the single particles continue to diffuse and join together to form larger clusters. Clusters grow and become larger and larger until eventually a single large cluster remains in a finite system.

There are various factors influencing the aggregation processes. One important physical parameter is the cluster mobility.[7,8] In order to take the cluster mobility into account we assume[7,8] that the diffusion coefficient $D_s(\gamma)$ of a cluster of size s is given by

$$D_s(\gamma) = D_0 s^\gamma, \tag{1}$$

where D_0 and γ are constants. Various types of cluster mobility can be studied by varying the index γ.

Other factors such as chemical reactivity, kinetic energy, mass, etc. of the clusters also influence the aggregation process.[9] Depending on these parameters the coagulation of two clusters may or may not take place during a collision. In the simulations these effects can be accounted for by assuming that the sticking probability $P_{ij}(\sigma)$ of two clusters of size i and j is given by[9]

$$P_{ij}(\sigma) = P_0 (ij)^\sigma, \tag{2}$$

where P_0 and σ are constants. By varying P_0 and σ one can study chemical reactivity and other effects on the aggregation process.

The dynamics of irreversible aggregation processes can be described through the cluster size distribution function $n_s(t)$ which is the total number of clusters of size s, $N_s(t)$, divided by the volume of the system L^d. We begin with the assumption[3,7] that for sufficiently large s and t the cluster size distribution has the scaling form

$$n_s(t) \sim s^{-\theta} f(s/t^z), \tag{3}$$

where z is the dynamic exponent describing the divergence of the characteristic cluster size in CCA. The normalization condition

$$\rho = \sum_s s n_s(t) \sim \int s n_s(t) ds \tag{4}$$

implies* that $\theta \equiv 2$ and[3,7]

$$n_s(t) \sim s^{-2} f(s/t^z). \tag{5}$$

---

* A pedagogic discussion of this derivation is given by Stauffer in this book.

The scaling function $f(x)$ toward zero for both $x \ll 1$ and $x \gg 1$ and therefore $n_s(t) \sim s^{-2}$ only for $x \sim 0^1$. This implies that the envelope of the cluster size distribution functions for various times is a straight line having a slope equal to -2. This results has been verified in the studies[7,9] of both diffusion-limited CCA[7] and the chemically controlled CCA.[9]

The mass dependent diffusion coefficient $D_s(\gamma)$ and the sticking probability $P_{ij}(\sigma)$ have a similar effect on the general shape of the cluster size distribution function.[7,9] For $\gamma \gg 0$ or $\sigma \gg 0$, relatively many small clusters remain in the system because the formation of large clusters is favored by having a higher diffusivity, or a higher sticking probability, respectively. On the other hand, for $\gamma \ll 0$ or $\sigma \ll 0$, small clusters die out rapidly by joining together and forming larger clusters. In this case the distribution of clusters is monodispersed and the peak in the distribution shifts to larger s with time. Consequently, as γ (or σ) is decreased from $\gamma \gg 0$ to $\gamma \ll 0$ (or $\sigma \gg 0$ to $\sigma \ll 0$) at a value of $\gamma = \gamma_c$ (or $\sigma = \sigma_c$) the shape of the cluster size distribution function changes from a monotonically decreasing function to a bell-shaped function.[7,9] Figure 1 shows a schematic log-log plot of $n_s(t)$ versus s in the two different regimes. These results imply that the form of the scaling function $f(x)$ in (5) depends on the dynamics of the aggregation process and is different in the two regimes.

For $\gamma > \gamma_c$, or $\sigma > \sigma_c$, the cluster size distribution decays as $s^{-\tau}$ where the exponent τ depends on γ and σ. In this regime $f(x)$ must be of the form

$$f(x) = x^{2-\tau} g(x) \qquad (\gamma > \gamma_c \text{ or } \sigma > \sigma_c), \tag{6}$$

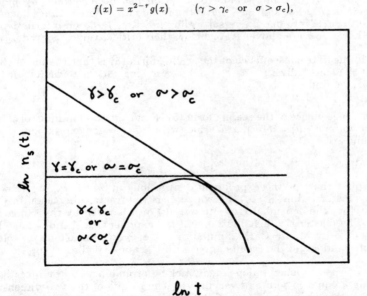

FIG. 1: Schematic plot of $n_s(t)$ versus t in a log-log plot. The characteristic shape of $n_s(t)$ depends on the dynamics of aggregation.

with $g(x) \simeq 1$ for $x \ll 1$ and $g(x) \ll 1$ for $x \gg 1$. Substituting (6) in (5) we find [3]

$$n_s(t) \sim t^{-w} s^{-\tau} g(x) \qquad (\gamma > \gamma_c \text{ or } \sigma > \sigma_c), \qquad (7)$$

with the scaling relation [3]

$$w = (2 - \tau)z. \qquad (8)$$

On the other hand, for $\gamma < \gamma_c$, or $\sigma < \sigma_c$, the cluster size distribution has no power law decay in s. In this regime $f(x)$ must have a form which would cancel the s^{-2} factor in (5). Therefore, $f(x)$ has the form

$$f(x) = x^2 h(x) \qquad (\gamma < \gamma_c \text{ or } \sigma < \sigma_c), \qquad (9)$$

with $h(x)$ decaying faster than any power of x in both $x \ll 1$ and $x \gg 1$ regimes. Substituting (9) in (5) we find

$$n_s(t) \sim t^{-w} h(x) \qquad (\gamma < \gamma_c \text{ or } \sigma < \sigma_c), \qquad (10)$$

with the scaling relation

$$w = 2z. \qquad (11)$$

Since $h(x)$ goes to zero for both small and large x, the t^{-w} dependence of $n_s(t)$ occurs only in the small scaling range of $x \sim 0^1$, and therefore $-w$ is the slope of the straight line which is tangent to a set of $n_s(t)$ curves for various values of s. The above dynamic scaling description has been verified by plotting $s^2 n_s(t)$ versus s/t^z. For a given γ and σ the results for different times have been shown[7,9] to scale into one universal scaling function $f(x)$ whose form depends on whether γ or σ are above, at, or below their critical values, γ_c and σ_c.

An immediate consequence of the scaling form (5) is that the mean cluster size $S(t) = \Sigma s^2 n_s(t)$ diverges as $t \to \infty$. Expressing $S(t)$ in terms of $n_s(t)$ and using (5), we find [3]

$$S(t) \sim t^z. \qquad (12)$$

Another consequence of the scaling form (5) is that the total number of clusters in a unit volume, $n(t) = \Sigma n_s(t)$, also scales with time. Using expressions (5) and (6) we find [10]

$$n(t) \sim \begin{cases} t^{-z} & \tau < 1 \\ t^{-w} & \tau > 1. \end{cases} \qquad (13)$$

This implies that for the experimental situations in which $D_s \sim 1/R$, where R is the cluster radius, $\gamma = -1/d_f$, where d_f is the fractal dimension in CCA, and therefore the dependence of $s(t)$ and $n(t)$ on t is given by the same exponent z. The cluster size distribution and the exponents z, τ and w have been measured[3,7,9] by Monte Carlo simulations in $d = 1$-3 dimensions for various values of γ and σ and the results agree with the above scaling theory.

The effects of a mass independent sticking probability can be investigated by setting σ equal to zero and varying P_0. The results of the two-dimensional simulations are shown in Fig. 2 of Ref. 9, where the total number of clusters $N(t)$ is plotted against t in a log-log plot. The results indicate that initially $N(t)$ decays slowly for small P_0, but asymptotically it decays with the same exponent independent of P_0. This implies that perhaps for all finite P_0 there is a crossover time, which depends on P_0, at which the scaling behavior becomes the same as CCA with $P_0 = 1$. As $P_0 \to 0$, the crossover time goes to infinity. For $P_0 \equiv 0$, the scaling behavior becomes independent of the diffusion coefficient and crosses over to a new universality class. It has been argued[11] that this universality class is characterized by an exponentially diverging $S(t)$ and a $n_s(t)$ which decays as $s^{-3/2}$ for all γ.

STEADY-STATE AGGREGATION

In many processes of both practical and theoretical interest a steady-state cluster size distribution develops in the system. This is in contrast to irreversible aggregation discussed in the previous section which has a permanent evolution in time with a decreasing number of clusters in the system.

A steady-state in the distribution of clusters can be achieved by feeding single particles (or monomers) into the system and removing the larger clusters according to some rule.[12] In the model that we have investigated,[12] at every unit time k particles are added at k different sites selected randomly. In addition, a cluster is discarded as soon as it becomes larger than a previously fixed number s_r. This is an extreme version of the situation in which the larger clusters leave the system with a higher probability.

In the steady-state coagulation process, both the total number of clusters $N(t)$ and the number of particles $M(t)$ in the system go to a constant value (N_∞ and M_∞) for long times. It is easy to show that in the Smoluchowski equation approach[13] both N_∞ and the characteristic time τ needed for relaxing to the final state scale with the feed rate k, with the exponent $1/2$. In the more general case one can assume that $N_\infty \sim k^\alpha$ and $\tau \sim k^{-\beta}$. In this case α and β satisfy the relation $\alpha + \beta = 1$ independent of the dynamics.

The results[12] for the steady-state values of the number of clusters N_∞ and for the relaxation time τ indicate scaling of these quantities as a function of the feed rate k. The values of the exponents α and β can be determined from the slopes of the straight lines drawn through the data on the log-log plots of N_∞ and τ versus k. The numbers we obtained for α and β depend on the dimension of the space in which the CCA takes place but they seem to be insensitive to the mass dependance of the diffusion coefficient or the parameter s_r.

In one dimension we found that $\alpha = 0.33 \pm 0.02$ and $\beta = 0.65 \pm 0.03$. In order to explain these results we solved the one-dimensional case exactly[12] and found $\alpha = 1/3$ and $\beta = 2/3$ in excellent agreement with the simulations. In $d = 3$ the simulations gave $\alpha = 0.47 \pm 0.05$ and $\beta = 0.54 \pm 0.05$, in agreement with the mean-field theory prediction $\alpha = \beta = 1/2$. In two dimensions we found that with logarithmic corrections we find $\alpha = 0.52 \pm 0.04$ and $\beta = 0.46 \pm 0.05$.

The results for α and β in $d = 1, 2$ and 3 enable us to discuss the relevance of the Smoluchowski equation (SE) for describing steady-state CCA. One notable feature of the results is that $\alpha + \beta \simeq 1$ in the simulations, in agreement with the SE prediction. On the other hand, the SE result of $\alpha = \beta = 1/2$ is in agreement with the simulations in $d = 2$ only with taking logarithmic corrections-to-scaling into account and it disagrees with the results in $d = 1$. This implies that the upper critical dimension for the kinetics of the steady-state coagulation is perhaps two.

The scaling of N_∞ and τ with k can be used to develop a general scaling description for $N(t)$. For the case when $k/L^d \ll 1$ and the initial number of particles is very small we can write

$$N(t) \sim k^\alpha f(k^\beta t), \qquad (14)$$

where $f(x)$ is a scaling function with $f(x) \sim x$ for $x \ll L^d$ and $f(x) = 1$ for $x \gg L^d$. The actual shape of $f(x)$ may depend on the parameters γ or s_r but for a fixed set of these numbers $N(t)$ can be expressed through the scaling form (14). The results for $N(t)$ were found[12] to collapse into a single scaling form as predicted by Eq. (14) when $N(t)k^{-\alpha}$ was plotted versus $k^{1-\alpha}t$.

I would like to thank Paul Meakin and Tamàs Vicsek with whom most of the researches reviewed here were carried out. This research was supported by NSF grant No. DMR-82-08051.

[1] See, e.g., *Kinetics of Aggregation and Gelation*, eds. F. Family and D. P. Landau (North-Holland, Amsterdam, 1984).
[2] S. K. Friedlander, *Smoke, Dust and Haze: Fundamentals of Aerosol Behavior* (Wiley, New York, 1977).
[3] T. Vicsek and F. Family, Phys. Rev. Lett. **52**, 1669 (1984).
[4] P. Meakin, Phys. Rev. Lett. **51**, 1119 (1983).
[5] M. Kolb, R. Botet and R. Jullien, Phys. Rev. Lett. **51**, 1123 (1983).
[6] M. Kolb and H. J. Herrmann (unpublished, 1984).
[7] P. Meakin, T. Vicsek and F. Family, Phys. Rev. B **31**, 564 (1985).
[8] M. Kolb, Phys. Rev. Lett. **53**, 1653 (1984).
[9] F. Family, P. Meakin and T. Vicsek, J. Chem. Phys. (October 1985).
[10] T. Vicsek and F. Family, Ref. 1, p. 101.
[11] W. D. Brown and R. C. Ball, J. Phys. A **18**, L517 (1985).
[12] T. Vicsek, P. Meakin and F. Family, Phys. Rev. A **32**, 1122 (1985).
[13] M. von Smoluchowski, Z. Phys. Chem. **92**, 129 (1918); Physik Z. **17**, 585 (1916).

FRACTAL PROPERTIES OF CLUSTERS DURING SPINODAL DECOMPOSITION

Rashmi C. Desai and Alan R. Denton

Department of Physics, University of Toronto
Toronto, Ontario CANADA M5S 1A7

Recently a molecular dynamics (MD) simulation of the liquid-vapor phase separation of a 2 - d Lennard-Jones fluid has been performed by Koch, Desai and Abraham. Here we analyse their constant energy simulation data for fractal properties during the phase separation process. We find that the capacity dimension of the largest cluster (i) is nonEuclidean, (ii) is rather insensitive to important dynamical phenomena, and (iii) has a very slow time variation.

BACKGROUND. During the past decade, a variety of macroscopic phenomena occurring in nature have been analysed in terms of fractal dimensions and other fractal properties.[1-2] Often such phenomena are related to growth in complex systems. In order to enhance and unify our understanding of many diverse systems, a variety[2] of growth models have been introduced and detailed behavior explored via numerical simulations. For instance, Monte Carlo simulations have been used to explore the Eden, DLA (diffusion limited aggregation) and ballistic models, and recently a biased diffusion model has been introduced[3] which includes the effects of both diffusive and ballistic aspects in growth. Surface effects[4,5] are also expected to play an important role in growth and in determining the extent of the non-Euclidean nature of resulting aggregates. It is also known that instabilities[5] are often among the key causes for the observed fractal aggregates. It appears that irreversible growth far from equilibrium often gives rise to scale-invariant fractal objects. One of the fundamental querstions is how this scale invariance arises in systems far from equilibrium.

Here we describe an analysis of fractal properties of growing clusters in a far-from-equilibrium situation. Our study differs in two main respects from others mentioned above. First, in contrast to the Monte Carlo technique used in many growth models, our simulation uses the well-known molecular dynam-

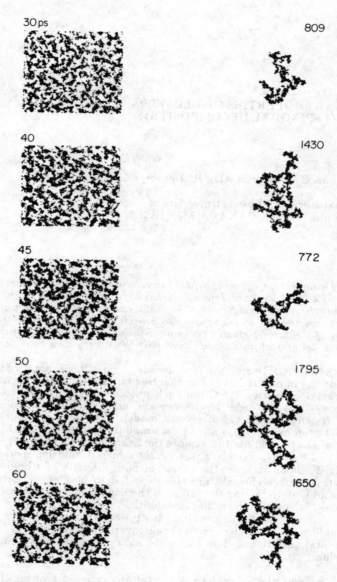

FIG. 1a: The left half shows the snapshot picture at the time indicated and the right half shows the largest cluster at that time with the number of atoms in it indicated at top right (see text).

ics (MD) technique, in which one models models many body dynamics by soplving the classical equations of motion. For a recent Monte Carlo study of spinodal decomposition in systems with long range forces, see Ref. 6. Second, our analysis of the MD data yields the time dependence of the fractal nature of the growing clusters.

The system is a very simple one: a one-component monatomic fluid of 5041 (71x71) atoms interacting via a Lennard-Jones potential and constrained to move in a two dimensional Euclidean geometry. The far-from-equilibrium situation is created by a "quench" of the homogeneous equilibrium fluid to an initial thermodynamically unstable state. Following the quench, the system beings to phase separate and evolves irreversibly towards an evential equilibrium state consisting of coexisting liquid and vapor phase separateed by an interface. Further details of the simulations are contained in Ref. 7. our analysis here is restricted to the "experiment C" of Ref. 7, which which the total number, total area and total energy of the system are held fixed (a non-equilibrium analog of the microcanonical ensemble).

The nonequilbrium phenomenon of phase separation in an initially unstable system is a first order phase transition and is referred to as spinodal decomposition. It is studied extensively[8] in binary mixtures (of metals, fluids, polymers, etc.) both in the laboratorya and via Monte Carlo simulation studies. In binary mixtures, the order parameter is the local concentratiorn fluctuation and the long wavelength hydrodymnamics description involves on conservation law. In contrast, in the case of liquid-vapor phase separation in a one component fluid, the local density fluctuation plays the role of the order parameter and the conserved quantities are the mass, (linear) momentum and energy densities. In a classical mean field desciption of the phenomenon,[9] the intial instability arises due to a negative thermodynamics derivative (of the chemical potential with respect to the local concentration in binary mixtures and of the pressure with respect to the local density in one-component fluids). Long wavelength fluctuations in the order parameter are responsible for driving the system to phase separate and nonlinearities are important in the time evolution. A linear stability analysis using fluctuating hydrodynamics (see Refs. 7,10) shows that the region of validity of a linear theory is at most a few correlation times. For the MD simulation discussed here, the initial quench astate is close to the triple point and for the parameters appropriate to argon, the correlation time is about 2 ps. During the spinodal decomposition and subsequent cluster growth, the broken translational symmetry[11] and the dynamic creation of interfaces with nonzero interfacial tension are clearly among the important physical effects that need detailed theoretical explanation.

RESULTS In the MD simulation,[7] it was found that the time evolution occurred in two main stages, each characterized by the prevailing mechanism of growth. The first stage was marked by the emergence of interconnected high-density and low-density regions forming a spatially periodic pattern with growing chateristic wavelength. Such behavior is characteristic of a spinodal decomposition and is seen in snapshot pictures of the system at early times (see Fig. la, left half). The second stage was characterized by the breakup of interconnected regions and subsequent growth of liquid clusters both by

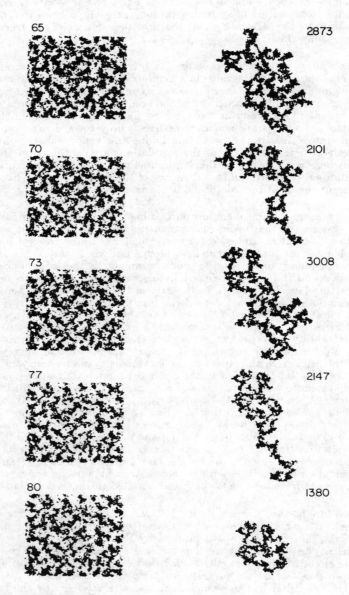

FIG 1b: Same caption as Fig. 1a.

monomer accretion from the low density vapor phase and by cluster coalescence. Figures lb and lc show (in the left half) respectively, the transition between the stages and the second stage. The right half in each of the figures la, lb, and lc shows the largest cluster at various times. Clusters are identified by the siple criterion that two atoms belong to the same cluster if they are separated by a distance less than some specified cutoff distance. Periodic boundary conditions are used to construct the cluster pictures. Note the dramatic necks, bulges, and voids in these clusters. Note also the occurrence of nick pinching in the transition stage of Fig. lb.

In order to investigate the possible fractal nature of hte system during its dynamic evolution, we have attempted to compute the time dependence of a variety of fractal dimensions which are in vogue.[1,12] These include the capacity dimension d_c[12b] for which the MD data give the least statistical error.[13] Since the fluid is homogeneous in both the prequench state and the final equilibrium state, we expect the Euclidean value of 2.0 for the fractal dimension at $t = 0$ and ∞. In Fig. 2 we show the time variation in d_c of the largest cluster. It is non-Euclidean and changes slowly from a value of about 1.4 at 30 ps to about 1.67 at 300 ps. The inset in Fig. 2 shows a typical log-log plot, from the slope of which d_c is obtained. The unit of the caliper length epsilon was chosen to be 1/200 of the MD cell size. The increase of d_c at early times reflects the increasing density of interconnected high density regions during the first stage of spinodal decomposition. At later times the increase in d_c indicates a tendency, as time advances, for the liquid clusters t to become increasingly compact (or less ramified). This process of compactification and the march toward Euclidean dimension is slow. The MD data goes only up to about 150 correlation times. We plan to investigate this process further in the future. Note from Fig. l that during roughly the first 100 corrleatiorn times, the system undergoes qualitative and dramatic changes in its morphology. The capacity dimension d_c is however oblivious to these changes.

In an interesting paper, Klein[14] has discussed a modified droplet model that includes the effects of ramified fluctuations. For systems with Euclidean dimension $d > 8$, he justifies the use of lattice-animal[12c] statistics for which the cluster size distribution is known. It is also known[12c] from Montge Carlo results for $2d$ site random animal radii that $1/d_f$ is about 0.65 to 0.6, where d_f is the fractal dimension obtained from the "mass-radius of gyration relation." If and d_c are equal, then our MD results for fractal dimension are consistent with this Monte Carlo result. Unfortunately from the MD data, we have not been able to determine d_f directly to a reasonable accuracy.

However, one may determine d_f indirectly from the density correlation function.[12d] In Ref. 7, a good scaling of the equal time correlation function was found (Fig. 13, Ref. 7). This scaled functionleads to a value of 1.7 ± 0.1 for d_f. If the scaling were exact, it would imply a time-independent fractal dimension. Thus the slow time variation of the fractal dimension (which eventually must approach the Euclidean value) signals a violation of the scaling of the equal-time density correlation function (Sec. VIII, Ref. 8) and its Fourier transform, the time-dependent structure factor which is commonly studied in laboratory experiments on spinodal decomposition.

One of us (RCD) would like to thank Farid Abraham, Stephan Koch and Jim Gunton for many discussions.

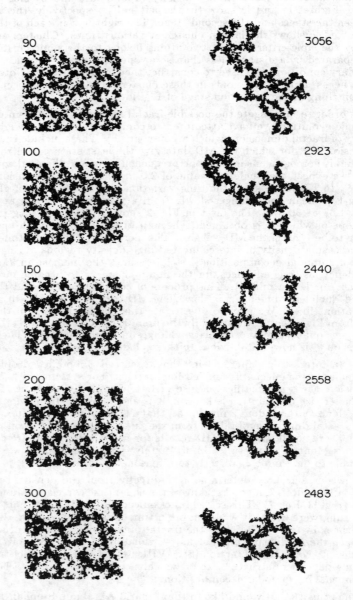

FIG. 1c: Same caption as Fig. 1a.

[1] B. Mandelbrot, The Fractal Geometry of Nature (Freeman, San Francisco, 1977).
[2] See other papers in these proceedings.
[3] R. Kapral, S. G. Whittington and R. C. Desai, 1985 preprint
[4] P. A. Rikvold, Phys. Rev. A 26, 647 (1982).
[5] T. Vicsek, Phys. Rev. Lett. 53, 2281 (1984).
[6] D. W. Heermann, Z. Phys. B 55, 309 (1984)
[7] S. W. Koch, R. C. Desai and F. F. Abraham, Phys. Rev. A 27, 2152 (1983).
[8] J. D. Gunton, M. San Miguel and P. S. Sahni, in *Phase Transitions and Critical Phenomena* (Academic, New York, 1983) eds. C. Domb and J. L. Lebowitz, Vol. 8, p. 267.
[9] An excellent review of limitations and validity of the mean field description is given in K. Binder, Phys. Rev. A 219, 341 (1984).
[10] S. W. Kaoch, R. C. Desai and F. F. Abraham, Phys. Rev. A 26, 1015 (1982).
[11] M. S. Jhon, J. D. Dahler and R. C. Desai, Adv. Chem. Phys. 46, 279 (1981); D. Jasnow Rep. Prog. Phys. 47, 1059 (1984); J. D. Gunton J. Stat. Phys. 34, 1019 (1984).
[12] (a) H. E. Stanley, J. Stat. Phys. 36, 843 (1984); (b) J. D. Farmer, E. Ott and J. A. Yorke, Physica 7D, 153 (1983); (c) D. Stauffer, Phys. Rep. 54, 1 (1979); (d) T. A. Witten and L. M. Sander, Phys. Rev. Lett. 47, 1400 (1981).
[13] Our algorithm for the capacity dimension was tested against two regular geometrical fractals, the Sierpinski gasket and Sierpinski carpet, and against a homogeneous $2d$ distribution. Compared to the exact values of 1.58, 1.89 and 2.00, our numerical results were 1.56 ± 0.05, 1.83 ± 0.01, and 1.99 ± 0.02 respectively. The error bars, here and in Fig. 2, correspond to the statistical errors in fitting a straight line to the numerical values.
[14] W. Klein, Phys. Rev. Lett. 47, 1569 (1981).

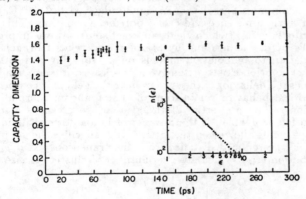

FIG. 2: Variation of capacity dimension d_c with time during the spinodal decomposition of a $2d$ Lennard Jones fluid. Inset shows a log-log plot of $n(\epsilon)$ vs. ϵ for the largest cluster at 90 ps. Capacity dimension is the negative of the slope.

KINETIC GELATION

David P. Landau

Department of Physics and Astronomy, University of Georgia
Athens, Georgia 30602

We describe a model for radical initiated, irreversible gelation in which kinetics are explicitly included. This model has been studied using computer simulation methods which allow determination of the cluster size distribution as well as the bulk properties. Although certain features of the behavior are percolation-like, other properties are quite different. We shall review the behavior of this model and relate the results to those for other models.

The phase transition from a sol (a collection of monomers which form only finite clusters) to a gel (a collection of monomers which includes an "infinite" cluster) has been studied for several decades. Theoretical descriptions have been provided from varying points of view. Early treatments modeled this transition by examining percolation on an unphysical lattice (Cayley tree) which ignored steric hindrence and cyclic polymer formation. This approach by Flory and Stockmayer[1] (which yields an exact solution) was improved upon in the mid 70's by de Gennes and by Stauffer who related branched macro-molecule formation to percolation models on a real, physical lattice.[2] (Although no exact solution exists, extensive numerical results are available.[3]) In both approaches the behavior of the system near the gel point can be described in terms of critical behavior reminiscent of that found in thermally-induced phase transitions. The properties of the system are given as a function of the conversion factor p which is the number of bonds formed divided by the total number of bonds allowed on the lattice. If p_c specifies the gel point, then $\varphi = (p - p_c)/p_c$ is the reduced distance from the transition. Near p_c the cluster size distribution $n_s(p)$, which give the number of clusters of size s, has the general form

$$n_s(p) = s^{-\tau} f(x), \tag{1}$$

where f is a monotonic function of $x = \epsilon s^\sigma$.

However $\tau = 2.5$ according to Flory-Stockmayer theory and is 2.2 for percolation on three-dimensional lattices. Various moments of the cluster size distribution can be related to the bulk properties of the model. For example, the second moment

$$\chi = \sum_s s^2 n_s \tag{2}$$

(where the "infinite cluster" is excluded from the sum) is closely related to the weight-average degree of polymerization. As $p \to p_c$,

$$\chi = c^{\pm} \epsilon^{-\gamma}, \tag{3}$$

where $\gamma = 1$ for Flory-Stockmayer theory and $\gamma \simeq 1.8$ for percolation. In addition, the amplitude ratio $R = c^-/c^+$, where c^- is obtained from $p < p_c$ and c^+ for $p > p_c$, is quite different for the two approaches being 1 for Flory-Stockmayer theory and ~ 10 for percolation in three dimensions. The fraction G of monomers which are in the "infinite" cluster may be defined by

$$G = 1 - \sum_s n_s s \tag{4}$$

(with the largest cluster excluded from the sum) and goes to zero as p_c is approached from above:

$$G = B\epsilon^{\beta}. \tag{5}$$

Here Flory-Stockmayer theory predicts $\beta = 1$ and three-dimensional percolation yields $\beta \simeq 0.44$.

Both theoretical approaches describe the behavior near the gel point in qualitatively similar terms but with very different critical exponents. Both theories also have shortcomings with regard to radical initiated gelation in that they allow bonds to be formed any place as a function of time whereas in some physical systems (e.g., mono-, bisacrylamide polymerization) bonds can only be formed at certain places due to a kinetic (chemical) process. Kinetic gelation models have thus been introduced which explicitely mimic this kinetic growth process and our goal is to determine how the gelation behavior compares with that obtained using other approaches.

We have studied[4] a model similar to that first proposed by Manneville and de Seze[5] which explicitly includes kinetic features which might be expected to be important for radical initiated irreversible gelation. In this model we consider $L \times L \times L$ simple cubic lattices with periodic boundary conditions. Individual lattice sites are then randomly occupied with either tetrafunctional monomers (with concentration c_4), bifunctional monomers (concentration c_2), or zerofunctional solvent monomers (concentration c_s) such that

$$c_4 + c_2 + c_s = 1. \tag{6}$$

The growth process begins with the random distribution of radicals (which we call initiators) on the lattice. The radicals dissociate and attack to the monomers by breaking a carbon double bond and leaving a free bond. Growth takes place with the random attachment of a free bond ("active center") to a nearest neighbor site with the result that the active center is now transferred to this newly bonded neighbor site. At each time step in the simulation an active center is randomly chosen and an attempt is made to form a bond to a nearest-neighbor site. If the neighbor is fully saturated however, (i.e., the number of bonds already formed equals the functionality of the monomer), no bond can be formed. If all the nearest neighbors of an active center are fully saturated, the active center becomes "trapped" and cannot participate in the growth process further. If an active center bonds to a site which contains another active center, then both are "annihilated." As bonds are added, clusters grow and merge and at p_c a cluster which spans the lattice emerges. The system is allowed to grow well beyond the gel point, and when the growth is terminated a new growth sample with a different distribution of initiators is begun. We have studied[4,9-12,15,20] this kinetic gelation model on lattices as large as $100 \times 100 \times 100$ with up to 2×10^4 growth samples over four orders of mangitude in initiator concentration c_I. In two dimensions[13] we have considered square lattices as large as 1600×1600 using in some cases infinite functional monomers to study the effects of eliminating trapping. This and similar models have now been studied extensively.[4-21]

The results of simulations reveal an interesting pattern of behavior. Finite size effects are quite important and become even more pronounced as c_I decreases. In comparison with simulation studies of thermally induced phase transitions we find that much larger lattices must be used for kinetic gelation investigations. We have therefore not included data from lattices smaller than $20 \times 20 \times 20$ in three dimensions and 200×200 in two dimensions. (For "small" c_I only larger lattice data were useful.) Using finite size scaling analyses of the bulk properties, i.e., gel fraction G and susceptibility χ, we find critical exponents β, γ which are, within our errors, the same as those found for percolation. Aharony[22] has suggested, however, that the amplitude ratio $R = c^-/c^+$ should be universal for models in the percolation universality class, while we find[4,20] values of R which are not only smaller than the percolation value but which also very with c_I. (An intriguing recent result[14] shows that if the initiators are placed periodically on a simple cubic lattice, then essentially the same critical exponents are obtained but the amplitude ratio R changes and becomes quite close to the percolation value.)

In addition, in three dimensions the cluster size distribution n_s does not show the monotonic behavior predicted by Eq. (1) but instead has very pronounced oscillations[10,20] with increasing cluster size which cannot be described by the droplet scaling form which appears to work so well for percolation. Maxima appear in n_s at integer multiples of a characteristic size s^*. (However, if the rate at which the initiators are created, i.e., the rate with which radicals dissociate, is slow compared to the bond formation rate, then the variation of n_s becomes monotonic.[11] The simple droplet scaling form still does not apply, however.) As bonds are added s^* grows linearly with p and there is no obvious change in n_s at the gel point. Finite size effects are also evident and tend to produce a spurious increase in the n_s for large s.[20] As c_I decreases p_c decreases as $c_I^{1/3}$ in three dimensions as long as no solvent and/or only moderate concentrations of bifunctional monomers are present. Once (quenched) solvent is introduced trapping becomes more important, and for sufficiently large c_s the system becomes completely trapped before a gel is ever formed.[9] The general role of trapping is still a question of great importance particularly for the case

$c_4 = 1$ (i.e., tetrafunctional units only) when the nature of the critical behavior as $c_1 \to 0$ (e.g., new universality class or merely crossover effects) is not unambiguously determined.

We have also recently studied[20] the fractal dimensionality d_f^{lc} of the largest (gel) cluster at p_c, its backbone d_f^{bb}, and its elastic backbone d_f^{eb} in three dimensions using the method of "burning."[23] Although it is difficult to obtain a large number of "well-formed" gel clusters and the range of lattice size which we could study was limited to $L \leq 60$, our results strongly suggest that d_f^{lc} and d_f^{eb} are the same for kinetic gelation as for percolation but that d_f^{bb} is different. This result again emphasizes the subtle difference between kinetic gelation and percolation.

This research has been carried out in collaboration with H. J. Herrmann, D. Stauffer, D. Matthews-Morgan, A. Chhabra, R. B. Pandey, S. Agarwal and N. Bahadur and has been supported by the National Science Foundation and the Program in Simulational Physics at the University of Georgia.

[1] P. J. Flory, J. Am. Chem. Soc. **63**, 3083,3091,3096 (1941); W. H. Stockmayer, J. Chem. Phys. **11**, 45 (1942); **12**, 125 (1944).
[2] P. G. de Gennes, J. Physique Lett. **37**, L1 (1976); D. Stauffer J. Chem. Soc. Faraday Trans. II **72**, 1354 (1976); Phys. Rev. Lett. **41**, 1333 (1978).
[3] D. Stauffer, *Introduction to Percolation Theory* (Taylor and Francis, London, 1985).
[4] H. J. Herrmann, D. P. Landau, and D. Stauffer, Phys. Rev. Lett. **49**, 412 (1982); H. J. Herrmann, D. Stauffer, and D. P. Landau, J. Phys. A **16**, 1221 (1983).
[5] P. Manneville and L. de Seze, *Numerical Methods in the Study of Critical Phenomena*, ed. I. Della Dora, J. Demongeat, and B. Lacolle (Berlin: Springer, 1981).
[6] N. Jan, T. Lookman, and D. Stauffer, J. Phys. A **16**, L117 (1983).
[7] A. Rushton and F. Family (unpublished).
[8] R. Bansil, H. J. Herrmann, and D. Stauffer, Macromolecules **17**, xxx (1984).
[9] D. Matthews-Morgan, D. P. Landau, and H. J. Herrmann, Phys. Rev. B **29**, 6328 (1984).
[10] A. Chhabra, D. Matthews-Morgan, D. P. Landau, and H. J. Herrmann, *Kinetics of Aggregations and Gelation*, ed. F. Family and D. P. Landau (North Holland, Amsterdam, 1984); J. Phys. A (in press).
[11] D. Matthews-Morgan and D. P. Landau, *Kinetics of Aggregations and Gelation*, ed. F. Family and D. P. Landau (North Holland, Amsterdam, 1984); and (to be published).
[12] R. B. Pandey and D. P. Landau, J. Phys. A **18**, L399 (1985).
[13] S. Agarwal, A. Chhabra, and D. P. Landau, Bull. Am. Phys. Soc. **30**, 486 (1985); and (to be published).
[14] N. Bahadur, D. P. Landau, and H. J. Herrmann, Bull. Am. Phys. Soc. **30**, 486 (1985); and (to be published).
[15] R. B. Pandey, J. Stat. Phys. **34**, 163 (1984).

[16] F. Family, Phys. Rev. Lett. **51**, 2112 (1983).
[17] T. Lookman, R. B. Pandey, N. Jan, D. Stauffer, L. L. Moseley, and H. E. Stanley, Phys. Rev. B **29**, 2805 (1984).
[18] D. Hong, H. E. Stanley, and N. Jan, Phys. Rev. Lett. **53**, 509 (1984); D. Hong, N. Jan, H. E. Stanley, T. Lookman, and D. Pink, J. Phys. A **17**, L433 (1984).
[19] H. J. Herrmann, D. C. Hong, and H. E. Stanley, J. Phys. A **17**, L261 (1984).
[20] A. Chhabra, D. Matthews-Morgan, D. P. Landau and H. J. Herrmann (to be published).
[21] A. F. Rushton, F. Family, and H. J. Herrmann, Proc. Conf. Macromolecular Dynamics 1982 (to be published).
[22] A. Aharony, Phys. Rev. B **22**, 400 (1980). This is also discussed by D. Stauffer in these Proceedings.
[23] H. J. Herrmann, D. C. Hong, and H. E. Stanley, J. Phys. A **17**, L261 (1984).

Sara Solla

DENDRITIC GROWTH BY MONTE CARLO

Janos Kertèsz, Jenö Szèp and Jòzsef Cserti

Institute of Theoretical Physics
Cologne University, D-5000 Koln, FRG
Eotvos University, Solid State Department
H-l088 Budapest Hungary

There is a class of pattern forming phenomena which can be described by the Laplace equation where the nonlinearity comes into the problem because of the moving boundary (which is the pattern itself). Many examples have been mentioned at this school; without seeking completeness we give a few references: the Saffman-Taylor instability,[1] dielectric breakdown,[2] flow through porous media[3] or dendritic crystal growth.[4] Here we want to deal with the last problem, but it should be emphasized that-mutatis mutandis-our method can be applied to different problems too.

There are several open questions concerning dendritic crystal growth, like mode selection, the role of anisotropy and fluctuations, or time evolution. The equations describing the phenomenon are well known,[4] but they are very complicated and therefore different approaches can be taken to attack the problem. One possibility is to build mathematically treatable models and to hope that they reflect the essential physics[5] or to construct simulations where the physically important features are represented by some parameters of the model.[6] On the other hand one can try to solve the equations numerically.[7] Our Monte Carlo (MC) simulation[8] is based on a well known numerical method of solving the Laplace equation and on a discretized version of the corresponding equations and therefore it should be considered to belong to the latter class of approaches.

We want to describe two-dimensional dendritic solidification in an (x,y) frame as grown from the $y = 0$ line in the positive y direction. Let us consider the following set of equations:[4]

$$\Delta u = 0 \tag{1a}$$

$$u(\underline{r}_\xi) = -d_0 k(\underline{r}_\xi) \tag{1b}$$

$$v_n(\underline{r}_\xi) = -D(\nabla u)\underline{n}(\underline{r}_\xi) \tag{1c}$$

$$u(y_0) = gy_0. \tag{1d}$$

Here u is the dimensionlesss temperature in the liquid, D the diffusion constant, d_o is the capillary length, the vector $\underline{r}_\xi$ is a point on the boundary parametrised by (we restrict ourselves to two-dimensional problems), k is the curvature at such a point, $\underline{n}$ is the normal vector and v_n the normal velocity of the interface, y_∞ some large distance, and g the value of u there. We neglected the time derivative in Eq. (1) which means that the diffusive relaxation is much faster than the growth of the crystal. In order to get instabilities, perturbations to Eq. (1) should be added, but due to the MC noise, they are always present in our simulation.

The standard MC method[9] solving the Laplace equation on a region S when the function is prescribed on the boundary ∂S is the following. N random walkers are started at a point $r \in S$ and at some time they hit the boundary

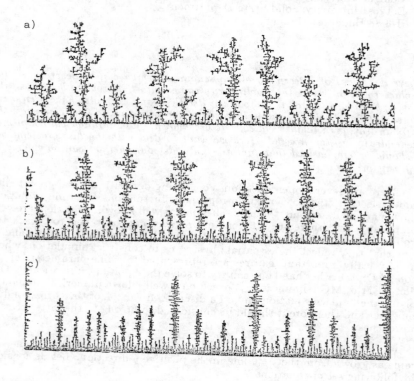

FIG. 1: Patterns generated in our model with d_o 0. Dg = (a) 0.07, (b) 0.01, (c) 0.001.

which is a sink for them. The average of the values of u at the terminating points will converge to the solution at r if $N \to \infty$. We apply this method to Eq. (1a) considering also Eqs. (1b-c) in a numerical way.[8] A similar approach was suggested by Ref. 10 to the Saffman-Taylor instability.

We discretise the problem by putting a square lattice mesh on the plane and occupied [empty] sites on it will belong to the solid [liquid] phase. Since we are interested in the shape of the interface which is governed by Eq. (1c) it is enough to determine u on the boundary, Eq. (1b), and next to it, in order to calculate the gradient in the rhs of Eq. (1c). In the discretised version of Eq. (1b) the curvature is determined from the occupation ratio of a 7×7 square.[6] Random walkers are started from the sites neighboring the boundary. One sweep along these sites means one MC step/site (MCS/s) which is the time unit. The information about the gradient is stored in an array F: if the walker hits the boundary the quantity D times

$$u(\text{on boundary next to the start}) - u(\text{at terminal}) \qquad (2a)$$

is added to it. After N MCS/s we have an approximation of

$$F = DN\nabla u. \qquad (2b)$$

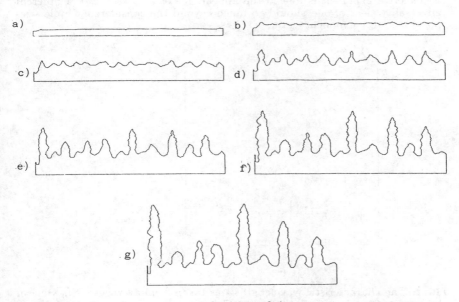

FIG. 2: Evolution of a pattern as a function of time: $d_o/g = 40$, $gD = 0.036$.

If at some site F becomes unity, this point is occupied. I.e., it freezes out. The velocity v of the boundary at this point is then

$$v = 1/N \qquad (3)$$

(distances measured in lattice units). For more details on the algorithm, see Ref. 8.

Let us first see the connection of this and another growth model, diffusion limited aggregation (DLA).[11] DLA is often referred to as the solution of Eq. (1) in the limit of vanishing surface tension $(d_o = 0)$.[12] But in the case of finite heat diffusion coefficient, we have an averaging effect due to the fact that the time N between two motions at the interface will be finite, and this can lead to a relevant smoothing of the pattern. For $D \to \infty, N \to 0$, i.e., the crystal grows infinitely fast [Eq. (3)]. Of course in the computer simulation $D \to \infty$ means that it is large enough to reach $F = 1$ whenever the difference in Eq. (2a) is positive. Only in this limit will our model coincide with DLA. All the patters in Fig. 1 were generated under the condition of vanishing surface tension. Fig. 1a (large D limit) has the same properties as DLA grown on a line substrate (diffusion controlled deposition[14,15]) and this is reflected also in the measured fractal dimension, $d_f \sim 1.7$. If smaller values of D are taken, the pattern and probably also the fractal dimension changes (Figs. 1b,c). The reason why the "clusters" are much less ramified and more elongated is that the finiteness of the diffusion constant slows the procedure and the attractive power of the gradient can better act. At the moment we do not know if the change in the effective exponent is due to the appearance of a crossover to a different universality class or one should go much beyond the considered sample sizes to see the DLA exponent again.

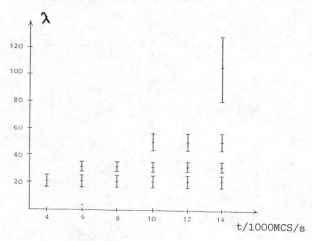

FIG. 3: The characteristic wavelengths λ as taken from averages of maxima in the Fourier transform of patterns like Fig. 2.

The nonzero surface tension is a stabilizing factor. Linear stability analysis tells us immediately what is the critical value q_o separating the small unsable wavenumbers from the large stable ones. Suppose there is a sinusoidal perturbabion of the solidification front with the small amplitude A and wavenumber q. If at a maximum of the wave, the value of u determined by Eq. (1b) is less than gA [g is the gradient, see Eq. (1d)], then the gradient is supporting the growth. This gives the instability criterion:

$$Ad_0 q^2 < Ag, \qquad (4)$$

leading to $q_0 = (g/d_0)^{1/2}$. For $q < q_0$ the systen is unstable and there is a roughly parabolic dispersion relation showing how fast is the initial exponential growth of the amplitude for a given wavenumber. One expects that at the beginning, the mode with the maximal speed will dominate if we start from a linear sample but apply a broad spectrum of noise (MC noise). In fact, our simulations confirmed this assumption: The measured wavelength which appeared first (cellular structure) was $q_0/2$ with an accuracy of 10 – 20%.

Figure 2 shows the time evolution of a dendritic pattern grown in our model, where the time between two snapshots was always 2000 MCS/s. Due to the MC noise fluctuations occur, then after some time a characteristic length is observable (Figs. 2c,d), corresponding to $q_0/2$. During further growth some branches become eventually larger and they screen the smaller ones in a "struggle for life" which leads to the appearance of longer characteristic wavelength (Figs. 2d,e). Figure 3 shows how the sequence of wavelengths appears as a function of time.

Further work is underway to get more accurate data, to study different geometries and the influence of diffusion in the solid and on its surface.

One of us (JK) thanks the Sonderforschungsbereich 125 for support, and D. Stauffer for reading the manuscript.

[1] J. Nittmann, G. Daccord, and H. E. Stanley, Nature **314**, 141 (1985).
[2] L. Niemeyer, L. Pietronero, and H. J. Wiesmann, Phys. Rev. Lett. **52**, 1033. (1984).
[4] L. Paterson, Phys. Rev. Lett. **52**, 1621 (1984).
[5] J. S. Langer, Rev. Mod. Phys. **52**, 1 (1980).
[5] E. Ben-Jacob, N. Goldenfeld, J. S. Langer, and G. Schon, Phys. Rev. Lett. **51**, 1930 (1983).
[6] T. Vicsek, Phys. Rev. Lett. **53**, 2281 (1984).
[7] D. A. Kessler, J. Koplik and H. Levine, Phys. Rev. **B30**, 2820 (1984).
[8] J. Szep, J. Cserti and J. Kertesz, J. Phys. A **18**, L413 (1985).
[9] P. R. Garabedian, Partial Differential Equations (Wiley, NY, 1964) p 483.
[10] L. P. Kadanoff, J. Stat. Phys. **39**, 267 (1985).
[11] T. A. Witten and L. M. Sander, Phys Rev. Lett. **47**, 1400 (1981).
[12] T. A. Witten and L. M. Sander, Phys. Rev. B **27**, 5686 (1983).
[13] P. Meakin, Phys. Rev. A **27**, 2616 (1983).
[14] Z. Racz and T. Vicsek, Phys Rev. Lett. **51**, 2382 (1984).

FLOW THROUGH POROUS MATERIALS

Jorge Willemsen

Schlumberger-Doll Research
Old Quarry Road
Ridgefield, CT 06877

My primary aim in this lecture is to bring to your attention a deterministic dynamical system which grows many of the patterns discussed at this school. Daccord and Guyon have discussed the current experimental situation, and I have nothing to add here except to note that the role of randomness of the porous material has been studied experimentally by J. D. Chen. I will start by reviewing some theoretical models which treat small-scale flow through porous media. These models are robust in the sense that they apply to a broad range of flow scenarios ranging from capillary dominance to viscous fingering in the case of unfavorable viscosity ratios. Next, I will discuss how special limiting situations can be treated through the use of simplifying assumptions. Invasion percolation is one example. Emergence of DLA-like viscous patterns is another.

Starting from ingredients scattered throughout the chemical engineering and petroleum engineering literature, the authors of Refs. (1) and (2) have put together a computational method appropriate for studying flow through porous media at the level of individual pores linked into a network. That is, first model a porous medium by a network which can be as random as you have patience for: put different sized pores, throats, lengths of throats, connectivities between pores, etc. For simplicity, my porous medium will be on a regular network, with no "pores" per se, but only pipes representing throats. These have radii chosen from some probability distribution.

Label the points where pipes intersect by node numbers "i,j,..". When a single fluid is flowing through the pipe linking i and j, the volume/sec of flowing fluid, J_{ij}, is governed by Poiseuille's law:

$$J_{ij} = \frac{A_{ij}^2}{\mu L_{ij}} (P_i - P_j) \qquad (1)$$

In this equation, P_i is the pressure at node i; A_{ij} is the area of the pipe linking i and j; L_{ij} is the length of the pipe; and μ is the viscosity of the fluid in the pipe, times a pipe-shape dependent constant.

Since fluid is conserved at every interior node i (those not source or sink nodes), $\sum_i J_{ij} = 0$. Denote the coefficient of $(P_i - P_j)$ as g_{ij}. Then current conservation implies that the pressures satisfy Kirchhoff's law at each interior node:

$$P_i = \frac{\sum_j g_{ij} P_j}{\sum_j g_{ij}} \tag{2}$$

Suppose next that there are two fluids in pipe ij at any given time. The simplest scenario is one in which the flow is piston-like, with the fluids separated by a meniscus. Introduce an additional pressure drop from i to j given by the Laplace-Young expression, $P_C \propto \gamma/R_{ij}$, where R is the pipe radius, and γ is the coefficient of surface tension times the cosine of the contact angle of the meniscus with the solid. Next, introduce a "permeability" which reflects that the resistance to flow will depend upon the viscosities of both the fluids in the pipe. I will return to this point later. This completes the setup of the problem, but keep in mind that there may be many situations in which the actual fluid configuration in the pipe is not so simple (3).

At this stage, there are node equations like Eq.(2) for all internal nodes joined by only a single fluid. More complicated expressions describe pipes containing two fluids. Finally, external conditions such as constant influx of fluid or pressure differential must be specified. The crank can then be turned. The calculation is very time consuming for two reasons. First, one does not "solve the circuit" once only -- at each time step, fluids move around, and so create a new circuit which must be re-solved. But more frustratingly, when capillary forces are important, the interface tends to slosh back and forth as capillary and viscous pressures adjust locally.

The above method has been used to calculate flow in the regime when capillary pressures are very large compared to the average viscous pressure drops due to the flow. These calculations are slow, and we have only little quantitative information relevant to the regime. Nevertheless, Koplik has verified that the following intuition is largely correct: the vast majority of fluid displacement moves occur where the capillary pressure is most favorable. This intuition forms the basis for the "Invasion Percolation" model we will discuss later (4,5). IP is thus a shortcut for simulating flow in the capillary dominated regime.

Very recently, Wilkinson (6) has applied the model to study flow in the limit of vanishingly small viscosity for the displacing fluid, and vanishingly small coefficient of surface tension. He has found flow patterns which resemble those observed experimentally, that is, "DLA-like" (7,8). In more detail, suppose that at time t fluid I has advanced into a pipe a distance x from its entry node. By analogy with Ohm's law for electricity, the hydraulic resistance to flow consists of two elements in series. Thus, if the resistor is of length L, of which x has R_1, while (L-x) has R_2, the resistance of the whole is $R = R_1 \frac{x}{L} + R_2(1-\frac{x}{L})$. That is,

$$-\frac{\Delta P}{L} = \frac{J}{A^2}\left[\mu_I \frac{x}{L} + \mu_{II}(1-\frac{x}{L})\right] \tag{3}$$

Next, because J_{ij} is a volume/time, one has directly $J_{ij} = A_{ij}\Delta X_{ij}/\Delta t$. That is: at time t, the pressure drop across the node is known. From this, one calculates the amount by which the interface advances through the pipe in the interval Δt. This alters the "permeability" of the pipe, and one must compute the pressures for $t+\Delta t$ afresh. This determines the next advance of the interface, etc. All of this is general. The simplifying elements in Wilkinson's calculation are setting the viscosity of the invading fluid equal to zero, and neglecting capillary forces altogether.

We have, then, two limiting regimes of considerable interest: invasion percolation in the limit of capillary dominance, and "DLA" in the limit of infinite viscosity ratio. In principle, the model is capable of handling the crossover regimes, though it runs extraordinarily slowly in these regimes. We now discuss each limiting region in turn.

Invasion percolation is a dynamic process through which a cluster grows into a sample through selection of paths of least resistance. The physical basis for the model is that when capillary forces totally dominate viscous forces, a displacing fluid tends to penetrate a sample saturated with another fluid at points where the capillary forces are instantaneously most favorable.

For the purpose of performing computer simulations of the above process, resistances to invasion are assigned randomly to the sites or bonds of a regular lattice, and are held fixed throughout the process. The invader grows by selectively moving to(through) the best uninvaded site(bond) as determined by the resistance distribution. The displaced fluid is treated as incompressible: once it has been surrounded the invader cannot penetrate it further.

Much has been learned about IP through computer simulations, but I must refer the interested student to Ref. (9) for details. In brief, the invading cluster is a fractal, with dimension different from that of static percolation in $d=2$, but equal in $d=3$ and above. In addition, the "trapped" clusters of defending fluid have a power-law size distribution as described by Stauffer in his lectures.

Since the IP growth cluster is fractal, the sample size sets the only natural length scale. This observation can be used to delineate the domain of applicability of the IP picture. For example, when buoyancy is taken into account, Wilkinson (10) and de Gennes (11) have independently estimated how a second length scale enters, effectively cutting off the size distribution of stranded oil blobs. The buoyancy estimates have been extrapolated to the case that viscous forces are reinstated into the dynamics, and although it is my opinion that these estimates are on less firm ground than those for buoyancy, the important thing is that IP supplies a point of departure for studying how typical sizes arise under realistic flow conditions.

Passing now to the viscous domain, a recurring theme in many of the tutorial lectures has been that DLA might be relevant to almost any growth phenomena involving solutions of the Laplace equation, and in which multiple time scales exist. We shall examine more carefully how the flow situation might be related to DLA.

First, write a master equation for diffusion which takes into account bond dependent hopping probabilities σ as follows:

$$n(i,t+\Delta t) = n(i,t) + \sum_{\{j\}} \sigma(j \rightarrow i) n_j(t) - \sum_{\{j\}} \sigma(i \rightarrow j) n_i(t) \qquad (4)$$

At steady state, Kirchhoff's law will be satisfied by the concentrations n_i, and this condition can be identified with Eq. (3) through the substitutions $P_i = n_i / \sum_{\{j\}} g_{ij}$, $\sigma(i \rightarrow j) = g_{ij} / \sum_{\{j\}} g_{ij}$. Thus, random walks may be used to solve for the pressures in the flow problem, provided that a steady state is achieved. Having done this, one may advance the interface according to the rules described earlier, and continue the calculation by performing more random walks. There is clearly nothing "wrong" with this - one may choose whatever method is appealing to solve the Laplace equation. But the advancement of the interface described above is not at all like the accretion of particles onto a DLA cluster. So what does the one process have to do with the other?

Heuristically, the key to understanding the connection between the two processes resides in $J \propto \Delta x / \Delta t$. According to this equation, the inviscid fluid advance is governed by both the pressure drop and the time step Δt. But there is no reason to identify this time step with the diffusional time step (one computer step). Suppose on the contrary that the flow time step corresponds to the random walk relaxation time (time to achieve steady state.) In that case the pressure drops will be approximated accurately. In addition, we can replace the deterministic growth process by a stochastic one, assigning a probability for a site to fill proportional to the pressure drop and to the pipe dimensions.

It would seem, then, that the flow will resemble DLA if DLA is not keenly dependent on diffusion providing the limiting time for the aggregation process. But that is already well-known (see Meakin's lecture.) In the large, the structure of the aggregate is the same whether adhesion occurs on the first encounter, or only after many.

To summarize these remarks: Assume that the detailed computational algorithm (1,2) adequately describes the physics of the fluid flow, as IP and now Wilkinson's results suggest. In the viscously unstable regime, a stochastic computation can be formulated utilizing random walks, which should approach the deterministic result if the aggregation probabilities are small and appropriately weighted. This should give results similar to DLA to the extent that DLA is robust against changes in the aggregation probabilities. No "proofs" here, only plausibility arguments.

Having said this, interesting questions remain: Is DLA on a lattice with random diffusion constants (RDLA) the same as DLA? (Most likely.) Further, for flow, the number of times walkers must encounter the already present cluster before adhering depends on properties of the bridge ij to the cluster. Will this "correlation" of the diffusion with the adhesion probability alter the details of the cluster?

A more serious problem is that DLA clusters are fractal on lattices with equal hopping probabilities, yet Wilkinson finds that the growth is dendritic rather than fractal on such lattices. The fact that the inviscid fluid actually grows "from inside out" suggests that screening might act differently in flow than in DLA: the correct physical interpretation of screening here is pressure build-up. But precisely how the randomness produces the right pressure field is not understood.

Furthermore, the introduction of an infinitesimal viscosity in the growing fluid introduces a cutoff length scale beyond which the flow will not be a DLA-like fractal. Exploiting further the analogy between Ohm's law and Darcy's law, we expect the pressure drop across a fractal to scale with the radial extent R of the fractal as $\Delta P \propto R^{2-d+d_T}$. Here d_T is a transport exponent which I will not dwell on further, but which depends on the properties of the fractal.

On the other hand the pressure drop for compact flow would have been dominated by $\mu_D R^{2-d}$. There is less work done if the fluid flows through the fractal provided $R \lesssim (\mu_D/\mu_I)^{1/d_T}$. This is an extremely crude estimate, which must certainly be refined. Qualitatively, however, I believe this reflects the mechanism which introduces a cutoff in the extent of fractal behaviour.

For future work: A key feature of DLA is that very little loopiness occurs. In the case of IP the same is true in three dimensions at the breakthrough point, but trapping does occur past that point. Is there an analogous second regime in the case of viscously unstable fluid flows, given experimentally viable boundary conditions? More generally, what are the interesting quantities one can examine theoretically and experimentally which characterize the crossover regimes?

Finally, can we use the microcopic flow equations as a "field theory" which can be studied mathematically to explore onset of fractal behaviour and crossovers? I believe so, though nothing has been done in this area as of yet. Certainly one direction is to use random walk computer methods in order to speed up calculations. But it is also likely that theoretical analysis of the equations themselves will prove fruitful, perhaps along the lines discussed by Kadanoff. For the students: we have here a physical system rich in scientific possibilities which has only begun to be explored!

REFERENCES

1. J. Koplik and T. J. Lasseter, SPEJ 25, 85 (1985).
2. M. M. Dias and A. C. Payatakes, "Immiscible Microdisplacement of Non-wetting Fluids in Porous Media", Schlumberger-Doll preprint submitted to JFM, 1985.
3. R. Lenormand and C. Zarcone, "Two Phase Flow Experiments in a Two-dimensional Permeable Medium", Schlumberger-Doll Research Report, 5th Int. PCH Conf., Tel-Aviv, 1984; J. D. Chen, private communication.
4. R. Lenormand and S. Bories, C. R. Acad. Sci., Paris, 291, 279 (1980).
5. R. Chandler, J. Koplik, K. Lerman, and J. F. Willemsen, J. Fluid Mech. 119, 249 (1982).
6. D. Wilkinson and J. D. Chen, SDR preprint, submitted to Phys. Rev. Lett.
7. T. A. Witten and L. M. Sander, Phys. Rev. Lett. 47, 1400 (1981).
8. R. Lenormand and C. Zarcone, in "Kinetics of Aggregation and Gelation", Ed. F. Family and D. P. Landau, Elsevier (NY), 1984, pp. 177-180.
9. D. Wilkinson and J. F. Willemsen, J. Phys. A16, 3365 (1983).
10. D. Wilkinson, Phys. Rev. A30, 520 (1984).
11. P. G. de Gennes, private communication.

Alla Margolina

CRACK PROPAGATION AND ONSET OF FAILURE

Sara A. Solla

IBM Thomas J. Watson Research Center
Yorktown Heights, New York 10598, USA.

We introduce a statistical model to describe the propagation of cracks along a geological fault [1]. A scale invariant mechanism for growth and coalescence of cracks results in catastrophic failure at a critical value σ_c of the applied stress σ.

The model is based on the observation that a geological fault is essentially a plane separating two elastic media. Under the application of a sufficiently small shear stress σ the fault remains locked and there is no relative displacement of these two elastic media (stick regime). When the stress σ exceeds a threshold value σ_c the two elastic media move with respect to each other (slip regime). The propagation of a zone of failure is controlled by the existence of asperity barriers that lock the two elastic media together. The gradual breakup of strong asperities as σ_c is approached from below leads to macroscopic failure and results in a transition from stick to slip.

A geological fault is therefore modelled as a two-dimensional array of asperities with a statistical distribution of failure strengths σ_f. If a stress σ is applied to the fault, the fraction p of broken asperities is given by

$$p = \text{Prob}(\sigma_f \leq \sigma). \tag{1}$$

Asperities are due to a variety of mechanisms, from bends in the fault to good pieces of rock going across it, and exhibit different shapes, sizes, and characteristic strengths. We model such variety through a Weibull distribution [2] of failure strengths,

$$P(\sigma) = \text{Prob}(\sigma_f \leq \sigma) = 1 - \exp[-(\sigma/\sigma_o)^n], \tag{2}$$

where σ_o is a reference strength and n is the order of the distribution.

The next stage in the construction of the model is the introduction of a scale invariant mechanism for transfer of stress. When an asperity fails, the stress σ on the failed asperity is transferred to one or more adjacent asperities. A renormalization group method is then used to investigate the properties of a hierarchical model for the stochastic growth of fault breaks through induced failure by stress transfer. We restrict this discussion to the case of a one-dimensional array, and refer the reader to the application of these ideas to the two-dimensional case corresponding to a geological fault [1].

Consider a one-dimensional array of asperities. An external stress σ results in a fraction p of broken asperities or microcracks, which are the seeds for macroscopic failure. A hierarchical mechanism for stress transfer is introduced by grouping pairs of asperities into cells and calculating the probability p' that a cell will fail as a function of the probability p for asperity failure. We obtain

$$p' = p^2 + 2p(1-p) P_{2,1} . \qquad (3)$$

The first term describes a direct process in which the cell fails due to the failure of the two asperities it contains. The second term describes induced failure and arises when only one of the two asperities in the cell fails. The stress σ that the failed asperity can no longer support is transferred within the cell to the unbroken asperity, which might now break under stress 2σ. $P_{2,1}$ is the probability that $\sigma_f \leq 2\sigma$ given that $\sigma_f > \sigma$. Such conditional probability is easily calculated for a Weibull distribution, leading to :

$$p' = 2p[1 - (1-p)^{2^n}] - p^2 . \qquad (4)$$

This recursion relation is the basis for the construction of a hierarchical model [3]: pairs of cells are grouped together into larger cells and the recursion relation (4) is iterated. The critical point at

$$p_c = 1 - (1/2)^{1/(2^n - 1)} \qquad (5)$$

describes the transition from a regime in which failed regions remain bound and the fault is locked ($p < p_c$) into a regime in which the failed regions grow indefinitely and the fault cannot support the external stress ($p > p_c$). Results for the critical stress σ_c and the critical exponent ν that characterizes the growth of the finite failed regions as the threshold is approached from below depend only weakly on the order n of the Weibull distribution [3].

Specific features of catastrophic failure have been incorporated into generalizations of this model. An interesting example is the introduction of a mechanism for blocking crack propagation by using a Weibull distribution with two characteristic failure strengths σ_o and $\gamma\sigma_o$:

$$P(\sigma) = (1-a)\{1 - \exp[-(\sigma/\sigma_o)^n]\} + a\{1 - \exp[-(\sigma/\gamma\sigma_o)^n]\} . \qquad (6)$$

The investigation of the resulting recursion relation as a function of n, a , and γ reveals a regime characterized by the existence of two critical points. The lower critical stress $\sigma_{c,w}$ is associated with the macroscopic failure of the weak asperities. The propagation of cracks is contained by the presence of intact strong asperities. A higher threshold at $\sigma_{c,s}$ has to be reached for the catastrophic failure of the whole system. This closing remark illustrates one of the possible extensions of the model to incorporate further features of the physical systems to which these ideas apply.

Acknowledgment

This work was done in collaboration with R.F. Smalley, Jr. and D.L. Turcotte, Department of Geological Sciences, Cornell University, Ithaca, New York 14853, USA. I thank them both for a very enjoyable interaction.

References

1. R.F. Smalley, Jr., D.L. Turcotte, and S.A. Solla, J. Geophys. Res. 90 , 1894 (1985).
2. D.G. Harlow and S.L. Phoenix, Adv. Appl. Probability 14 , 68 (1982).
3. D.L. Turcotte, R.F. Smalley, Jr., and S.A. Solla, Nature 313 , 671 (1985).

Fereydoon Family

THE THETA POINT

Naeem Jan,* Antonio Coniglio, Imtiaz Majid and H. Eugene Stanley

Center for Polymer Studies and Department of Physics
Boston University, Boston, MA 02215 USA

We exploit the relationship between the limit of the $n \to 0$ of the n-vector model and self-avoiding walks (SAW) to relate the number of closed polygons, N of $N + 1$ links to the radius of gyration R_N of SAW's of N steps, i.e., $N(N + 1) \sim R_N^{-d}$ where d is the dimensionality of the space. The relationship also holds at the Theta point: $N_w(N + 1) \sim R_\theta^{-d}$ where R_θ is the radius of gyration of the interacting SAW's at the θ-temperature and N_w is the appropriately weighted polygon number. We show that a walk on the hull of the percolation clusters at the critical threshold P_c of the triangular lattice is identical to an interacting SAW and the critical properties of this walk are the θ-point critical properties.

The interacting polymer chain has received considerable attention[1-8] but as yet there has not been a satisfactory resolution of the problem for $d = 2$. The weakly interacting polymer chain is well understood both from the theoretical and experimental standpoint. The elegant argu;ment of Flory[1,2] where the excluded volume repulsion energy is equated to the entropic elastic energy leads to the relationship

$$\nu = 3/(2 + d), \tag{1}$$

where the exponent ν is defined by $R \sim N^\nu$, R is the end-to-end distance of the polymer or its radius of gyration and N the number of links. The relationship is exact in $d = 1$, 2 and 4 (the upper critical dimensionality) and agrees to within 2% of the best numerical results in $d = 3$. As the strength of the interaction is increased by lowering the temperature or, equivalently, by immersion of the polymer in a poor solvent the stable conformations are the collapsed states where ν is simply $1/d$. However the θ-point, which may be considered a tricritical point,[1] is the termination of a line of 2nd order phase transitions which occur for $T > \theta$. The upper critical dimensionality is therefore 3 and here mean-field exponents with logarithmic corrections are exact, i.e., $\nu = 1/2$ and $\gamma = 1$ where γ is defined by $C_N \sim \mu^N N^{\gamma-1}$ (C_N is the number of distinct walks of N steps and μ is the connectivity constant). The non-trivial case, therefore, is d equal to 2.

There is a well-defined relationship, through the $n \to 0$ limit of the n-vector model, between the radius of gyration of SAW and the number of SAW's that return to the origin.[1] The number of such polygons of $N + 1$ links is

$$\mathcal{N}(N+1) \sim R_N^{-d}, \tag{2}$$

where R_N is the radius of gyration of the N step walk. This relationship is also applicable to the case of interacting SAW's with the polygons weighted in the appropriate manner, but R_N is now given by

$$R_N(T) \sim N^{\nu(T)}, \tag{3}$$

where $\nu(T)$ is expected from general considerations to belong to one of three universality classes, i.e., for $T > \theta$, $\nu(T) = \nu_{SAW}$; for $T = \theta$, $\nu(T) = \nu_\theta$ and for $T < \theta$, $\nu(T) = 1/d$, the collapsed phase.

Now consider the set of clusters at the percolation threshold P_c of the triangular lattice. It is possible to show that a walk on the hull of these clusters is equivalent to an interacting SAW on the honeycomb lattice with nearest-neighbor and next-nearest neighbor couplings.[9] The clusters may be considered as closed polygons and on the general grounds of universality we expect that $\nu(T)$ for this walk belongs to one of the three classes enumerated above. Since it is a walk on the perimeter of percolation clusters at P_c, then $\nu > 1/d$ and there is strong numerical evidence[10,11] that ν for this walk is 0.57 ± 0.01. This walk on the perimeter of the incipient infinite cluster is in the same universality class as the indefinitely-growing self-avoiding walk (IGSAW) which also has $\nu = 0.57$.[12]

We have shown that walks on the hull of percolation clusters at P_c is identical to an interacting SAW. These walks are in a universality class distinct from ν_{SAW} and $\nu_{collapsed}$ of the interacting polymer chain and thus describes the critical properties at the θ-point. We thank L. Peliti and J. W. Lyklema for useful discussions.

[1] P. G. de Gennes, *Scaling Concepts in Polymer Physics* (Cornell University Press, Ithaca, 1953).
[2] P. J. Flory, *Principles of Polymer Chemistry* (Cornell University Press, Ithaca, 1953).
[3] A. Baumgartner, J. de Physique **43**, 1407 (1982).
[4] K. Kremer, A. Baumgartner and K. Binder, J. Phys. A **15**, 2879 (1982).
[5] A. L. Kholodenko and K. F. Freed, J. Chem. Phys. **80**, 900 (1984).
[6] M. J. Stephen, Phys. Lett. A **43**, 363 (1975).
[7] R. Viłlanove and F. Rondelez, Phys. Rev. Lett. **45**, 1502 (1980).
[8] J. A. Marqusee and J. M. Deutch, J. Chem. Phys. **75**, 5179 (1981).
[9] A. Coniglio, N. Jan, I. Majid and H. E. Stanley, to be published.
[10] R. F. Voss, J. Phys. A **17**, L373 (1984).
[11] A. Weinrib and S. A. Trugman, Phys. Rev. B **31**, 2993 (1985).
[12] K. Kremer and J. W. Lyklema, Phys. Rev. Lett. **54**, 267 (1985).

FIELD THEORIES OF WALKS AND EPIDEMICS

Luca Peliti

Dipartimento di Fisica, Università "La Sapienza"
Piazzale Aldo Moro 2, I-001895 ROMA, Italy
and GNSM-CNR, Unità di Roma

Field theories of random walks with memory and of epidemic processes are reviewed and their implications for the asymptotic behavior of such processes are briefly discussed. Stress is laid upon difficulties in extending this approach to more general growth processes.

Several processes of aggregation and growth produce fractal objects, which look the same, if they are observed at a larger or smaller scale. It is this property that makes it at all possible to describe them by means of a set of fractal dimensions such as those discussed by Stanley[1] in this School. Self-similarity is a quite striking property, which is also prominent in other phenomena, such as in critical fluctuations. It has been possible in this case to trace the origin of self-similarity to the existence of a fixed point of a renormalization group transformation: a transformation of the Hamiltonian describing the system that acts as a simple rescaling of its observable properties, such as correlation functions. This justifies the hope to find such a satisfactory explanation, along with a successful computational scheme, also for the problems which concern us here.

We are nevertheless still quite far from this goal. The purpose of my lecture is to present a few cases in which the renormalization group approach to aggregation phenomena has met with a success comparable with that found in critical phenomena. I shall also mention attempts to extend the method to different cases where such a success has not yet been found. I am still confident that the roots of self-similarity will one day be exposed by one formulation or another of a renormalization group transformation, but I think that some radically new ideas are needed to arrive at such a formulation.

I shall illustrate the application of the renormalization group by means of its field theoretical formulation, admittedly a quite esoteric method, but useful for its computational power and reliability. I sketch here the steps of its application to our problems. One first writes down a field theoretical model descibing the process. This can be done quite generally by means of the powerful Fock space technique for classical objects, introduced by Doi[2] and by Grassberger and Scheunert[3] (for a pedagogicaly introduction, see Ref. 4). One then makes an attempt to solve this model by means of perturbation theory. This is sensible for problems such as random walks with memory which have a regime which is adequately described by a noninteracting field theoretical model. This is the point at which this approach finds itself in difficulties when it is applied to diffusion-limited aggregation and related models. Random walk and epidemic models, on the contrary, behave essentially like free diffusion when space dimension is sufficiently high. There is an upper critical dimension d_c beyond which perturbative corrections to the asymptotic behavior are finite. Renormalization methods are introduced to control divergent corrections for space dimensions lower than d_c. They allow one to obtain divergence free expressions of observable quantities, by expressing them as functions of observable properties of the system at a certain renormalization wave number κ. By exploiting the arbitrariness of κ, one succeeds in expressing the properties of the system observed at larger and larger length scales by means of flow equations in parameter space: self-similarity is recovered when a fixed point of these flow equations is reached. Since details of these procedures may be found in standard textbooks, such as Ref. 5, I shall not dwell on the renormalization program, but only introduce the field theoretical models and sketch the results.

In Sec. 2 are sketched the Doi-Grassberger-Scheunert formalism and its application to the derivation of a field theoretical description of aggregation phenomena. The results of this approach to random walks with memory are reported in Sec. 3, whereas Sec. 4 is dedicated to epidemic models, of a class which encompasses some forms of time-dependent percolation. Diffusion limited aggregation and the Eden model are briefly mentioned in Sec. 5.

FOCK SPACE FORMALISM FOR CLASSICAL OBJECTS

This formalism[2-4] makes it possible to write down systematically field theoretical descriptions of birth death processes on a lattice. Let the microscopic state $|\underline{n}\rangle$ of the system be identified by the set $\underline{n} = \{n_{r,\alpha}\}$ of occupation numbers of particles of species α at the site r of a d-dimensional lattice. The macroscopic state $|\phi\rangle$ is then defined by the probability $\phi(\underline{n})$ of finding the system in the microscopic state $|\underline{n}\rangle$. All states are considered to be elements of a Hilbert space, and in particular:

$$|\phi\rangle = \sum_{\underline{n}} \phi(\underline{n})|\underline{n}\rangle. \tag{1}$$

The evolution of the state $|\phi\rangle$ is given by a master equation of the form

$$\frac{\partial |\phi\rangle}{\partial t} = L|\phi\rangle, \tag{2}$$

where the Liouvillian operator L is expressed in terms of the annihilation $a_{r,\alpha}$ and creation $\pi_{r,\alpha}$ operators defined by

$$a_{r,\alpha}|\ldots, N_{r,\alpha}, \ldots\rangle = n_{r,\alpha}|\ldots, N_{r,\alpha} - 1, \ldots\rangle, \tag{3}$$

$$\pi_{r,\alpha}|\ldots,n_{r,\alpha},\ldots\rangle = |\ldots,n_{r,\alpha}+1,\ldots\rangle. \tag{4}$$

These operators satisfy the usual commutation relations,

$$[a_{r,\alpha}, \pi_{r',\alpha'}] = \delta_{rr'}\delta_{\alpha\alpha'}. \tag{5}$$

The Hilbert space structure is introduced in such a way that the Hermitean conjugate $a_{r,\alpha}^+$ of the annihilation operator is given by

$$a_{r,\alpha}^+ = \pi_{r,\alpha} - 1, \tag{6}$$

corresponding to the "inclusive" scalar product of Grassberger and Scheunert.[3] One can then apply standard manipulations to solve Eq. (2) by means of path integrals (see, e.g., Ref. 6). One thus obtains a field theoretical model of the Martin-Siggia-Rose[7] type, identified by the Lagrangian

$$\mathcal{L} = \sum_{r,\alpha} \overline{\psi}_{r,\alpha}\dot{\psi}_{r,\alpha} - L[\overline{\psi},\psi], \tag{7}$$

where the classical fields $\overline{\psi}$ (imaginary), ψ (real) have taken the place of the operators a^+, a respectively. In going from Eq. (2) to Eq. (7) one should take care to express the Liouvillian as a normal product, with all annihilation operators on the right of the creation ones. The Lagrangian (7) is the starting point of perturbation theory. One separates its "free" part, bilinear in $\overline{\psi}$, ψ, from the rest, which is then treated as a perturbation. it is usually possible to simplify drastically the model by going at once to the continuum limit.

FIELD THEORIES OF RANDOM WALKS

It is easy to write down the Liouvillian of simple diffusion by considering that the basic process is one in which a particle is annihilated at site r of the lattice and created at one of its neighboring sites $r+e$. Denoting by λ the jump rate, we have

$$L = \sum_r [\lambda \sum_e (\pi_{r+e} - \pi_r)a_r], \tag{8}$$

leading, in the continuum limit, to the Lagrangian

$$\mathcal{L} = \int d^dr [\overline{\psi}\frac{\partial}{\partial t}\psi - D(\Delta\overline{\psi})\psi], \tag{9}$$

where $D = \lambda|e|^2$ is the diffusion constant and Δ is the Laplacian. This model is of course exactly soluble: in our context it is sufficient to integrate by parts the second term and to observe that the functional integration over the imaginary field $\overline{\psi}$ compels ψ to satisfy the diffusion equation. As a consequence, the Laplace(time) - Fourier (space) transform of the endpoint probability distribution function $G(r,t)$ for a walk that leaves the origin at time $t=0$ is given by

$$\tilde{G}(p,z) = \int_0^\infty dt e^{-zt} \int d^d r e^{ip\cdot r} G(r,t)$$
$$= [z + Dp^2]^{-1} \tag{10}$$

Let us now consider a kinetic version of the self-avoiding walk (SAW),[8] where the walker is suppressed (with a rate μ) if it steps on a site which it has already visited. One might suppose that the walker drops a poison on each site it is about to leave.[9] Indicating by W the walker and by P the poison, the Liouvillian is

$$L = \sum_r [\lambda \sum_e (\pi_{r+e,w} \pi_{r,p} - \pi_{r,w}) a_{r,w} + \mu(1 - \pi_{r,w}) \pi_{r,p} a_{r,w} a_{r,p}], \qquad (11)$$

and the corresponding Lagrangian, in the continuum limit, is given by

$$\mathcal{L} = \int d^d r [\overline{\psi}_w \dot{\psi}_w + \overline{\psi}_p \dot{\psi}_p - D(\Delta \overline{\psi}_w) \psi_w - \lambda(\overline{\psi}_w + e^2 \Delta \overline{\psi}_w) \overline{\psi}_p \psi_w + \mu \overline{\psi}_w (1 + \overline{\psi}_p) \psi_w \psi_p]. \qquad (12)$$

The second line represents the interaction terms. Power counting shows that (i) they do not produce infrared divergent contributions (and therefore do not modify the asymptotic behavior of a free random walk) if the dimensionality of the space is larger than four, (ii) that the leading contributions are due to the two cubic terms that do not involve derivatives. The asymptotic behavior of the walk is thus determined by the simplified Lagrangian

$$\mathcal{L} = \int d^d r [\overline{\psi}_w \dot{\psi}_w + \overline{\psi}_p \dot{\psi}_p - D(\Delta \overline{\psi}_w) \psi_w - \lambda \overline{\psi}_w \overline{\psi}_p \psi_w + \mu \overline{\psi}_w \psi_w \psi_p]. \qquad (13)$$

When one expands the correlation function of this model by means of Feynan graphs, one notices that the diagrams and their contributions correspond to those of the static theory of polymer statistics in a good solvent.[8,10] The structure and the results of the static renormalization group can be carried over without modification to this case. One obtains a swollen form for the coil, which is characterized by a fractal dimension $d_f = 1/\nu$, where ν is the exponent of the gyration radius, and is accurately given by the Flory expression $\nu = 3/(d+2)$. The expansion of the exponents in powers of $\epsilon = d_c - d$ is less reliable than the Flory expression, for reasons that are not yet fully understood.

A different situation is found for the so-called "true" self-avoiding walk, where the poison, instead of suppressing the walker, turns it away from visited. The interaction Lagrangian now reads

$$\mathcal{L}_I = \int d^d r [-\lambda \overline{\psi}_w \overline{\psi}_p \psi_w - \mu(\nabla \overline{\psi}_w) \cdot (\nabla \psi_p) \psi_w]. \qquad (14)$$

Because of the two extra gradients the supper critical dimsion is now shifted to two. The renormalization of this model is fairly complicated,[11] involving the introduction of a renormalized diffusion constant and of three coupling constants. One thus obtains logarithmic corrections to the asymptotic behavior in two dimensions. One can also set up an expansion of the fractal dimension of the coil, $d_f = 1/\nu$, in powers of $\epsilon = 2 - d$. The result of the ϵ-expansion is different from the Flory result (which however seems to reproduce the correct exponent in one dimension) and from numerical results on fractal lattices.[12] One may remark that adding also a very small probability of the walker being killed by the poison brings the model back, in general, to the univesrality class of the self-avoiding walk. By the same method, one may investigate the effects of quenched disorder, of long-range interactions with the poison, of poison diffusion, etc.

FIELD THEORIES OF EPIDEMIC PROCESSES

A very interesting epidemic process, introduced by Cardy,[13] has been shown by Janssen[14] and by Cardy and Grassberger[15] to represent a form of dynamic percolation. We consider units on a d-dimensional lattice that can be in one of three states: healthy (H), affected (A) or immune (I). An A site may affect one of its neighbors, with a rate λ if it is healthy or μ if it is immune and ma to immune with a rate ξ. The Liouvillian is then given by

$$L = \sum_r \sum_e \{\lambda(\pi_{r+e,A} - \pi_{r+e,H})\pi_{r,A} a_{r+e,H} a_{r,A}$$
$$+ \mu(\pi_{r+e,A} - \pi_{r+e,I})\pi_{r,A} a_{r+e,I} a_{r,A}]$$
$$+ \xi(\pi_{r,I} - \pi_{r,A}) a_{r,A} \}. \tag{15}$$

We may consider the extreme case in which the contagion rate μ of immunes vanishes. In a path integral formulation, one can then integrate over $\overline{\psi}_H$ obtaining:

$$\dot{\psi}_{r,H} = -\lambda \psi_{r,H} \sum_e (1 + \overline{\psi}_{r+e,A}) \psi_{r+e,A}. \tag{16}$$

In the same way one obtains a law of evolution of the immune field

$$\dot{\psi}_{r,I} = \xi \psi_{r,A}. \tag{17}$$

Inserting Eq. (16) in the Lagrangian one obtains a model that only involves the affected field. Going to the continuum limit, expanding in powers of the field and stopping at the lowest nontrivial order, we obtain, up to the redefinition of some constants, the Lagrangian,

$$\mathcal{L} = \int d^d r [\overline{\psi}_A \dot{\psi}_A - D\Delta \overline{\psi}_A \psi_A + \nu \overline{\psi}_A \psi_A - \gamma \overline{\psi}_A^2 \psi_A + g \overline{\psi}_A \psi_A \int^t d\tau \psi_A(\tau)]. \tag{18}$$

One can partially integrate with respect to time to express the action in terms of $\overline{\psi}_A$ and of the variable

$$\Phi = \int^t d\tau \psi_A(\tau). \tag{19}$$

The result reads

$$S = \int dt \mathcal{L} = \int dt \int d^d r [-\partial_t \overline{\psi}_A \partial_t \Phi + D \partial_t \overline{\psi}_A \Delta \Phi - \nu \partial_t \overline{\psi}_A \Phi$$
$$- \gamma \overline{\psi}_A^2 \partial_t \Phi - \frac{1}{2} g \partial_t \overline{\psi}_A \Phi^2]. \tag{20}$$

This action is symmetrical with respect to the transformation

$$\gamma \overline{\psi}_A(t) \longleftrightarrow \frac{1}{2} g \Phi(-t). \tag{21}$$

This symmetry is broken when the contagioon rate is sufficiently high. One interesting feature of this model is that it reproduces the statistics of percolation clusters in the static limit. One may check that the upper critical dimension is $d = 6$ and that the static exponents (in particular the fractal dimension $d_f = d - \beta/\nu$, where β is the exponent expressing the vanishing of the mass of the percolating cluster at the percolation threshold, and the fractal dimension of links, $d_{red} = 1/\nu$) are reproduced in an expansion in powers of $\epsilon = 6 - d$. The dynamical description also allows us to introduce an exponent z characterizing the vanishing of the contagion speed at the percolation threshold. Sites at a distance $\xi \sim |\xi - \xi_c|^{-\nu}$ are affected after a time $\tau \sim |\xi - \xi_c|^{-\nu z}$. This time may have the interpretation of a chemical distance between sites if we assume that the contagion tends to follow the shortest available path (and the exponent z is the same as that called d_{min} in Stanley's lectures). One obtains thereby the following expression for the spreading dimension d_s (called d_ℓ in Stanley) of percolation clusters,

$$d_s = d_f/z. \tag{22}$$

To first order in $\epsilon = 6 - d$ we obtain

$$d_s = 2 - \frac{\epsilon}{14} + O(\epsilon^2). \tag{23}$$

Janssen[14] has also shown how the same formalism allows one to understand on a new basis the relation of animal statistics to the Lee-Yang singularity edge problem.

EDEN MODEL AND DIFFUSION-LIMITED AGGREGATION

While in the original Eden model[16] exactly one particle is added at each time step to the aggregate, we define it by postulating that each member of the aggregate may add one particle at one of tits neighboring sites, if it is empty, with a certain rate λ. The "time" of the original Eden model is therefore proportional to the aggregate mass of this model. Let us stipulate that the sites can be in two states: empty (E) or full (F). The Liouvillian of the process therefore read

$$L = \sum_r \left\{ \sum_e \lambda(\pi_{r,F} - \pi_{r,E})\pi_{r+e,F} a_{r,E} a_{r+e,F} \right\}. \tag{24}$$

The Lagrangian of the process reads, in the continuum limit,

$$\mathcal{L} = \int d^d r [\overline{\psi}_E \dot{\psi}_E + \overline{\psi}_F \dot{\psi}_F - \Lambda(\overline{\psi}_F - \overline{\psi}_E)\psi_E(\psi_F + e^2 \Delta \psi_F)$$
$$- \mu(\overline{\psi}_F - \overline{\psi}_E)\overline{\psi}_F \psi_E \psi_F + \text{higher order terms}]. \tag{25}$$

Some constants have been redefined. Integrating out the fields we obtain

$$\dot{\psi}_E = -\Lambda \psi_E(\psi_F + D\Delta \psi_F). \tag{26}$$

Integrating this differential equation and substituting its solution into Eq. (26), we obtain a Lagrangian

$$\mathcal{L} = \int d^d r [\overline{\psi}_F \dot{\psi}_F - \tau \overline{\psi}_F \psi_F - D \overline{\psi}_F \Delta \psi_F$$
$$- g \overline{\psi}_F \psi_F \int^t dt' \psi_F(t') - \mu \overline{\psi}_F^2 \psi_F + \text{higher order terms}]. \tag{27}$$

with a suitable redefinition of the constants. This Lagrangian coincides with that of Cardy's model [Eq. (18)], but with a different sign of the memory term: it corresponds, thereforpe, to epidemics with sensibilization rather than immunization. Cardy[13] has shown that in this case the "free" fixed point, where the model behaves like ordinary diffusion, is always instable, and that renormalization group equations show a runaway towards higher values of the coupling constant. As a consequence, perturbation theory is not reliable: information about the asymptotic behavior must be obtained by nonperturbabtive methods.

A similar phenomenon happens when one tries to write down a field theoretical model for diffusion-limited aggregation.[17] A diffusion particle D may turn into an aggregated one A, and stop if it is at an otherwise empty site neighboring an aggregated particle. one can thus write down the Liouvillian

$$L = \sum_r \left[\lambda \sum_e (\pi_{r+e,D} - \pi_{r,D}) a_{r,D} \right.$$
$$\left. + \mu \sum_{k=0}^{\infty} \frac{(-1)^k}{k!} \pi_{r+e,A}(\pi_{r,A} - \pi_{r,D}) \pi_{r,A}^k a_{r,A}^k a_{r+e,A} a_{r,D} \right]. \tag{28}$$

We have not written down explicitly the source terms at infinite which ensure the steady incoming of D particles. The model can then be handled in the same way as the previous ones. It is possible to integrate out the fields $\overline{\psi}_D$, ψ_D, obtaining the equation

$$\dot{\psi}_D = D\Delta\psi_D + \mu u(r)\psi_D, \tag{29}$$

where the potential $u(r)$ is given by

$$u(r) = \sum_e \sum_{k=0}^{\infty} \frac{(-1)^k}{k!} (1 + \overline{\psi}_{r+e,A}) \overline{\psi}_{r,A} \overline{\psi}_{r,A}^k \psi_{r,A}^k \psi_{r+e,A}. \tag{30}$$

In the limit $\mu \gg D \gg 1$, one recovers a Laplacian description of DLA similar to the one used to describe dielectric breakdown.[19] On the other hand, one easily sees that pure diffusive behavior is never recovered, and therefore that perturbation theory is useless.

We have seen that field theory is a useful method for identifying universality classes of growth processes, and to investigate their asymptotic behavior. Nevertheless, being linked to the existence of a reliable perturbaation theory, it fails when the phenomenon of interest is never approximated by a free diffusion process.

I wish to thank J. L. Cardy, B. Derrida, P. Grassberger, Y. Shapir and Y. C. Zhang who have contributed in various ways to shape my ideas about aggregation processes. I am also grateful to G. Paladin for a critical reading of the manuscript.

[1] H. E. Stanley, this book.
[2] M. Doi, J. Phys. A **9**, 1465,1479 (1976).
[3] P. Grassberger and M. Scheunert, Forts. der Physik **28** 547 (1980).
[4] L. Peliti, J. Physique (Paris), to be published.
[5] D. J. Amit, *Field Theory, the Renormalization Group, and Critical Phenomena* (McGraw Hill, NY, 1978).
[6] L. S. Schulman,Techniques and Applications of Path Integrals (J. Wiley, NY,1981).
[7] P. C. Martin, E. D. Siggia and H. A. Rose, Phys. Rev. **8**, 423 (1973).
[8] P-G de Gennes, *Scaling Concepts in Polymer Physics* (Cornell Univ Press, Ithaca, 1979).
[9] P. Grassberger, unpublished.
[10] J. W. Essam, Rep. Prog. Phys. **43**, 834 (1980).
[11] S. P. Obukhov and L. Peliti, J. Phys. A **16**, L147(1983).
[12] A review of this and related results may be found in L. Peliti and L. Pietrtonero, "Random walks with memory," La Rivista del Nuovo Cimento, to be published.
[13] J. L. Cardy, J. Phys. A **16**, L709 (1983).
[14] H. K. Janssen, Z. Physik B **58**, 311 (1985).
[15] J. L. Cardy and P. Grassberger, J. Phys. A **15**, L267 (1985).
[16] M. Eden, in J. Neyman (ed.), "Proc. 4th Berkeley Symp. on Math. Statistics and Probability" (Univ. Calif. Press, Berkeley, 1961).
[17] T. A. Witten and L. M. Sander, Phys. Rev. Lett. **47**, 1400 (1981); Phys. Rev. B **27**, 5686 (1983).
[18] L. Peliti and Y-C Zhang, to be published.
[19] L. Niemeyer, L. Pietronero and H. Weismann, Phys. Rev. Lett. **52**, 1083 (1984).

Unidentified student

TRANSPORT EXPONENTS IN PERCOLATION

Stéphane Roux and Etienne Guyon

L.H.M.P.–E.S.P.C.I.
10 rue Vauquelin 75231 PARIS Cedex 05, France

This meeting has been devoted to "Growth and Form" Here, we would like to investigate how forms and, more specifically, randomness in geometry as found in percolative structures, govern the macroscopic transport behavior. In the first part, we show how in the node-link-and-blob picture of the percolation infinite cluster, one can derive bounds for critical exponents. This method can be used for scalar transport, for elasticity, or for the specific problem (discussed by P. N. Sen in this book) of "continuous percolation." In the second part, we give an upper bound for the elastic critical exponent which may well be an exact relation (conjecture).

BOUNDS FOR THE CRITICAL EXPONENT IN ELASTICITY[1]

Let us consider the elastic problem formulated by Kantor and Webman.[2] Bond percolation is performed on a lattice where each bond is a spring (elastic constant β) and where, at each site, a set of angular springs (elastic constant γ) connects any pair of bounds. The Hamiltonian can be written as

$$H = \frac{1}{2}\beta \sum_{i,j} A_{ij}\left[(\vec{u}_i - \vec{u}_j)\cdot\vec{r}_{ij}\right]^2 + \frac{1}{2}\gamma \sum_{i,j,k} A_{ij}A_{jk}(\theta_{ijk})^2, \tag{1}$$

where
$A_{ij} = 1$ if the bond (i,j) is present; 0 if not
$\vec{u}_i = $ the displacement of site i
$\vec{r}_{ij} = $ unit vector along (i,j)
$\theta_{ijk} = $ angle between (i,j) and (j,k).

Any other model (e.g., a beam network[3]) which contains angular or flexion elasticity so as to recover the usual percolation threshold is relevant. On the other hand, the problem studied by S. Feng and P. N. Sen[4] where bonds are free to rotate around the sites has a mechanical threshold larger than the electrical one and probably belongs to a different universality class.

A macro-link, of length ξ being given, we can overestimate the elastic constant k of this link by considering that the blobs are infinitely rigid. Conversely, we can underestimate the link stiffness by neglecting the elasticity of some bonds. More precisely, we will cut any bond which does not belong to the shortest path connecting the two ends of the chain. Now if the effort applied at the ends are pure forces F (Fig. 1) (as in Kantor and Webman's work[2]) we find
$$k_{SP} < k < k_{SCB}, \tag{2}$$
where
$$k_{SCB} = \left(\frac{\beta a}{N_{SCB}} + \frac{\gamma}{S^2_{SCB}} \right)$$
$$k_{SP} = \left(\frac{\beta a'}{N_{SP}} + \frac{\gamma}{S^2_{SP}} \right) \tag{3}$$

$N_{SCB}[N_{SP}]$ = number of singly-connected bonds SCB [of bonds belonging to the shortest path SP]

$a[a']$ = average along the set of SCB (i,j) [S.P. bonds] of $\cos(\delta)^2$ where δ is the angle between $\vec{F}$ and $\vec{N}_{ij}$; this factor $0 < a < 1$ plays no significant role

$S^2_{SCB}[S^2_{SP}]$ = total squared radius of gyration of the set of SCB sites [SP sites] projected onto an hyperplane perpendicular to $\vec{F}$. This factor arises naturally because at a specific site i (Fig. 2) the force F creates a moment $M = F \cdot r_i$ which will produce a rotation $\gamma F r_i$ which in turn will give rise to a displacement $\gamma F r_i^2$. This has to be summed over the set of relevant sites ($S^2 = \sum_i r_i^2$).

Kantor[5] has shown numerically that $S^2_{SCB} \sim N_{SCB}\xi^2$ and $S^2_{SP} \sim N_{SP}\xi^2$. The two terms in the r.h.s. of (3) do not scale the same way, and for $\xi^2 \gg \gamma/\beta$ (near threshold) only the second term remains. Now, using the well-known results (lecture by Stanley) one obtains $N_{SCB} \sim (\Delta p)^{-1}$ and $N_{SP} \sim (\Delta p)^{-\nu d_{min}}$
$$k_{SCB} \sim (\Delta p)^{1+2\nu} \qquad k_{SP} \sim (\Delta p)^{\nu(d_{min}+2)}. \tag{5}$$

To go from the microscopic k over scale ξ to the macroscopic coefficient K, we can use the general scaling relation $K \sim k(\Delta p)^{\nu(d-2)}$ so that, as $K_{SP} < K < K_{SCB}$ and $K \sim (\Delta p)^T$ one finally gets

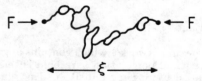

FIG. 1: Macro-link submitted to a force F.

$$1 + \nu d < T < \nu(d_{min} + d). \tag{6}$$

The analysis of result (4) shows that *moments created locally* inside the macro-link by pure forces applied at the ends are dominant at threshold. But, in a macroscopic percolative structure, torques are not only *local*. They assist up to a ξ length-scale. Therefore any macro-link is globally subject to a moment M and a force F with a mean value $M \sim F\xi$. We have seen previously that the effects of forces are by themselves irrelevant insider a link of $\xi^2 \gg \gamma/\beta$ (would they be tensile or shear forces). This important notion was missing in the preceeding analysis (reproduced from Kantor and Webman work). We have now to investigate the case of pure moments $M = F\xi$ applied at the end of the chain (Fig. 3). This will allow us to recover the previous bounds without any need for numerical simulation. The reason why the bounds are equal comes fsrom a coincidence in the value of S^2. It is straightforward in our two-limit cases to compute the effective elastic constants

$$\begin{aligned} k_{SCB} &\sim \Delta p^{1+2\nu} \\ k_{SP} &\sim \Delta p^{\nu(d_{min}+2)}, \end{aligned} \tag{7}$$

which give again (6).

A similar and somewhat simpler treatment can be used in the scalar transport case

$$1 + \nu(d - 2) < t < \nu(d_{min} + d - 2),$$

if t is the critical index.

FIG. 2: Effect of a localized rotation.

FIG. 3: Macro-link submitted to torque M.

Halperin et al (see P. N. Sen lecture) have obtained an important result in the case of "continuous percolation." Let us suppose that the transport coefficient varies like δ^m where δ is a parameter, the probability density of which tends to a finite positive value as δ goes to 0. Then, using the S.C.B. approximation, Halperin et al showed that the critical exponents could be modified. More precisely, their results were the following:

scalar transport   *elasticity*

$\bar{t} \geq 1 + \nu(d-2)$ if $m \leq 1$ $\bar{T} \geq 1 + \nu d$ if $m \leq 1$

$\bar{t} \geq m + \nu(d-2)$ if $m \geq 1$ $\bar{T} \geq m + \nu d$ if $m \geq 1$

If we apply the scalar transport approximation we then can prove that

$\bar{t} \leq \nu(d_{min} + d - 2)$ if $m \leq \nu d_{min}$ $\bar{T} \leq \nu(d_{min} + d)$ if $m \leq \nu d_{min}$

$\bar{t} \leq m + \nu(d-2)$ if $m \geq \nu d_{min}$ $\bar{T} \leq m + \nu d$ if $m \geq \nu d_{min}$

These results lead to the following conclusion (Fig. 4) for $m > \nu d_{min}$; the indices given are exact within the node-link-and-blob scheme.

ANOTHER BOUND FOR T.[1]

We have seen that *forces do not propagate scalarly* ($T \neq t$). They create (or annihilate) torques. But, in a $1-D$ medium, *torques do propagate scalarly!* This can be used to construct easily an upper bound for T. The essential point is that our structure is, at length scale up to ξ, *topologically* a 1-dimensional body. Therefore, we can develop the following correspondances:

electric transport		*torques transport*
current	⟷	torque (moment)
potential	⟷	rotation
conductance	⟷	angular elastic constant

Knowing the solution of the electric problem allows us to construct a field of of admissible stress (boundary conditions and equilibrium equations satisfied everywhere) for one macro-link.

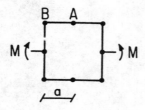

FIG. 4: Torque applied to a loop.

The reason why this field is only admissible and not *the* solution is that there exist loops in the backbone which impell a perfect scalar transport. In a loop, tensile and compression forces will arise (Fig. 5) to counterbalance partially the torque. In any event, if one computes the elastic energy of this admissible field of stress E_s, then the result is an overestimate of the elastic energy E. So in a macro-link

$$E_s \sim M^2(\Delta p)^{-t+\nu(d-2)} \sim F^2 \Delta p^{-(t+2\nu)+\nu(d-2)},$$

or

$$k_s \sim (\Delta p)^{t+2\nu-\nu(d-2)}.$$

At the macroscopic level, one finally obtains

$$K > K_s \sim \Delta p^{t+2\nu}.$$

So

$$T \leq t + 2\nu.$$

The table gives estimates of the bounds and the comparisons with known values.

We emphasize that the same type of singularity seems to govern both scalar and elastic transport through the *scalarity of torque propagation*. An additional exponent 2ν has to be taken into account to "translate" forces into moments and back rotations into displacement.

d	2	3	4	5	6
ν	1.33	0.88	0.72	0.58	0.5
d_{min}	1.13	1.35	1.50	1.75	2
$1 + \nu(d-2) \leq$	1	1.88	2.44	2.74	3
$t \leq$	1.29±0.01	2.02±0.02	2.5	2.76	3
$\nu(d_{min} + d - 2)$	1.5	2.07	2.52	2.76	3
$1 + \nu d \leq$	3.66	3.64	3.88	3.90	4
$\delta \leq$	3.96±0.04	3.8±0.5			
$t + 2\nu \leq$	3.96	3.78	3.95	3.92	4
$\nu(d_{min} + d)$	4.16	3.83	3.96	3.92	4

TABLE 1: Estimates of the bounds and comparison with known values for $d = 2$ to 6.

[1] S. Roux, Compte-Rendus de l'Ac. des Sci. [Paris], accepted for publication.
[2] Y. Kantor and I. Webman, Phys. Rev. Lett. **52**, 1891 (1984).
[3] S. Roux and E. Guyon, preprint.
[4] S. Feng and P. N. Sen, Phys. Rev. Lett. **52**, 216 (1984).
[5] Y. Kantor, J. Phys. A **17**, L843 (1984).

NON-UNIVERSAL CRITICAL EXPONENTS FOR TRANSPORT IN PERCOLATING SYSTEMS

Pabitra N. Sen,* James N. Roberts* and Bertrand I. Halperin+

*Schlumberger-Doll Research
Ridgefield, Connecticut 06877-4108

+Department of Physics, Harvard University
Cambridge, Massachusetts 02138

When there is a distribution of bond strengths, the transport exponent depends on this distribution.[1-4] A distribution of bond strengths arises in the "Swiss cheese" models,[2] where circular or spherical holes are randomly placed in a uniform transport medium. For example, in the $2D$ case a bond is present if two neighboring holes (radius a) do not overlap. If we approximate the ith neck by a thin rectangle of width δ_i and length $\ell \sim \sqrt{\delta_i a}$, the electrical conductance is given by $g_i \approx \sqrt{\delta_i/a}$. The fluid flow permeability through the space in between the circles varies as $\delta_i^{5/2}/a^{1/2}$. In $3D$, the smallest cross-section is roughly triangular and $g_i \sim \delta_i^{3/2}$, fluid flow permeability $\sim \delta_i^{7/2}/a^{1/2}$. The force constant γ_i for bond-bending is proportional to $\delta_i^{7/2}$ in $3D$ and to $\delta_i^{5/2}$ in $2D$.

In all these cases we have $g \approx \delta^m$. We also find that the probability distribution $P(\delta)$ goes to a finite constant for $\delta \to 0$. It follows that for a uniform $P(\delta)$, the bond conductance is a random variable with probability density function

$$P(g) = (1-\alpha)g^{-\alpha} \quad 0 < g \leq 1$$
$$= 0 \quad g \geq 1,$$

where $\alpha = 1 - 1/m$. We plot the conductivity Σ of a $2D$ square lattice as a function of size L, with half of the bonds missing. The bonds present have random values of g as above. We find that the conductivity Σ varies as $\Sigma \sim L^{-\bar{t}/\nu}$, where $\bar{t}$ is the transport exponent and ν the correlation length exponent. $\bar{t}$ is the same as t, the standard conductivity exponent for $\alpha < 0$, $m < 1$, but $\bar{t} \approx t + \alpha/(1-\alpha)$ for $m > 1$, $0 < \alpha$; ν is independent of bond strength.

[1] P. N. Sen, J. N. Roberts and B. I. Halperin (unpublished).
[2] H. I. Halperin, S. Feng and P. N. Sen, Phys. Rev. Lett. 54, 2391 (1985).
[3] P. M. Kogut and J. P. Straley, J. Phys. C 12, 2151 (1979).
[4] A. Ben-Mizrahi and D. J. Bergman, J. Phys. C 14, 909 (1981).

LÉVY WALKS VERSUS LÉVY FLIGHTS

Michael F. Shlesinger

Office of Naval Research/Physics Division
800 North Quincy Street
Arlington, Virginia 22217-5000 USA

and

Joseph Klafter

Corporate Research Science Laboratory
Exxon Research and Engineering Company
Annandale, New Jersey 08801 USA

We explore the behavior of random walkers that fly instantaneously between successive sites, however distant, and those that must walk between these sites. The latter case is related to intermittent behavior in Josephson junctions and to turbulent diffusion.

INTRODUCTION

When the vogue in probability theory was to prove the Central Limit Theorem under the weakest possible conditions, Paul Lévy in the 1920's searched for exceptions to it. He asked questions about scaling for probability distributions and was led to working with random variables with infinite second moments. A finite second moment would describe a characteristic scale and gaussian behavior for the probability density. An infinite second moment implies the absence of a definite scale, which is the paradigm of fractals. Lévy discovered the class of probability distributions governing the sum of these random variables. They describe a random walk with an infinite mean squared displacement for each jump. Today this is called a Lévy flight. Each flight is parameterized by an exponent describing how the first infinite moment diverges.

We begin by discussing the Lévy flight in the continuum and then restrict ourselves to a lattice. Attention is focused on the fractal dimension of the set of points visited by a walker that instantaneously flies over the intervening points separating successively visited sites. Next we introduce the Lévy walker who can only take one step of unit length at a time, but may take many correlated steps in the same direction.

LÉVY FLIGHTS[1,2]

Lévy considered a set $\{X_i\}$ of identically-distributed random variables each governed by a probability density $f_i(x)$ whose Fourier transform is denoted by $\tilde{f}_i(k)$. He asked the question of when the new random variable X_0, given by

$$c^\alpha X_0 = c_1^\alpha X_1 + \cdots + c_N^\alpha X_N, \tag{1}$$

with the c's being constants related by an auxillary condition

$$c^\alpha = c_1^\alpha + \cdots + c_N^\alpha, \qquad 0 < \alpha \leq 2,$$

has the same probability density as the X_i? Choose for $i > 1$ all the $< X_i > = 0$, $c_i = 1$ and $\alpha = 2$. Then $X_0 = N^{-1/2} \sum_{i=1}^{N} X_i$. This sum of N random variables properly normalized by the $N^{-1/2}$ has the gaussian density, or $f_0(k) = \exp(-Dk^2)$ with D a constant, as an asymptotic solution for small k. For $\alpha < 2$, Lévy showed the solutions

$$f_0(k) = \exp(-D|k|^\alpha) \tag{2}$$

exist. These cases correspond to $f_0(x) \sim x^{-1-\alpha}$ for $x \to \infty$, i.e., the moment $< X^\alpha >$ diverges. These processes represent a random walker (with step i represented by X_i) that visits a disconnected set of self-similar clusters of fractal dimension α. This will become clearer when we restrict the walk to a lattice, as in the next section.

WEIERSTRASS FLIGHTS: A DISCRETE SPACE LÉVY FLIGHT

Consider a random walk on a lattice. For simplicity we can use a one-dimensional lattice, but the results can be easily generalized to higher dimensional cubic type lattices. Let $p(\ell)$ be the probability for a random walk jump of vector displacement ℓ. We choose[2]

$$p(\ell) = \frac{n-1}{2n} \sum_{j=0}^{\infty} n^{-j} \left(\delta_{\ell, b^j} + \delta_{\ell, -b^j} \right), \tag{3}$$

with $n, b > 1$. Jumps of all orders of magnitude can occur in base b, but each successive order of magnitude displacement occurs with an order of magnitude less probability in base n. The walker makes about n jumps of unit length before a jump of length b occurs, and a new cluster of sites visited is begun to be generated. Eventually a fractal set of points is visited. To better see this, consider

$$\tilde{p}(k) = \frac{n-1}{n} \sum_{j=0}^{\infty} n^{-j} \cos(b^j k), \tag{4}$$

which is Weierstrass' example of an everywhere non-differentiable function for $b > n$. For small k

$$\tilde{p}(k) \sim 1 - \frac{1}{2}k^2\overline{\ell^2}, \qquad (5)$$

if $\overline{\ell^2} = \Sigma \ell^2 p(\ell)$ is finite. If $b^2 > n$, then $\overline{\ell^2} \propto \sum_{j=0}^{\infty}(b^2/n)^j$ diverges and Eq. (5) cannot be used. Instead note that

$$\tilde{p}(bk) = n\tilde{p}(k) - (n-1)\cos k, \qquad (6)$$

which has the following small k behavior

$$\tilde{p}(k) \sim 1 - k^\alpha \sim e^{-k^\alpha}, \quad \text{with} \quad \alpha = \ln n / \ln b. \qquad (7)$$

The exponent α now appears in the form of a fractal dimension. For $\alpha = 2$, as with Eq. (5), gaussian behavior results and for $\alpha < 2$, $\overline{\ell^2} = \infty$ and the Lévy flights are obtained. The trail of the flight is of dimension α. Note for a homogeneous space the upper critical dimension of the random walk trail is 2, i.e., brownian motion covers a two-dimensional space.

LÉVY WALKS[3,4]

Let $\Psi(\vec{\ell}, t)$ be the probability that a walker makes $\vec{\ell}$ correlated steps in the same direction and that it takes a time t. We write Ψ as

$$\Psi(\vec{\ell}, t) = \psi(t|\vec{\ell})p(\vec{\ell}), \qquad (8)$$

where $p(\vec{\ell})$ is the probability that ℓ correlated steps take place (this is the analog of a jump of length ℓ in the Lévy flight) and $\psi(t|\ell)$ is the probability density that these ℓ steps take at time t. We choose for $|\vec{\ell}| \gg 1$,

$$p(\vec{\ell}) \sim |\vec{\ell}|^{-2-\beta}, \qquad (9)$$

and

$$\psi(t|\vec{\ell}) = \delta(|\vec{\ell}| - t), \qquad (10)$$

so the correlated motions occur with a constant velocity. The probability density that a jump begins at time t is

$$\psi(t) = \sum_{\ell} \Psi(\vec{\ell}, t). \qquad (11)$$

The mean squared displacement $<R^2(t)> = \sum_{\vec{\ell}} |\ell^2| P(\vec{\ell}, t)$, where $P(\vec{\ell}, t)$ is the probability of being at site $\vec{\ell}$ at time t, can be shown to be given by[2,3] in Laplace space,

$$\int_0^\infty e^{-st} <R^2(t)> dt = [s - s\psi^*(s)]^{-1} \sum_i \partial^2 \tilde{\Psi}^*(\vec{k}=0, s)/\partial k_i^2, \qquad (12)$$

where * denotes the Laplace transformed function. For $0 < \beta \leq 1$, and $s \to 0$ and $\bar{t}$ being the average time spent for a completed jump,

$$\psi^*(s) \sim 1 - s\bar{t} + const.\ s^{1+\beta}. \tag{13}$$

For $-1 < \beta < 0$, $\bar{t} = \infty$ and

$$\psi^*(s) \sim 1 - const.\ s^{1+\beta}. \tag{14}$$

For either case $\partial^2 \tilde{\Psi}^*(\vec{k} = 0, s)/\partial k^2 \sim s^{\beta-1}$. This leads to

$$<R^2(t)> \sim \begin{cases} t^2, & -1 < \beta \leq 0, \ \bar{t} = \infty \\ t^{2-\beta}, & 0 < \beta < 1, \ \bar{t} < \infty, \ \overline{t^2} = \infty \\ t \ln t, & \beta = 1, \ \bar{t} < \infty, \ \overline{t^2} \text{ log divergent} \\ t, & \beta > 1, \ \overline{t^2} < \infty. \end{cases}$$

These same results exactly have been derived[5] by Geisel et al for a nonlinear mapping governed by a universality exponent z, if we set $\beta = (2-z)/(z-1)$. Their mapping describes the intermittent change between clockwise and counterclockwise voltage phase rotations in a Josephson junction. The number of cycles in any direction is the equivalent of $p(\ell)$ for the number of steps of a walker in a given direction.

One can imagine other scaling relationships. If, say, a walker has a larger velocity when it walks coherently for a larger distance, then an expression of the form

$$\psi(\ell, t) = \delta(|\ell| - t^\gamma), \qquad \gamma > 1 \tag{16}$$

can apply. For example, in turbulent diffusion a particle caught in a big vortex can have this type of behavior, i.e., move a larger correlated distance at a higher velocity than a particle in a small vortex which moves smaller distances at lower velocities. If we choose $\gamma = 3/2$,

$$<R^2(t)> \sim t^3 \tag{17}$$

is derived, a well-known relation in turbulent diffusion.

THE TRAPPING PROBLEM

Finally we consider the lifetime of a Lévy walk in the presence of a random distribution of traps. For brownian motion this is an old problem.[7] The probability $\phi(t)$ that the walker has not been trapped by time t develops a long tail because the walker may start in a large trapless region of radius R. This occurs with a probability $\propto e^{-V}$ where in d-dimensions $V \sim R^d$. The first passage time distribution to leave this region is $\exp(-const.\ t/R^2)$. For $t \to \infty$ this lead to $\phi(t) \sim \exp(-t^x)$ with $x = d/(d+2)$. For Lévy walks with $<R^2(t)> \sim t^2$

$$\phi(t) \sim \int_0^\infty e^{-V/V_0} \exp(-const.\ t/r) dr \sim \exp(-t^{d/(d+1)}), \tag{18}$$

which is a more rapid decay law. In effect if we fix a time T, then the brownian walker searches in two dimensions a circle of radius $T^{1/2}$, while the Lévy walker searches (albeit fractally) a circle of radius T. The effect of large trapless regions is felt less acutely by the Lévy walker. It has been suggested[8] that certain animals such as ants perform Lévy walks when searching for food in a new area. The above analysis may imply that starving Lévy walk ants possess a slight evolutionary advantage over ants performing other walks, such as even the SAW. Flying ants can be considered by the reader.

REFERENCES

[1] B. B. Mandelbrot, *The Fractal Geometry of Nature* (Freeman, San Francisco, 1982).

[2] E. W. Montroll and M. F. Shlesinger, in *Studies in Statistical Mechanics*, Vol. 11, "From Stochastics to Hydrodynamics," eds. J. L. Lebowitz and E. W. Montroll (North-Holland, Amsterdam, 1984), pp. 1-121.

[3] M. F. Shlesinger, J. Klafter and Y. M. Wong, J. Stat. Phys. **27**, 499 (1982).

[4] M. F. Shlesinger and J. Klafter, Phys. Rev. Lett. **54**, 2551 (1985).

[5] T. Geisel, J. Nierwetberg and A. Zacherl, Phys. Rev. Lett. **54**, 616 (1985).

[6] F. Wegner and S. Grossman, Zeits. für Physik B **59**, 197 (1985); S. Grossman and I. Procaccia, Phys. Rev. A **29**, 1358 (1984).

[7] B. Ya Balagorov and V. G. Vaks, Sov. Phys. JETP **38**, 968 (1974).

[8] B. Ninham, private communication.

Preparing the Grand Mechoui

GROWTH PERIMETERS GENERATED BY A KINETIC WALK: BUTTERFLIES, ANTS AND CATERPILLARS

Alla E. Margolina

E. I. du Pont de Nemours Company
Central Research and Development
Experimental Station
Wilmington, Delaware 19898 USA

Diffusion processes can be represented by a random walk on a fractal substrate. One might also study a walk that creates its own substrate by certain rules while walking: The trace left by such a walk forms a cluster of visited sites S. E.g., a walker on a lattice, which at each step chooses its direction at random and occupies the next site with probability p_c (or blocks it with probability $1 - p_c$), creates a percolation cluster at the percolation threshold p_c. How does this cluster grow in time? The growth occurs, apparently, through the sites that are nearest neighbors to visited sites but which were not tested before by the walk (not yet blocked or visited). These sites are called growth sites[1] and the set of these sites we call a growth perimeter G. One can appreciate that the direct study of growth perimeters leads to an interesting problem that has two aspects: pure growth patterns and the types of diffusion leading to these patterns. Here we will discuss a new butterfly walk[2] that visits only the growth sites and, therefore, concentrates on a pure growth phenomena. In the end we will return to the diffusion aspect of the problem and compare butterfly diffusion to the random diffusion on percolation (the "ant").

Consider spreading of an infection on a square lattice started by an initial sick site S. Imagine the infection spreading by an infected butterfly who flies from one growth site G to another, randomly choosing the next from the probability distribution

$$P(r) \sim 1/r^\alpha \tag{1}$$

where r is the distance from the most recently added sick particle to the chosen growth site and α is a parameter that governs the effective repulsion or attraction between G sites. The chosen G site is converted into an S site. The typical cluster of S sites with a growth perimeter G is shown in Fig. 1. If this procedure is performed on a percolation cluster, the additional choice has to be made: we convert G sites into S sites with probability p_c or into immune I sites with probability $1 - p_c$. The sick and immune sites stay as such forever while the growth perimeter G constantly changes its identity each time the sick site is added. All its not yet determined nearest neighbors become G's only when growth occurs, i.e., when the S site is added, and therefore the "butterfly" moves in an artificial growth time $t = S$. After t time steps a large ramified cluster of S sites has been formed. The fractal dimension d_f of the grown cluster is the same as for percolation cluster at p_c ($d_f = 91/48$ in $d = 2$) since the ultimate connectivity of the S sites is identical to the connectivity of occupied sites in percolation. The main property of interest for this type of a walk is the fractal dimension d_G of the growth perimeter G

$$G \sim R^{d_G} \sim S^{d_G/d_f} \tag{2}$$

where R is the average cluster radius.

CROSSOVER FROM LONG-RANGE TO SHORT-RANGE BUTTERFLY

Let us concentrate now on the effect of the tuning parameter α from Eq.(1) on the growth patterns. The case of $\alpha = 0$ reduces to the Eden model on a percolation substrate and G sites are chosen at random (with probability $1/G$). One can distinguish between the two well-defined limits in the butterfly behaviour: the long-range limit when the butterfly tends to fly far away (but is limited by the extremities of the cluster) and the short-range limit when the butterfly tends to fly close by (mostly to the nearest neighbors for large positive α). The limiting value α_c can be roughly estimated by considering the mean length of a flight, $<r> = \int rP(r)dr$, which implies $\alpha_c = 2$. Namely, the long-range behavior is observed for $\alpha < 2$ and it breaks down for $\alpha > 2$.

FIG. 1: The cluster of 1500 sites formed by an $\alpha = -3$ butterfly on a Euclidean lattice with $p = 1$. Only the growth sites * are shown. The resemblance to a butterfly is coincidental.

Moreover, we find that for $\alpha > 8$ the short-range behavior established and the growth perimeter assumes roughly the same value for all $\alpha > 12$.[2] How is this crossover behavior reflected on growth? The main result is that d_G changes continuously from the smaller long-range value $d_G = 0.76V(\alpha < 2)$ to the larger short-range value $d_G = 1.04(\alpha > 8)$. Therefore, the spatial growth sequence can enhance or hinder the growth of G. We find that the the kinetic exponent d_G can be continuously tuned while the static d_G does not change. Thus the dynamic universality classes are quite independent of the static ones. The butterfly turns out to be the first one-cluster growth model having such a feature and thereby yields insight into a generic feature for growth models.

THE INTERPRETATION OF THE LONG-RANGE LIMIT

A conjecture was proposed[4] for a long-range value of d_G based on the assumption that the scaling form[5] for the finished cluster perimeters P and the perimeters of still growing clusters is the same if the clusters are large enough

$$P = (1 - p_c)/p_c + AS^\sigma \qquad (3)$$

Here $\sigma = 1/(\nu d_f)$ where ν is a correlation length exponent A is a constant. The first term in Eq.(3) can be easiliy identified with the number of blocked sites in a growing cluster of size S. Therefore, the growth perimeter G, which is equal to the number of all nearest neighbors to cluster sites S minus the identifed blocked sites, should be proportional to the "excess" perimeter

$$G \sim S^\sigma \sim R^{1/\nu} \qquad (4)$$

Comparing Eq.(2) and Eq.(4), we find

$$d_G = 1/\nu \qquad (5)$$

This conjecture is in very good agreement with the numerical data for the long-range limit in two and three dimensions, and is exact for the Cayley tree. Note that the fractal dimension d_G of the growth perimeter in the long-range limit appears to be the same as the fractal dimension of singly-connected bonds[6] which might lead to some interesting insights into the connectivity dynamics of the growing percolation clusters. We also suggest[4] a new scaling relation for the chemical dimension d_ℓ [defined as $S \sim \ell^{d_\ell}$, where ℓ is a chemical length: $d_\ell = 1/(1 - \sigma) = d_f/(d_f - d_G)$, predicted earlier by numerical observation[7].

THE INTERPRETATION OF THE SHORT-RANGE LIMIT

The short-range limit value $d_G = 1.04 \pm 0.04$ appears to be in a different kinetic universality class than the "ant" for which we find[8] $d_G = 0.93$. Let us see if this statement holds for a short range butterfly and a random walk performed on a Euclidean lattice ($p = 1$). For $\alpha > 16$ the butterfly in $d = 2$ is a well defined space-filling walk (see Fig.2). The resulting behavior turns out to be the same for butterfly and random walk growth perimeters: $G \sim S/lnS$. This makes the discrepancy between the short-range butterfly and ant dynamic behavior on percolation clusters even more intriguing. At present we do not have a satisfactory explanation of this phenomenon.

THE DIFFUSION ASPECT OF THE BUTTERFLY MODEL: THE CATERPILLAR

Let us rescale the growth time $t = S$ back to usual $t \sim S^{d_w/d_f}$, thus allowing for revisiting of already visited sites [here d_w is the dimension of the walk, defined as $t \sim R^{d_w}$. It is as if the random walk proceeds normally until it meets a growth site. At this moment the decision to occupy it or not is made according to a butterfly probability, Eq. (1). If the growth site is not occupied, the revisiting continues until the next growth site. I suggest to call this type of walk a grow-limited diffusion process or "caterpillar". If we assume the relation for the ant[9] $d_w/d_f = 2 - d_G/d_f$ to hold for all α, we get a crossover for $d = 2$ percolation in a spectral dimension $d_s = 2d_f/d_w$ from $d_s = 1.25$ (long range) to $d_s = 1.38$ (short range) with the value for the ant $d_s = 1.32$ in between the two. Thus the random diffusion process corresponds to a moderately short-range butterfly with $\alpha=4$. This is in accord with our understanding that the random walk must have a finite probability to visit the growth sites that are relatively far away.

This work was done in collaboration with A. Bunde, H. J. Herrmann, I. Majid and H. E. Stanley. I wish to also thank A. Aharony, S. Havlin, F. Leyvraz, R. Rubin and D. Stauffer for stimulating discussions.

[1] F. Leyvraz and H. E. Stanley, Phys. Rev. Lett. 51, 2048 (1983).
[2] A. Bunde, H. J. Herrmann, A. Margolina, and H. E.Stanley Phys. Rev. Lett. 55, 653 (1985).
[3] F. Family and T. Viczek, J. Phys A 18, L75 (1985).
[4] A. Margolina, J. Phys. A 18, xxx (1985).
[5] D. Stauffer, Phys. Rep. 54, 1 (1979).
[6] A. Coniglio, J. Phys A 15, 3829 (1982).
[7] S. Havlin and R. Nossal, J. Phys A 17, L427 (1984).
[8] H. E. Stanley, I. Majid, A. Margolina and A. Bunde, Phys. Rev. Lett. 53, 1706 (1984).
[9] H. E. Stanley, J. Stat. Phys. 36, 843 (1984).

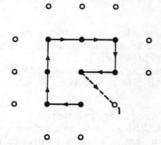

FIG. 2: Typical 7-step trap for a short-range kinetic walk. Butterfly finds a way out by flying to one of the growth sites (o) according to Eq. (1). Growth site 1 is the most probable choice.

ASYMPTOTIC SHAPE OF EDEN CLUSTERS

Deepak Dhar

Laboratoire Physique Theorique et Hautes Energies,
Universite Pierre et Marie Curie
4, Place Jussieu, 75230 PARIS Cedex, France

There has been a lot of interest in the study of the Eden model in recent years. It is known that Eden clusters are compact in any dimension.[1] The model can be solved exactly on the Bethe lattice,[2] and a systematic expansion in powers of $(1/d)$ about the $d = \infty$ solution can be developed.[3] The scaling properties of the average width of the active region have been studied by simulations[4] and also in a theoretical model,[5] but are not well understood. In addition, the vibrational spectra of Eden clusters without loops have been studied.[6]

An important unsettled question concerns the asymptotic shape of Eden clusters.[7] Eden noticed the fairly circular shape of clusters in his early two dimensionsonal simulation. This is suggestive of a greater symmetry, similar to that observed in equilibrium models near phase transitions (the critical correlations are isotropic, e.g., in the Ising model on a scaling lattice). In the following, I prove that in large enough dimensions, the Eden clusters are not spherical. Non-sphericity has been observed in simulations of other growth models such as Witten-Sanders aggregates also.

The Eden model is equivalent to the epidemic model defined as follows[1]: At time $t = 0$, all sites of a d-dimensional hypercubical lattice are 'healthy,' except the origin which is 'infected.' A healthy site having at least one infected neighbor has a probability dt of becoming infected in any subsequent small interval of time dt. An infected site never recovers.

As time increases, more and more sites get infected, and the boundary separating the infected from healthy sites moves outwards. For large times, the extent of a cluster in any direction is proportional to the average velocity of the boundary in that direction. It is shown below that for large d, the velocity along the axes is strictly greater than the velocity along the principal diagonal direction $(1, 1, 1)$ (in the Euclidean norm), and hence the clusters are non-spherical.

UPPER BOUND ON THE INFECTION VELOCITY ALONG THE DIAGONAL

This is an elementary refinement of an argument originally due to Hammersley.[8] Let $Prob(X, T)$ denote the probability that the site $X \equiv (L, L, \ldots, L)$ of a d-dimensional hypercubical lattice is infected at time T. For each infected site, we define an infection path to the origin by connecting a site to its earliest infected neighbor.

Then $Prob(X, T)$ is certainly less than the sum over all possible infection paths W of the probability that the infection would reach X along W in time T. Now an infection path $(L+n_i)$ steps in the $+X_i$-direction and n_i steps in the $-X_i$-direction $(i = 1$ to $d)$ has a total length $\omega = \sum_{i=1}^{d}(L + 2n_i)$. The total number of possible infection paths corresponding to a given set $\{n_i\}$ is less than the number of all random walks with length ω having some restriction on the number of steps in different directions. The latter is given by

$$N(L, \{n_i\}) = \omega! / \prod_{i=1}^{d}[(L + n_i)! n_i!]. \tag{1}$$

The probability that the infection covers a distance ω (measured along bonds) in time T is given by

$$P = e^{-T} T^{\omega} / \omega!. \tag{2}$$

Then from (1) and (2) we get that

$$Prob(X, T) \leq \sum_{\{n_i\}} N(L, \{n_i\}) \cdot P \tag{3}$$

$$= e^{-T} \left[\sum_{n=0}^{\infty} T^{L+2n} / [(L + n)! n!] \right]^{d}. \tag{4}$$

For large L and T, the summation over n in Eq. (4) can be evaluated by the method of steepest descent, and we get

$$\log Prob(X, T) \leq -T + d[\sqrt{L^2 + 4T^2} - L \log((L + \sqrt{L^2 + 4T^2})/2T)]. \tag{5}$$

Let ϑ be a solution to the equation

$$-1/d + \sqrt{\vartheta^2 + 4} - \vartheta \log(\vartheta/2 + \sqrt{1 + \vartheta^2/4}) = 0. \tag{6}$$

If $L > \vartheta T$, and T is large, the right hand side of Eq. (5) is large and negative so that $Prob(X, T)$ is exponentially small. This implies that an upper bound on the infection front velocity is given by $\vartheta\sqrt{d}$, where ϑ is determined by Eq. (6).

LOWER BOUND ON THE INFECTION VELOCITY ALONG THE AXES

Let P_m denote the hyperplane with the equation $x_1 = m$. The infected site is introduced at the origin at time $t = 0$ when all other sites are healthy. Let $<T>$ denote the average time elapsed before the infection reaches P_1. The average time elapsed before the infection reaches P_m is strictly less than $m <T>$, and $1/<T>$ provides a lower bound to the velocity of infection along the axes.

Consider the infection process restricted to P_0. This is a $(d-1)$-dimensional growth process. Let S_n be a (random) number of perimeter sites when the number of infected sites is n, and τ_n be the interval between the time when the nth and $(n+1)^{th}$ sites get infected. The time-history of this process is characterized by the function $n(t)$ giving the number of infected sites at time t. Clearly

$$n(t) = r, \quad if \quad \sum_{i=1}^{r-1} \tau_i \leq t < \sum_{i=1}^{r} \tau_i. \tag{7}$$

Consider now the infection of sites in P_1 by those in P_0. We shall disallow infection of sites in P_0 by those in P_1. This does not affect the validity of the bound derived. Let $P(t)$ be the probability that none of the sites in P_1 are infected at time (t) for a given history of infection $n(t)$ in P_0. Then $P(t)$ satisfies the differential equation

$$\frac{d}{dt}P(t) = -n(t)P(t), \tag{8}$$

which has the solution

$$P(t) = exp[-\int_0^t dt' n(t')]. \tag{9}$$

The average waiting time T for a fixed $n(t)$ is given by

$$T = \int_0^\infty P(t)dt$$

$$= \int_0^\infty dt \exp[-\int_0^t dt' n(t')]. \tag{10}$$

Using Eq. (7) this may be reexpressed in terms of $\{\tau_i\}$

$$= \int_0^{\tau_1} dt' \exp(-t') + \int_{\tau_1}^{\tau_1+\tau_2} dt' \exp(-\tau_1 - 2t') + \int_{\tau_1+\tau_2}^{\tau_1+\tau_2+\tau_3} dt' \exp(-\tau_1 - 2\tau_2 - 3t') + \cdots$$

$$= \frac{(1-e^{-\tau_1})}{1} + e^{-\tau_1}\frac{(1-e^{-\tau_2})}{2} + e^{-\tau_1-2\tau_2}\frac{(1-e^{-\tau_3})}{3} + \cdots \tag{11}$$

For a fixed sequence $\{s_i\}$, different r_i's are mutually independent exponentially distributed random variables with

$$<e^{-n r_n}> = S_n/(n+S_n). \qquad (12)$$

For a $(d-1)$-dimensional hypercubical lattice in all configurations

$$S_n \geq 2(d-1)n^{(d-2)/(d-1)} = C_n \qquad (13)$$

Taking the expectation value of Eq. (11) over all possible histories, and using the estimate Eq. (13), we get

$$<T> \leq (1+C_1)^{-1} + C_1(1+C_1)^{-1}[(2+C_2)^{-1} + C_2(2+C_2)^{-1}[(3+C_3)^{-1} + \cdots \qquad (14)$$

For any given value of d, Eq. (14) with C_n's given by Eq. (13) is a convergent series, and may be summed numerically to get an upper bound on $<T>$. For large d, the asymptotic behavior of $<T>$ is determined as follows: For $n << (d-1)^{d-1}$ we can approximate $C_n(n+C_n)^{-1}$ by $\exp[-n^{1/(d-1)}/2(d-1)]$ and

$$\exp[-\sum_{i=1}^{n} i^{1/(d-1)}/2(d-1)] \approx \exp[-n^{d/(d-1)}/2d].$$

The infinite series (14) then becomes

$$\sum_{n=1}^{\infty} \exp[-n^{d/(d-1)}/2d] n^{-1+1/(d-1)}/2d.$$

Changing the summation to integration, and changing variables from n to $x = n^{1/(d-1)}$ we get

$$<T> \sim \int_{1}^{\infty} \exp[-x^d/2d] dx/(2d). \qquad (15)$$

Eq. (15) shows that the infection velocity along the x_1-axis in d dimensions, given by $1/<T>$, varies as $d^{1-1/d}$ for large d. Since the upper bound on the velocity along the main diagonal in the Euclidean norm varies as $\sqrt{d}$, For large enough d the velocity along axes is greater than the velocity along diagonal and the clusters are diamond-like and non-spherical.

Numerically, these bounds are not very good. Only for $d \geq 54$ do we find the lower bound on the velocity along the axes exceeding the upper bound along the main diagonal. Some improvements are clearly possible, but the more interesting cases of $d = 2$ or 3 would perhaps require other techniques.

[1] D. Richardson, Proc. Camb. Phil. Soc. **74**, 515 (1973); D. Dhar, Phys. Rev. Lett. **54**, 2058 (1985).

[2] J. Vannimenus, B. Nickel and V. Hakim, Phys. Rev. B **30**, 391 (1984).

[3] G. Parisi and Y. C. Zhang, Phys. Rev. Lett. **53**, 1791 (1984).
[4] Z. Racz and M. Plischke, Phys. Rev. A **31**, 985 (1985); R. Jullien and R. Botet, preprint (1985).
[5] S. Edwards and D. Wilkinson, Proc. Roy. Soc. Lon. A **381**, 17 (1982).
[6] D. Dhar and R. Ramaswamy, Phys. Rev. Lett. **54**, 1346 (1985); D. Dhar, J. Phys. A **18**, xxx (1985).
[7] D. Mollison, J. Roy. Stat. Soc. B **39**, 325 (1977).
[8] J. M. Hammersley, J. Roy. Stat. Soc. B **28**, 491 (1966).

Mike Kosterlitz

OCCUPATION PROBABILITY SCALING IN DLA

Leonid A. Turkevich

The Standard Oil Company/Corporate Research Center
4440 Warrensville Center Road
Cleveland, Ohio 44128 USA

Scaling relates the DLA Hausdorff dimension D to the perimeter occupancy probabilities of the cluster tips. On a 2d square lattice, we find $D = 5/3$ for DLA, $D = 2$ for the Eden model, and $D = 4/3$ for the Brown-Boveri dielectric breakdown model. D is not universal: $D = 7/4$ for the 2d triangular lattice. The results are extended to $d \geq 2$. We find no upper critical dimension for DLA, although $D \to d-1$ for large d. Our values for D are fully consistent with Meakin's Cartesian lattice simulations for $2 \leq d \leq 6$.

In this talk, I want to convince you of the importance of occupation probabilites in diffusion-limited aggregation (DLA).[1] These are the probabilites for a random-walker to end up on the growing perimeter sites of a DLA cluster. In the first section, the occupancy probabilities for growth sites are related to the fractal dimension D of the cluster.[2] A simple scaling relation implements the physics that all the growth takes place at, and is driven by, the tips. This reduces DLA to classical electrostatics, with other models as innoccuous variants. In order to calculate D, we only need the scaling, with cluster mass N, of the occupancy probability, $P_{max}(N)$, of the maximally extending tips. In the next section, we utilize an *Ansatz*[2] that the large-scale structure of the cluster determines the strength of the tip cusps (wedges). This permits a simple electrostatic calculation of the scaling of $P_{max}(N)$. We implement the electrostatics in $d = 2$, using conformal transformations, and investigate the mild nonuniversality (lattice dependence) of D. In the last section, we extend these results to higher dimensions.[3] The theory yields no upper critical dimension for DLA on Cartesian lattices. The results are in good agreement with Meakin's large Cartesian lattice simulations.

RELATION OF OCCUPANCY PROBABILITIES TO D

First a few definitions. Suppose we have a DLA cluster of N particles. We send in the $N+1^{st}$ random-walker from infinity. This random walker may land on any one of the perimeter sites of the cluster (indexed by i) with probability p_i. These occupancy probabilities are not uniform, as they would be for the Eden model, but rather depend intimately on the diffusive (fractal) character of the random walk. In particular, there is a far greater probability for the random-walker to adhere to a tip perimeter site than to a perimeter site deep within the cluster. It is this physics that we wish to capture in our scaling treatment.

If the DLA cluster were fully developed to its extremities, it would have a nominal radius

$$r_0 = N^{1/D} a, \qquad (1)$$

where a is the lattice constant. However the interior of the cluster is screened (with screening length ξ): additional random walkers are only captured by perimeter sites in the "active zone." Thus the DLA cluster is fully developed (i.e., with the number of particles within radius r scaling as $N(r) \sim r^D$) only up to a radius $r_- = r_0 - \xi$. In order to still accommodate all the N particles at the cluster, the cluster must extend further, to a radius $r_+ > r_0$. Self-similarity ensures that all the lengths, r_0, r_+, ξ, scale similarly with N.

We now consider DLA clusters specifically in $d = 2$. The occupancy probabilites may be expressed as a probability density, namely, $p_N(r)dr$ is the probability that the $N+1^{st}$ random walker lands in the annulus of width dr at radius r. Clearly $p_N(r)$ is maximal at the tips; in fact, $p_N(r)$ diverges as $r \to r_+$. The probability that the $N+1^{st}$ random walker lands at the tips is

$$P_{max}(N) = \int_{r_+ - a}^{r_+} p_N(r)dr, \qquad (2)$$

in which case the cluster grows: $r_+^{(N+1)} = r_+^{(N)} + a$. There is zero probability for this walker to land within the fully developed region—this is just the effect of screening. The probability that the $N+1^{st}$ random walker lands within the active zone (but not at the tips) is $1 - P_{max}$, but, in this case, the cluster does not grow: $r_+^{(N+1)} = r_+^{(N)}$. Combining these possibilities

$$r_+^{(N+1)} = P_{max}(r_+^{(N)} + a) + (1 - P_{max})r_+^{(N)}, \qquad (3)$$

which may be expressed differentially,

$$\frac{dr_+^{(N)}}{dN} = P_{max}(N)a. \qquad (4)$$

Equation (4) quantifies the physics that all the growth of a DLA cluster is driven by the occupancy probability at the tips. Concomitantly, Eq. (4) also expresses the fact that with growth $p_{N+1}(r)$ (the occupancy probability density for the $N+2^{nd}$ random walker) is altered from $p_N(r)$, while without growth

$p_{N+1}(r) \approx p_N(r)$. Finally, since $r_+ \sim N^{1/D}a$, the singularity in the occupancy probability density $p_N(r \to r_+)$, as the tip is approached, completely determines the fractal dimension D. In the infinite time limit, the random walk diffusion reduces[4] to $\nabla^2 u = 0$, where u is the concentration field of the random walker. As the perimeter sites are perfect traps, we have the boundary condition $u = 0$ along the perimeter of the cluster. Since we are isotropically launching random walkers from infinity, we also have the boundary condition that $u = 1$ for $r = R_\infty$. The occupancy probability density $p(\vec{r})$ is just proportional to the flux $\vec{\nabla} u$ of random walkers at perimeter site $\vec{r}$. Thus, in order to solve for the probability density $p(\vec{r})$, we merely solve an electrostatic problem $\nabla^2 \phi = 0$ for a conducting cluster ($\phi = 0$) inside an infinite radius conducting circle ($\phi = 1$)

$$p(\vec{r}) = \frac{|\vec{E}(\vec{r})|}{\int_\pi |\vec{E}(\vec{r})| d_s}, \qquad (5)$$

where the electric field $\vec{E} = -\vec{\nabla}\phi$, and where the probability density is normalized over the perimeter π of the cluster. One may consider[4] a whole class of models, where (5) is generalized

$$p(\vec{r}) = \frac{|\vec{E}(\vec{r})|^\eta}{\int_\pi |\vec{E}(\vec{r})|^\eta ds}. \qquad (6)$$

For $\eta = 0$ we recover the Eden model,[5] $\eta = 1$ DLA and $\eta > 1$ a class of dielectric breakdown models considered by the Brown-Boveri group.[4]

THE LARGE-SCALE STRUCTURE ANSATZ

The electrostatic problem of a randomly-branched conducting cluster within a conducting cylinder is too difficult to solve exactly. Our scaling relation (4), however, only requires the scaling behavior of the probability density *at the tips* of the cluster. The interior of the cluster is screened, and thus the scaling of P_{max} should be insensitive to the details of the ramified, random interior. In order to obtain the scaling of P_{max}, it suffices to consider any object *with the same cusp structure* as the DLA cluster.

For a cluster grown on a $2d$ square lattice, the easy growth directions are precisely oriented along the lattice axes. The cusp structure of the cluster should thus reflect the symmetry of the lattice. In fact, for the large square-lattice simulations of Meakin[6] and of Ball[7], the shapes of the clusters take on the definite appearance of a diamond (a square rotated $\pi/4$ from the Cartesian axes). We thus propose, as an ansatz, that the cusp structure of a DLA cluster grown on a square lattice is identical to the cusp structure of a diamond outline of sites on the lattice.

This ansatz enables us to calculate, via conformal transformation, the electrostatic potential near the tip of a DLA cluster. Near the tip of a square of length L,

$$E \sim L^{-2/3} \delta s^{-1/3}, \qquad (7)$$

where δs measures the distance to a tip. Normalizing with (5), we obtain the probability density

$$p_N(r) \sim \xi^{-2/3}(r_+ - r)^{-1/3}. \qquad (8)$$

Using (2), we obtain the occupancy probability at a tip, $P_{max} \sim (a/\xi)^{2/3}$. The scaling equation of growth (4) yields

$$N \sim (r/a)^{5/3}, \qquad (9)$$

i.e., $D = 5/3$, where we have used the self-similarity assumption that r_+, ξ, r_0 all scale similarly. This result (9) is consistent with the largest square lattice simulations ($D_\alpha = 1.71 \pm 0.02$, $D_\beta = 1.72 \pm 0.06$) by Meakin.[6]

We can verify the consistency of the diamond cusp structure ansatz by examining the numerical scaling of $P_{max}(N)$. Substituting $\xi \sim N^{3/5}$ into $P_{max} \sim \xi^{-2/3}$ gives $P_{max} \sim N^{-2/5}$, which we have verified[2] with small realizations of DLA clusters grown using a CTRW formalism. This has also been checked by Meakin,[8] with large simulations, who finds $P_{max} \sim N^{-0.39}$.

The extension of these 2d square lattice results to the more general class of models (6) is straightforward. Normalizing (7) according to (6),

$$p_N(r) \sim \xi^{\eta/3-1}(r_+ - r)^{-\eta/3}, \qquad (10)$$

whence (2) yields $P_{max} \sim (a/\xi)^{1-\eta/3}$. Integrating the scaling equation of growth (4),

$$N \sim (r/a)^{2-\eta/3}, \qquad (11)$$

i.e., $D = 2 - \eta/3$. This trivially yields $D = 2$ for the Eden model, $D = 11/6$ for $\eta = 1/2$ and $D = 4/3$ for $\eta = 2$. These last results are in excellent agreement with the simulation results of Meakin[5] ($D_\alpha = 1.86 \pm 0.02$, $D_\beta = 1.92 \pm 0.05$, for $\eta = 1/2$ and $D_\alpha = 1.44 \pm 0.02$, $D_\beta = 1.39 \pm 0.10$ for $\eta = 2$).

Our *Ansatz* of the large-scale structure of the cluster determining the cusp structure may be applied to DLA grown on other lattices. For the 2d triangular lattice we would expect a DLA cluster "in the large" to resemble a hexagon, with cusp angle $2\pi/3$. An argument identical to that used for the square lattice yields $D = 7/4$. The latest numerical simulations of Meakin[6] on a hexagonal lattice obtain $D_\beta = 1.71$, with stringent enough error bars to statistically (but not systematically) rule out $D = 7/4$. However, no discernable hexagonal structure has emerged in these clusters, so (i) the clusters may not be large enough to properly identify D, or (ii) a hexagon, in fact, never emerges, indicating that the stronger singularity present for off-lattice DLA dominates that generated by the triangular lattice (while the still stronger square-lattice singularity drives 2d square-lattice DLA to the diamond shape).

We remark that the off-lattice results[6] $D = 1.71$ may be qualitatively understood as an averaging over coordination of the accreted random-walkers. The two possible 2d isotropic lattices bracket the off-lattice result: $D = 5/3$ (4-fold coordination) and $D = 7/4$ (6-fold coordination). It is amusing to note that a hypothetical 5-fold coordinated structure yields $D = 12/7 = 1.7143\ldots$ Expressed in this language, a theory for off-lattice DLA must somehow be a theory for the local coordination of the particles in the cluster.

Finally, the above results may be extended to anisotropic (but still 4-fold coordinated) lattices, with oblique angle β. A straightforward exercise yields

$$D = \frac{3\pi - \beta}{2\pi - \beta}, \tag{12}$$

which is bounded $1.50 \leq D \leq 1.67$. The (weak) non-universality of this result may be understood by "unskewing" the oblique lattice. Naively, we should recover isotropic Cartesian lattice DLA, but the "unskewing" tampers with the distribution of sources at infinity, uniaxially concentrating the source of random-walkers. The prediction (12) of this theory is sufficiently dramatic as to warrant testing by simulation.

HIGHER DIMENSIONS

We now turn to an application of these geometric ideas to higher dimensions.[3] We consider for simplicity only Cartesian lattices. In $d = 3$ we must integrate the probability distribution away from the tips to obtain

$$P_{max} = \int_{r_+ - a}^{r_+} p_N(r) r\, dr. \tag{13}$$

The scaling equation of growth (4) is unchanged. The large-scale structure *Ansatz* motivates the solution of the ectrostatics of a cube held at $\phi = 0$ inside a conducting sphere (at R_∞) held at $\phi = 1$. As the divergence of the electric field near the cube edges is weaker than at the cube corners, we solve for the dominant singular behavior near the tip of a rectilinear cone. With azimuthal symmetry, $\phi = R(r)\Theta(\theta)$; the radial part $R \sim r^\nu$, i.e, $E \sim r^{\nu-1}$, whence (13) yields $D = 2 + \nu$. The separation constant ν appears in the polar (Legendre) equation

$$\frac{1}{\sin\theta}\frac{d}{d\theta}(\sin\theta\Theta') + \nu(\nu+1)\Theta = 0, \tag{14}$$

and is determined by the boundary condition that the cone be at constant potential,[9]

$$P_\nu(\cos\beta) \equiv F(-\nu, \nu+1; 1; z) = 0, \tag{15}$$

where $P_\nu(\cos\theta)$ is the Legendre function of order ν, $F(\alpha, \beta; \gamma; z)$ is the hypergeometric function, and where

$$z = \frac{1}{2}(1 - \cos\beta). \tag{16}$$

We remark that the cone angle β is measured exterior to the cone from its axis (i.e., $\beta = 3\pi/4$ for a rectilinear cone).

The extension to higher dimensions is straightforward. To obtain P_{max}, the probability distribution must be integrated away from the tips of a hypercone

$$P_{max} \sim \int_{r_+ - a}^{r_+} p_N(r) d^{d-1} r. \tag{17}$$

Using hyperspherical coordinates and assuming hyperazimuthal symmetry, $\phi = R(r)\Theta(\theta)$; the radial part $R \sim r^{\nu-d+3}$, i.e., $E \sim r^{\nu-d+2}$, whence (17) yields $D = 2+\nu$. The separation constant ν appears in the polar equation

$$\frac{1}{\sin^{d-2}\theta}\frac{d}{d\theta}(\sin^{d-2}\Theta') + (\nu - d + 3)(\nu + 1)\Theta = 0, \tag{18}$$

and is determined by the boundary condition that the hypercone be at constant potential

$$F(d - 3 - \nu, 1 + \nu; \frac{d-1}{2}; z) = 0. \tag{19}$$

The results of this analysis are shown in the table below for Cartesian lattices (the radius of gyration exponent $\beta = 1/D$). In the last column are given the simulation results of Meakin.[10] The agreement of theory with experiment is quite remarkable.

d	D	β	β(expt)
2	1.667	0.600	0.592±0.017
3	2.463	0.406	0.401±0.009
4	3.333	0.300	0.303±0.011
5	4.245	0.236	0.238±0.009
6	5.182	0.193	0.182,0.193,0.186 (0.239,0.202)
7	6.137	0.163	
8	7.103	0.141	
9	8.077	0.124	
10	9.058	0.110	

The experimental results for $d = 6$ are the measured radius of gyration exponents for five individual clusters; the last two would imply $D < d - 1$, the mean-field result! Theoretically, (19) indicates there is to be no upper critical dimension for DLA, but rather that $D \to d - 1$ for large d.

I thank Harvey Scher for numerous invaluable discussions of this work.

[1] T. A. Witten, Jr. and L. M. Sander, Phys. Rev. Lett. **47**, 1400 (1981), Phys. Rev. B **27**, 5686 (1983); P. Meakin, Phys. Rev. A **27**, 604 (1983).
[2] L. A. Turkevich and H. Scher, Phys. Rev. Lett. **55**, 1026 (1985).
[3] L. A. Turkevich and H. Scher, Phys. Rev. A (submitted).
[4] L. Niemeyer, L. Pietronero and H. J. Wiesmann, Phys. Rev. Lett. **52**, 1033 (1984).
[5] M. Eden, in *Proc. Fourth Berkeley Symp. on Mathematical Statistics and Probability*, ed. J. Neyman (Univ. of Calif. Press, Berkeley, 1961), Vol. 4, p. 223.
[6] P. Meakin, this book.
[7] R. C. Ball and R. M. Brady, to be published.
[8] P. Meakin, private communication.
[9] J. D. Jackson, *Classical Electrodynamics*, 2nd Ed. (Wiley, New York, 1975), pp. 94-98.
[10] P. Meakin, Phys. Rev. A **27**, 1495 (1983).

FRACTAL SINGULARITIES IN A MEASURE AND "HOW TO MEASURE SINGULARITIES ON A FRACTAL"

Leo P. Kadanoff

The James Franck Institute
The University of Chicago
5640 South Ellis Avenue
Chicago, Illinois 60615

This is a preliminary report upon a piece of research being carried out by T. C. Halsey, Mogens Jensen, Leo Kadanoff, Itamar Procaccia and Boris Shraiman. It is an outgrowth of the thinking reflected in the work of Hentschel and Procaccia[1] and of Halsey, Meakin and Procaccia.[2] A fuller report on this work will appear later.

SINGULARITIES

In many different areas of physics one is interested in describing how an object, perhaps fractal, may be covered by a measure. A measure is just a probability assigned to each piece of the object. For example, in dynamical systems theory, the object may be a strange attractor and the measure the probability for visiting a given piece of the attractor. In a random resistor network, the object might be a percolating cluster of resistors and the measure might be proportional to the magnitude of the current through each resistor.

In any case, imagine some object, perhaps fractal, sitting in an ordinary Euclidean space. The object is divided into N pieces with the jth piece having size ℓ_j. Associated with the jth piece is M_j, the measure of that piece.

This measure may have singularities. For example if the object is the line between -1 and 1 $(x \in [-1,1])$ and the measure between x and $x + dx$ is equal to $M(x)dx$ with $M(x) = |x|^{-1/2}$, then the interval $[-\ell, \ell]$ has measure $\int_{-\ell}^{\ell} dx M(x) \sim \ell^{1/2}$. Thus, there is a singularity with index 1/2 at the origin. In general if, for small ℓ_j, we find

$$M_j \sim \ell_j^\alpha, \tag{1}$$

we say there is a singularity of type α.

If the measure were uniformly distributed over a d-dimensional space, α would be d. Hence α is a kind of dimension for the singularity. In ancient times in critical phenomena, α was denoted by the symbol y.

COUNTING SINGULARITIES

One example of interest is a DLA aggregate with the measure being the probability that a walker will land at a given point.[3] In this and other examples, there are infinitely many singularities (in the limit where the cluster becomes infinitely large) and a whole range of different possible values of α.

To characterize what happens, let $N(\alpha)d\alpha$ be the number of singularities of type α' for all α' lying between α and $\alpha + d\alpha$. As we divide the object more and more finely we get more and more singularities. Let ℓ be a typical size of a subdivision and define an index for the number of singularities, $f(\alpha)$, via

$$N(\alpha)d\alpha = \rho(\alpha)d\alpha \ell^{-f(\alpha)}. \qquad (2)$$

The higher $f(\alpha)$ is, the larger the density of singularities of type α. In critical phenomena, $f = d$. Notice that f may be interpreted as the dimension of the set of singularities of type α.

MEASURING $f(\alpha)$ AND α

Now I wish to know ("measure") the possible values of α and the function $f(\alpha)$ for some particular set. To do this, I follow an approach based upon a partition function

$$Z_N(q,\tau) = \sum_{j=1}^{N} M_j^q \ell_j^{-\tau}. \qquad (3)$$

I define $\tau_N(q)$ as the value of τ which makes this partition function equal to 1:

$$Z_N(q, \tau_N(q)) = 1. \qquad (4)$$

I find that the resulting value of $\tau_N(q)$ is not very sensitive to how the splitting into pieces is done and that the limit $N \to \infty$ exists. Hence define

$$\tau(q) = \lim_{N \to \infty} \tau_N(q). \qquad (5)$$

Given a data set, $\tau(q)$ can be directly measured.

What is the thing that has been measured? Say that in splitting up the set we obtain a piece with a given ℓ-value with a probability density $\rho(\ell)$. Then from Eqs. (1), (2) and (3) we can write the partition function as

$$Z_N(q,\tau) = \int d\ell p(\ell) \int d\alpha \rho(\alpha) \ell^{-f(\alpha)} \ell^{q\alpha} \ell^{-\tau}. \qquad (6)$$

The α-integral will contribute almost entirely from the region in which the exponent of ℓ, $q\alpha - f - \tau$, is a minimum. This will occur when

$$q = f'(\alpha), \qquad (7a)$$

or else when α reaches an endpoint of a region in which $\rho(\alpha)$ is nonzero. (Here we neglect the latter possibility.) If the result is not to go to zero or infinity as $N \to \infty$, we must have the total exponent of ℓ be zero. (See the condition of Eq. (4).) Thus we also have

$$q\alpha = f(\alpha) + \tau, \qquad (7b)$$

as in Ref. 2.

Our measurement gave us τ as a function of q. Equations (7a,b) are implicit equations which will express α as a function of q and f as a function of q and will then finally give $f(\alpha)$. To make this conclusion more explicit notice that these equations are of the form of a Legendre transformation. We can rewrite them in the form

$$f(\alpha) = -\tau(q) + q\alpha, \qquad (8a)$$

where q is determined as a function of α via

$$\tau'(q) = \alpha. \qquad (8b)$$

WHAT IT MEANS

Varying q and τ is a trick for exploring the different regions of α. For large positive q we are looking at small values of α, i.e., places in which the measure is very highly concentrated. For q large and negative we instead study parts of the fractal for which the measure is very small, i.e., the larger values of α. At $q = 0$, the measure drops out and f is the Hausdorf dimension. This turns out to be the maximum possible value of f.

The research reported here was supported by the ONR.

[1] H. G. E. Hentschel and I. Procaccia, Physica 8D, 440 (1983).
[2] T. Halsey, P. Meakin and I. Procaccia, preprint.
[3] This example was studied in detail in Ref. 2 and also in the parallel and independent work by H. Scher and L. Turkevitch.
[4] A. For a mathematical basis for this type of approach, see R. Bowen, *Equilibrium States and the Ergodic Theory of Anisov Diffeomorphisms*, Lecture Notes in Math. No. 470 (Springer, Berlin, 1975). B. For a somewhat analogous attack upon random resistor networks, see L. de Arcangelis, S. Redner and A. Coniglio, Phys. Rev. B 31, 4725 (1985).

Breakfast

Lunch

LIST OF PARTICIPANTS

AMITRANO Concezione	Universita di Napoli	Italy
BALL Robin	Cavendish Laboratory, Cambridge	UK
BERVILLIER Claude	Service de Physique Théorique, CEN Saclay	France
BEYSENS Daniel	SPSRM, CEN Saclay	France
BLUMBERG Robin	Harvard University, Cambridge	USA
BOON JeanPiere	Université Libre de Bruxelles	Belgium
BRAK Richard	King's College, London	UK
BROOMHEAD David	Royal Signal Radar Establishment, Malvern	UK
CANNELL David	University of Santa Barbara	USA
CARIDE Anibal	Centro Brasileiro de Pesquisas Fisica, Rio de Janeiro	Brazil
CHHABRA Ashvin	Mason Laboratory, Yale Station	USA
CHOWDHURY Debashish	Universität zu Köln	W.Germany
CONDE Olinda	Universitade de Lisboa	Portugal
CONIGLIO Antonio	Universita di Napoli	Italy
COSCIA Vincenzo	Universita di Napoli	Italy
DACCORD Gerard	Dowell Schlumberger, St Etienne	France
DAOUD Mohammed	Laboratoire Leon Brillouin, CEN Saclay	France
DEARCANGELIS Lucilla	Boston University	USA
DESAI Rashmi	University of Toronto	Canada
DHAR Deepak	Université P. et M. Curie, Paris	France
DJORDJEVIC Zorica	Institute of Physics, Beograd	Yougoslavia
DUARTE Jose	Universitade da Porto	Portugal
DURAND Dominique	Université du Mans	France
ENGLAND Paul	Imperial College, London	UK
ERZAN Ayse	Universitade da Porto	Portugal
FAMILY Fereydoon	Emory University, Atlanta	USA
FERONE Vincenzo	Universita di Napoli	Italia
GOBRON Thierry	Ecole Polytechnique, Palaiseau	France
GONZALES FLO Augustin	Instituto di Fisica UNAM	Mexico
GUYON Etienne	E.S.P.C.I., Paris	France
HALSEY Thomas	University of Chicago	USA
HEERMANN Dieter	Universität zu Mainz	W.Germany
HERRMANN Hans	S.P.T., CEN Saclay	France
HOLDSWORTH Peter	Oxford University	UK
HONG Daniel	Boston University	USA
JAN Naeem	St Francis Xavier University, Antigonish	Canada
KADANOFF Leo	University of Chicago	USA
KANG Kiho	Boston University	USA
KERTESZ Janos	Hungarian Academy of Science, Budapest	Hungary
KOLB Max	Université de Paris sud, Orsay	France
KOSTERLITZ Mike	Brown University, Providence	USA
LAM Pui-Man	Chinese Academy of Sciences, Beijing	China
LANDAU Dave	University of Georgia, Athens	USA
LEYVRAZ François	Boston University	USA
MAGALHAES Aglae	Centro Brasileiro de Pesquisas Fisicas, Rio de Janeiro	Brazil
MARGOLINA Alla	Dupont de Nemours, Wilmington	USA
MATTHEWS MOR. David	University of Georgia, Athens	USA

MEAKIN Paul	Dupont de Nemours, Wilmington	USA
MISBAH Chaouqi	Physique des Solides ENS, Paris	France
MURAT Michael	Tel Aviv University	Israel
NOULLEZ Alain	Université libre de Bruxelles	Belgium
OSTROWSKY Nicole	Université de Nice	France
PALADIN Giovanni	Università "La Sapienza" di Roma	Italy
PELITI Luca	Università "La Sapienza" di Roma	Italy
PETIT Luc	E.S.P.C.I., Paris	France
RAMASWAMY Rama	Tata Institute, Bombay	India
RARITY John	Royal Signal Radar Establishment, Malvern	UK
REINA Juan	Boston University	USA
ROUCH Jacques	Université de Bordeaux I	France
ROUX Didier	Centre de Rech. Paul Pascal, Bordeaux	France
ROUX Stephane	E.S.P.C.I., Paris	France
RYS Franz	Fritz Haber Institute, Berlin	W.Germany
SCHAEFFER Dale	Sandia National Laboratory	USA
SCHLESINGER Michael	Office of Naval Research, Arlington	USA
SCHMITTMANN Beate	Universität Düsseldorf	W.Germany
SECO Fernando	Brown University, Providence	USA
SELINGER Jonathan	Harvard University, Cambridge	USA
SEN Pabitra	Schlumberger Doll Research, Ridgefield	USA
SIMIEVIC Avraham	Hebraic University of Jerusalem	Israel
SOLLA Sara	I.B.M. Research Center, Yorktown Heights	USA
SORNETTE Didier	Université de Nice	France
STANLEY Eugene	Boston University	USA
STAUFFER Dietrich	Universität zu Köln	W.Germany
STEPANEK Petr	Institute for Macromolecular Chemistry, Pragues	Czechoslovakia
TEXEIRA José	Laboratoire Leon Brillouin, CEN Saclay	France
TURKEVICH Leonid	Standard Oil Cie., Cleveland	USA
VAN DONGEN Peter	State University of Utrecht	Netherlands
VAROQUI Raphaël	C.R.M., Strassbourg	France
WILLEMSEN Jorge	Schlumberger Doll Research, Ridgefield	USA
WITTEN Tom	Exxon Res. and Dev., Annandale	USA

INDEX

Activator 175
Active zones 42,63,132,294
Aerosol 136,223
Aggregation 136,231,257
  2-dimensional 171
  ballistic 73, 115
  chemically limited 128, 194
  cluster-cluster 7,49, 117,136,188,201, 222,232
  colloidal 116,200,211,222
  diffusion limited (DLA) 5,60 69,107,117,146, 194, 222,251,255
  kinetics 136,191,231
  metal particles 128
  reversible 171,214,225
  under shear 171
Alexander /Orbach conjecture 29
Algorithm 120, 251,257
  Hoshen-Kopelman 86
Alumina 201
Angular correlation function 125
Anisotropy 63,78,221,225, 249
Ant (in the labyrinth),27, 86, 170, 284
Anticorrelation 221
Approximation to DLA 66
Automata cellular 4, 175

Backbone 46, 86 104,164,247
  exponent 104
Binary mixture 211,238
Blobs 49,104,256,274
Boundary layer model 77
Bridges 106
Brown-Boveri model 293
Brownian motion 117,282
  particles 142
Bruxesselator 178
Butterflies 90, 284

CTRW (Continuous-time-R.W.) 293
Cab-O-Sil 201
Cancer growth 113
Capacity dimension 237
Capillary force 74,171,255
  length 250
  number 203
Carbon black 201
Catastrophic failure 260

Caterpillar growth-limited-diffusion 288
Cayley tree 287
Cell colonies 112
Central limit theorem 279
Charge density109
Chemical fractal dimension 15,32, 287
  length 15
Classroom experiment 84
Cluster 79, 238,251,264, 284,288
  fractal dimension 142
  incipient infinite 80, 264
  number 85,105
  perimeter 286
  radius 95
Cluster size distribution 13,136,191,224 228, 244
Clustering of clusters: see aggregation
Coagulation (see aggregation)
Coexistence curve 212
Colloid(s) 171,199,223,231
Colloidal (see aggregation)
Computer 4,80 108 111, 244
Conductivity 50,72,105,164,275
Conformal transformations 293
Contrast 151
Convection 70,165
Copper sulphate 70
Correlation function 62,71,123,146,226
  length 8,95, 169, 286
Critical exponent 10,25, 92, 246,273
  phenomena 7,24,104,211,265
Curie point 92
Cutoff 70, 258

Darcy's law 73
Dendritic growth 71,117, 249 258
Deposition diffusion controlled 70,251
Deterministic dynamical system 254
Diamond shape 125,295
Dielectric breakdown 4, 75,107,114,146,293
Differentiation 178
Diffusion 57,117,257, 284
  anomalous 89
  constant 50,69, 130,142,166,175, 218,228,232, 250
  equations 56
  ionic 71
  rotational 117
  translational 117

Diffusive velocity 223
Dispersion 165
  hydrodynamical 170
Droplet model 98, 242,246
Dye 166
Dynamic configuration 8, 102
Dynamics of fractal 198

Eden model 5,61, 111,270,286, 288,294
  (invading) 115
Eden trees 115
Elastic modulus 166,274
  properties 48, 199, 260
Elasticity 167, 273
Electrical breakdown 3
Electrostatic analogy 108,293
Epidemic 5 45, 86,265, 288
  shell model 18
Euclidean dimensionality 21,112
Excluded volume 17, 168,227, 263
Exponent "gap" 105
  backbone 166
  critical 7,94, 273
  infinite set of 106
  of Fisher 84

Fallacy   97
Faraday's law 71
Field theory 258, 265
Finite size 15, 75,84, 246
Fixed point 265
Flocculation 222,231
Flory theory 227
Flow through porous material 254
Fock space   266
Forest fire 32, 86
Form factor 153
Fractal 4, 71, 167, 258, 284
  dimension 11,27, 95,105,123,146,
    191,205,218,223, 237,251,280
  dynamics 198
  exact 21
  random 23,55
  surface 107,167
  volatile 45
Fracton 29, 89,160
Fracture(s) 168,260
Fugacity 10

Gel swelling 229
Gel(s) 4, 224
Gelation 18, 140, 224 228, 244
  kinetic 12
Generating function 11
Geological fault 260
Gold aggregates 128, 194
Grain space 163
Growth 3,90,102,112,265,284
  droplet 57
  history 5
  irreversible 4,238
  needle 60
  non-equilibrium 130, 237
  of skin cancer (tumor) 5,111
  parabolic 58
  perimeter   284
  probability 103
  stochastic 55
Growth sites 12,90,101, 293
  probability distribution 101
Guinier approximation 154

Hele-Shaw cell 73, 203
Heterogeneity 163
Hierarchical model 105,127, 261
Hopping 169, 257
Hull 37, 264
Hydrodynamic radius 199, 218
Hydrodynamics 240
Hypercone 297
Hypercube 293
Hyperscaling 95

Infection velocity 289
Information dimension 108
Inhibitor 175
Instabilities 237
Integral 98
Intensity cross-correlations 219
Ising model 34

Josephson junction 282

Kinetic aspects   86,130
  gelation: see gelation
Kinetics 136, 231
Kirchhoff's law 255

Laplace equation 107,249 257
Latex suspension 204, 225

Lattice animals 9,90,225
  honeycomb 86
  triangular 96
Lennard-Jones fluid 237
Lichtenberg figure 3
Light scattering 151,187,198,211 218
Linear stability analysis 176, 251
Links ("red" bonds) 39,104,274
Liouvillian operator 266
Lévy flights 14, 280
Lévy walks 281

Mean field 18,224,227,235,238
Measure 299
Mechanical properties 163
Mesh 166
Minimum path 30
Mole in a labyrinth 19
Molecular Dynamics 237
  weight (of an aggregate) 190
Monte Carlo 82,237,249
Morphogens 174

Network 254
Neutron scattering 187
Noise (Monte Carlo) 251
Non- universality 293
Non-linear(ity) 174,249
Non-local(ity) 5
Number fluctuationn 218

Occupation probability 293

Packing 163
  appolonian 166
  random 163
Parasite 90
Partition function 8,300
Pattern 3, 75,163 174,238,249,254
Peclet number 166
Penetrable DLA 61
Percolation 11,24, 79,103,167,228, 273, 284
  cluster 83,106, 285
  conductance 163
  invasion 85, 255
  probability 94
  threshold 80, 164, 228, 264
Perimeter 38,101, 294
Periodic boundary conditions 84,126, 238

Permeability 164,255
Phase separation 238
  transition 8,91
Poiseuille's law 254
Polyacrylamide 19
Polydisperse systems 142,159,163, 201, 224
Polymer(s) 199, 227
Polymerization 136
  see gelation
Polysaccharide solution 204
Polystyrene 199, 218,229
Pore space 164
Pore(s) 254
Porosity 164
Probability occupancy 293

Quench 238

Radius (hydrodynamic) 199
  of gyration 10,117, 188,228, 263,274
Random bond breaking 130
  matter 163
  number 82
  packing 163
  resistor network 32,103
  superconducting network 32,106
Random walk 16,23,55,250,257, 265, 279,284,294
  Kinetic Growth Walk (KGW) 17
  Self Avoiding (SAW) 17,263
Rate equation 136
Rate constant 136
Rayleigh (forced) 166
Reaction diffusion 174
Renormalization 27,261,265
Reorganization 130
Reptation 85
Resistivity exponent 104
Reynolds number 165,205
Rheology 163,208
Rickvold model 115

Scale invariance 218,237,260
Scaling 12,56, 103,199,223,232,242,274, 293
  basic concepts 137
  corrections to 23, 91, 236
  laws 94
  theory 90

Scattering from fractal objects 155
  from fractal surfaces 158
  small angle X-ray 151
  small angle neutron 151
  techniques 149
  wave vector 150
Screened growth model 18,115
Screening 36,224,228, 258
  length 294
Self avoiding walk: see random walk
Self- similar(ity) 3,55,265
Shape (asymptotic) 288
  (diamond) 127
Shear rate 205
Shear stress 260
Shear thinning (non-Newtonian fluid) 75, 203
Sierpinski gasket 21, 104,166,243
Silica spheres 128, 187, 211
Simulation (see computer)
Simulation numerical 142, 237
Singularities 299
Smoluchowski equation 18,130,143,224,235
Sol-gel transition 18, 167, 244
Spinodal decomposition 237
Static configuration 6, 102
Statistical uncertainties 123
Steric hindrance 16
Sticking probability 18,112,187, 232
Stokes law 164
Structure factor 189
  large-scale 295
Superconducting bond(s) 105
Surface 12, 132
  chemical 15
  effect 237
  tension 74, 238, 255
Susceptibility 98
  ratio 101

Termite 32, 90
Theta point 263
Tip 293
  priority factor 114
Torque 275
Transport exponent 258,273
  properties 132

Trapping 17,246,256,282
Tricritical point 263
Turbidity 214
Turbulent diffusion 282

Universality 98,132
  classes 235, 251,264,273, 286
Upper critical dimensionality106,128,236,263,281,29

Van der Waals equation 98
Vapor aggregates 201
Viscosity 164,255
Viscous fingering 73, 203,254
  fluids 203
Voltage distribution 102

Water-lutidine mixture 211
Wedge 293
Weierstrass flights 280
Wetting 211
Witten-Sander (see Aggregation DLA)

X-ray scattering 187

Zinc metal leaves 71